The RADIATION BIOLOGY of the VASCULAR ENDOTHELIUM

The RADIATION BIOLOGY of the VASCULAR ENDOTHELIUM

Edited by
David B. Rubin, M.D.
Rush Medical College
Chicago, Illinois

CRC Press
Boca Raton Boston London New York Washington, D.C.

Library of Congress Cataloging-in-Publication Data

The radiation biology of the vascular endothelium / edited by David B. Rubin.
p. cm.
Includes bibliographical references and index.
ISBN 0-8493-4840-4 (alk. paper)
1. Vascular endothelium. 2. Blood-vessels--Effect of radiation on. 3. Tumors--Blood-vessels--Effect of radiation on. I. Rubin, David B.
[DNLM: 1. Endothelium, Vascular--radiation effects. 2. Endothelium, Vascular--physiopathology. WG 500 R1285 1997]
QP110.V34R33 1997
612.1′3--dc21
DNLM/DLC
for Library of Congress 97-34506
CIP

Direct all inquiries to CRC Press LLC, 2000 Corporate Blvd., N.W., Boca Raton, Florida 33431.

International Standard Book Number 0-8493-4840-4
Library of Congress Card Number 97-34506
Printed in the United States of America 1 2 3 4 5 6 7 8 9 0
Printed on acid-free paper

Contributors

A. William Blackstock, M.D.
Department of Radiation Oncology
University of North Carolina
School of Medicine
Chapel Hill, NC

Ed Robert Blazek, M.D.
Departments of Radiation Oncology and Medical Physics
Rush-Presbyterian-St. Luke's Medical Center
Chicago, IL

Clement A. Diglio, Ph.D.
Department of Pathology
Wayne State University
School of Medicine
Detroit, MI

Elizabeth A. Drab-Weiss, M.D.
Department of Medicine and Radiation Oncology
Rush-Presbyterian-St. Luke's Medical Center
Chicago, IL

Luis F. Fajardo L-G, M.D.
Stanford University School of Medicine
and
Veterans Affairs Health Care System
Palo Alto, CA

Mitchell Friedman, M.D.
Department of Medicine
Tulane University School of Medicine
New Orleans, LA

Zvi Fuks, M.D.
Department of Radiation Oncology
Memorial Sloan-Kettering Cancer Center
New York, NY

Itzhak D. Goldberg, M.D.
Department of Radiation Oncology
Long Island Jewish Medical Center
New Hyde Park, NY

Michael M. Graham, Ph.D., M.D.
Departments of Radiology and Radiation Oncology
University of Washington Medical Center
Seattle, WA

Melvin L. Griem, M.D.
Department of Radiation and Cellular Oncology
University of Chicago
Chicago, IL

Adriana Haimovitz-Friedman, Ph.D.
Department of Radiation Oncology
Memorial Sloan-Kettering Cancer Center
New York, NY

Seema S. Kantak, Ph.D.
Department of Radiation Oncology
Wayne State University
School of Medicine
Detroit, MI

Lester Kwock, Ph.D.
Department of Radiology
University of North Carolina
School of Medicine
Chapel Hill, NC

Luka Milas, M.D., Ph.D.
Department of Experimental Radiation Oncology
University of Texas
M.D. Anderson Cancer Center
Houston, TX

Mira Milas, M.D.
Department of Surgery
Emory University School of Medicine
Atlanta, GA

Agostino Molteni, M.D., Ph.D.
Department of Pathology
Northwestern University Medical School
Chicago, IL

James V. Moore, Ph.D., F.R.C. Path.
Laser Oncology Programme
Paterson Institute for Cancer Research
Christie Hospital
Manchester, UK

Cliff Murray, Ph.D.
Laboratory of Molecular Oncology
CRC Academic Department of Clinical Oncology
City Hospital
Nottingham, UK

James M. Onoda, Ph.D.
Department of Radiation Oncology
Wayne State University
School of Medicine
Detroit, MI

Lanell M. Peterson, B.A.
Department of Radiation Oncology
University of Washington Medical Center
Seattle, WA

Eliot M. Rosen, M.D., Ph.D.
Department of Radiation Oncology
Long Island Jewish Medical Center
New Hyde Park, NY

David B. Rubin, M.D.
Departments of Medicine and Radiation Oncology
Rush-Presbyterian-St. Luke's Medical Center
Chicago, IL

Chung-hsin Ts'ao, Ph.D.
Department of Pathology
Northwestern University Medical School
Chicago, IL

William F. Ward, Ph.D.
Department of Radiology
Northwestern University Medical School
Chicago, IL

David C. West, Ph.D.
Department of Immunology
University of Liverpool
Liverpool, UK

Preface

I enjoy challenging my students by asking them how they think a lung doctor came to spend his time irradiating aortic cow cells in a plastic dish. As a means of answering this question, but more important, as a way to acknowledge and thank my teachers, the following should be told. My odyssey in research, and the earliest ideas for this book, began when I was a medical student at Rush Medical College in Chicago in 1972 where I heard Dr. Gerald Moss, now Dean of the University of Illinois Medical School, present grand rounds about acute lung injury in our severely wounded soldiers in Vietnam. I was fascinated by the notion that lung endothelial cells would be damaged by hemorrhagic shock. Around this time, I read two articles that were pivotal to the development of my research interest. In the first, Ng and Vane reported that the lung vessels convert an inactive peptide, angiotensin I, into a potent vasoconstrictor, angiotensin II. Soon afterward, I read the report by Una and Jim Ryan who described how this process occurred in the lining of lung vessels, the endothelium which prominently expressed membrane-bound angiotensin converting enzyme (ACE). When I learned that the lungs contain more endothelial cells than all the other tissues combined, and that these cells possess a variety of biochemical functions that regulate circulation, my plan was set: to become a pulmonary endocrinologist and measure biochemical functions of lung endothelial cells.

These ideas were nurtured in my last year of medical school by Drs. Richard Hughes, Drummond Rennie, and Joseph Messer, my pulmonary, renal, and cardiology teachers at Rush. As an intern, I knew I was on the right track when I found a new book by Alain Junod, Professor at Geneva, entitled *Lung Metabolism.* Within a year, my plan as a pulmonary research fellow was to study endothelial functions in patients with lung disease; however, John Clements, Professor at UCSF/CVRI, told me that I could study patients for 20 years and still not know what's going wrong with them. He then directed me to Bob Mason and Leland Dobbs, who firmly but gently guided me into the realm of cell biology and academic medicine. They first sent me to Una and Jim Ryan at the University of Miami, who taught me how to culture endothelial cells from calf pulmonary arteries and measure ACE.

Back at UCSF, I found Denis Gospodarowicz, who had recently characterized fibroblast growth factor, and he too generously guided me through endothelial experiments. During my fellowship, I developed the bias that oxidative stress to endothelial cells was one of the most common events in lung injury; and during the last year of my research training, Dr. Junod, visiting as an alumnus of UCSF/CVRI, invited me to join him for a year in Switzerland to study hyperoxia and endothelial uptake of purines. Returning to Chicago as junior faculty in pulmonary medicine at Northwestern Medical School, I met Betty Drab-Weiss. We "hired each other" and have been research partners ever since. Betty has an uncanny ability to turn my cosmic ideas (as Dr. Junod would call some of them) into practical experiments. I also met Dr. William Ward, a Professor at Northwestern Medical School. "Uncle Bill" initiated Betty and me into the radiation biology club. Bill was irradiating rat lungs and examining markers of endothelial biochemical function. He taught me that therapeutic radiation was mostly an oxidative stress. He also convinced me that, as a clinical problem caused by an oxidative stress, radiation exposure was a frequent event whose onset and localization of injury was better defined than most other injuries suffered by our patients. Before leaving Northwestern and returning to Rush, I met Dr. Ken Bauer, now at Genentech, who taught Betty and me about flow cytometry and the cell cycle. At Rush, I credit Dr. J. P. Szidon, Professor of Medicine, for helping me mature my scientific outlook, and I am also deeply indebted to Dr. Henri Frischer, Professor of Medicine, who introduced me to the world of thiols. Last, I have been taught about the radiation biology of endothelial cells from investigations carried out by the contributors to this book, who are leaders in their field. Their patience for my too-slow compilation of this body of work is very much appreciated.

David B. Rubin, M.D.

Foreword

This book brings together a number of expert reviews on the biology and radiobiology of the vasculature, with special reference to the endothelial cell component of blood vessels. This is a research field in which there have been dramatic changes in recent years in technology and data collected, and therefore in the interpretation of radiation responses.

For the first 70 or 80 years of radiation biology, the blood vessels were either ignored or regarded simply as a problem in the etiology of normal tissue injury, but of no potential therapeutic benefit in the treatment of tumors. The negative impact of a poor microcirculation in tumors, especially in terms of radioresistance of hypoxic cells at a distance from each capillary, was well recognized. Diffusion distances for oxygen are often exceeded by the long intercapillary distances in solid tumors, but hardly ever in normal tissues. Therefore, tumor hypoxia was regarded as a disadvantage for the radiotherapist, and considerable energy has been expended on improving the oxygen delivery, or providing chemicals that can substitute for oxygen (e.g., the nitroimidazoles, or radiations [higher LET] whose effect is less oxygen dependent). The recognition that tumor blood vessels show inappropriate, intermittent closure or even reversal of flow more recently added the concept of acute, intermittent hypoxia to that of chronic diffusion hypoxia. This served to increase the efforts to find ways of overcoming tumor-specific hypoxic radioresistance. It has also increased the efforts to develop techniques that can measure hypoxia or blood flow in experimental and human tumors.

In the 1970s and early 1980s, several lines of study developed independently in vascular research and oncology research and changed the perception of blood vessels and especially of the contribution of endothelial cells to damage in tumors and normal tissues. These lines included the following:

- The recognition that normal tissue injury after irradiation could result from vascular damage, with a long latency. Late expression of damage in vessels could lead to secondary consequences of ischemic injury (e.g., in spinal cord, lung, and kidney), which could be distinguished from direct radiation damage to the parenchymal cells of the tissue.
- Hyperthermia re-emerged as an alternative cause of damage to tumors. It was shown *in vitro* to be more effective on cells in poor nutritional microenvironments, and therefore provided a reason for the greater effectiveness of moderate heat in damaging cells in solid tumors than in normal tissues. Studies of blood flow and other vascular parameters, however, showed that the vasculature could itself be a target, with a therapeutic window of time and temperature in which vascular occlusion could occur in solid tumors but not in normal tissue vessels.
- The process of angiogenesis in tumors was identified as a potential target for tumor therapy. Much effort is being poured into understanding the angiogenic factors that tumor cells secrete in order to induce new vessel formation, and many attempts are being devised to block these factors and thereby prevent the expansion of the network needed for continued tumor growth.
- The growth characteristics of normal and newly formed tumor vessel endothelium was compared and shown to differ markedly. The proliferating endothelial cell was

identified as a potential target for cytotoxic agents. Development of stains involving histochemical reactivity showed the proliferating endothelium to also have altered biochemical characteristics, with less expression of differentiation markers. This led to the additional concept of targeting based on immaturity and different functional status of the endothelial cells.

- The study of endothelial cell biology suddenly exploded. Its key role in controlling molecular traffic, fluid permeability and coagulation, and cytokine production and turnover have all been identified in recent years and intensively studied. The biochemical studies have been correlated with light and electron microscopy, and the newer techniques of molecular biology have allowed a greater depth of understanding. In particular, the endothelial cell involvement in vascular tone control (e.g., via NO) has attracted great interest.
- A number of novel approaches to cancer therapy have been shown to have a vascular component to their anti-tumor action in mice. These effects were usually not anticipated from the initial screening programs, whether those were *in vitro* or in rodent tumors or xenografts. For example, several cytotoxic drugs and the cytokine or biotherapy approaches (e.g., of interferon, the interleukins, and TNF) have been shown to have significant effects on endothelial cell function and on vascular sufficiency.
- New approaches targeted at tumor cells were found to be limited in their effectiveness because of the problems of access to all tumor cells in a solid tumor. This results from abnormal permeability of tumor vessels and the lack of draining capillaries. The normal driving pressure of intravascular–interstitial–intralymphatic drainage is lost, and diffusion of large molecules within the intersitium of tumors is grossly hindered.
- New radiotherapy approaches aimed at killing tumor cells using tumor-specific localization of radiosensitizers has been re-evaluated. Two examples are boron neutron capture therapy and photodynamic therapy. The sensitizers (borated compounds or hematoporphyrin derivatives) are difficult to deliver to every tumor cell, and may indeed localize within the neovascular endothelial cells. The applied neutron beam or laser light that activates the sensitizing drug may then localize the damage in the vessels, leading to secondary tumor cell starvation. The target cell is critical if an optimal timing for a differential effect after drug administration is to be achieved.
- Differential metabolism or activation of prodrugs may occur in the mature endothelial cells of normal vessels and the immature cells lining the neovasculature, leading to a differential delivery of the metabolite of the prodrug to tumor and normal tissues, as with the radio- and chemoprotective drug WR 2721 (Ethyol, Amifostine).
- The endothelium has been recognized as a barrier for the delivery of immunoglobulins aimed at delivering toxins to tumor cells. The potential of delivering the antibodies to the endothelial cells, because of proliferative, biochemical, or surface antigen differences, is now appreciated and several successful antibodies have been described.

Thus, research interest in the microcirculation, and in endothelial cells in particular, has grown dramatically in the past two decades. A specific subset of that work has focused on the interaction of radiation with the endothelial cells. Most of the chapters in this book focus on ionizing radiation, with little attention to ultraviolet radiation or the radio frequency, microwaves, or ultrasound that are often used to generate hyperthermia. Nevertheless, a kaleidoscope of effects is presented to the reader. The expertise of endothelial cell biologists and radiobiologists from widely differing backgrounds has been brought together. Possible mechanisms of cell damage are presented, and endogenous chemicals that can influence damage expression are illustrated. A range of techniques for assessing endothelial cell function, viability, and integrity are considered. The importance of the relationship of endothelial cells to each other and to adjacent cells, both at the luminal and abluminal surfaces, is elucidated.

The chapters cover both the basic aspects of endothelial cell biology and the practical consequences of this for the oncologist, with special reference to the radiation oncologist. After reading this book, no clinical or experimental oncologist should be able to ignore the intimate relationship and interpendence of the tumor and normal tissue cells with the endothelial cells that form the barrier to, and the bearer of, their essential nutrient supplies. Any new or old approach to cancer therapy should now include an assessment of whether the action of the cytotoxin occurs on the endothelial cell or on the tumor cell. If the endothelium is damaged to such an extent that vessels occlude or rupture, the consequence is likely to be much more severe for tumors than for normal tissues. Each cord of tumor cells is too far from any other adjacent capillary to be able to benefit from a collateral supply if the nearest vessel closes down. By contrast, in normal tissues, there is usually an abundance of vessels available to increase the flow in times of stress. The loss of a capillary can often be compensated therefore by collateral vessels opening up.

As any new field expands, there will be doubting Thomas's on the sidelines ready to criticize and raise objections. But the differences between quiescent normal endothelium and that in expanding neovasculature is gradually gaining recognition. After the identification in mice of a 50-fold difference in proliferation rates between tumor and normal endothelium, subsequent studies in man emphasized there was a smaller differential, but still about tenfold. However, this is still a much greater therapeutic window than the proliferative differential between parenchymal cells in bone marrow and intestinal crypts versus solid tumors, where the normal cells can be proliferating even more rapidly and therefore limit most current proliferation-based chemotherapy!

The important thing in oncology research is always to keep in mind that a differential is needed if an agent is to be more effective in damaging tumors than normal tissues. This applies as much to vascular effects as to direct cellular attack. That differential may occur at the level of intrinsic radio- or chemosensitivity of proliferating immature endothelium versus mature quiescent endothelium. Or it may result from a difference in the redundancy in the two systems. Mature and new vascular networks may also have a differential response to the biochemical and neurological signals for increasing or decreasing flow and thereby altering the delivery of radiomodifiers like oxygen. This may allow differential manipulation of radiosensitivity via microenvironmental changes, either of the endothelium itself or of the tumor cells it supplies. Finally, manipulation with chemical sensitizers or protectors, either at the time of irradiation or after the event, may also provide an opportunity for differential modification of injury. Most of these aspects are covered in the different chapters of this book.

Two other areas where the studies reported in this volume may have some relevance, are for the prognostic value of vascular density and vascular irradiation for restenosis of atherosclerotic vessels.

In the last few years, following the initial studies in Boston and Oxford, a correlation has been found between the probability of developing distant metastases and the density of vessels in hot spots within the primary tumor. No convincing biological explanation has been provided to explain why high vessel density should lead to worse prognosis. Indeed, it is contrary to the expectation based on poor vasculature and poor microenvironment being associated with resistance to radiation and to chemotherapy. Assuming that all endothelial cells are being recognized in the staining procedures, and if one accepts that "hot spot" density is more relevant than overall vascular density, it will be interesting to see if this ultimately relates to the natural history of these tumors or to their response to therapy of the primary tumor.

In the last two years, radiation has suddenly become popular as a possible way of overcoming the proliferative wound-healing response of ruptured layers in the coronary vessels after balloon angioplasty. The exposure of intimal cells to the bloodstream when the endothelial barrier and the elastic lamina are disrupted results in a neo-intimal proliferation that contributes to the approximately 50% restenosis rate. It is proposed that intravascular irradiation, via inserted radioactive stents or intracatheter delivery of radioactive seeds may arrest this proliferative response. A differential effect of radiation on endothelial cells required to divide to re-cover the denuded area, and on the undesirable smooth muscle proliferation, may be crucial. In this field, damaged and reactive proliferating endothelial and vascular smooth muscle cells are the target, in a host that has a tendency toward atherosclerotic plaque deposition. These factors may, of course, grossly influence the ability to extrapolate from existing experimental data from proliferative (in tumor) or quiescent (in normal) endothelial cell studies. This is an example where research in oncology may have great value to those working in an adjacent field, in this case cardiovascular medicine, and collaborative projects may advance the understanding in both fields.

I feel sure that many researchers, both clinical and experimental, will find this volume a treasure trove of new information. It is particularly helpful to have so many disparate approaches brought together in a single setting.

Juliana Denekamp

Contents

Dedicated to my daughters,
Ariel and Elena

CHAPTER 1

The Endothelial Cell Is a Unique Target of Radiation: An Overview

Luis F. Fajardo L-G, M.D.

This is a concise review of the other chapters that constitute the main substance of this book. Each of the chapters that follow is an important contribution to our understanding of the endothelial cell role in the radiation injury of mammals. The functions of the endothelial cells (EC) have become more and more obvious in latter years.[4] Decades ago, EC were seen simply as the lining of the blood vessels and the heart that provided transport for various substances from or into the bloodstream. Important developments have occurred since then, including the understanding of the interactions between platelets and EC, the establishment of continuous EC cultures, the identification of angiogenic factors, and the effects of cytokines on the endothelium. All of these observations led to a plethora of data indicating that the EC are far more complex than expected.[4] And many more discoveries are expected.

Aside from those common to all cells, the many functions of EC include: interaction with the soluble coagulation systems and with platelets; participation in the inflammatory response and the immune response, including adhesion to leukocytes; participation in the vascular tone regulation; synthesis of multiple cytokines; synthesis of the stromal components that surround blood vessels; unique metabolic functions such as those involving lipoproteins; and most importantly, the formation of new microvessels in the fully developed mammal: angiogenesis.[4] From reviewing the extensive information, one has to conclude that endothelial cells vary considerably from one site to another (e.g., from large arteries to large veins to microvessels) and from one animal species to another.[4] This variation affects proliferation kinetics, morphology, antigen expression, prostaglandin secretion, and collagen synthesis.[4,5]

As will be described in the other chapters, most of the functions of EC can be affected by ionizing radiation.[5] The contributors to this book have focused their sections on certain aspects of the interaction of EC with radiation, which reflect recent areas of investigation and important observations.

The *in vivo* morphologic alterations that occur in irradiated endothelium are reviewed by **Kwock, Blackstock, and Friedman**[11] in Chapter 7. In rat lungs, 6 weeks after exposure to 30 Gy in 5 fractions (FR), they observed denudation of endothelium and areas with enlarged EC. The latter coincide with the compensatory hypertrophy of endothelial cells studied by Rosen and Goldberg[15] (see below). Microradiographs of sections of such lungs at 4 to 6 weeks showed constriction points, giving a sausage-like appearance to the microvessels. At that time there was a 70% decrease in vascular lung perfusion (^{99m}Technetium MAA) (see also Graham,[8] below). Using similar doses, the authors have observed cell detachment in confluent monolayers of EC (from various tissues), leaving discontinuities that became larger with time. The exposed cells were "highly vacuolated." The latter coincides with our personal observations of extensive vacuolation (documented by transmission electron microscopy) in bovine aortic endothelial cells (BAEC) exposed to single doses (SD) of 2.5 to 20 Gy.[10] Our study, in 1982, also documented hypertrophy of remaining cells after detachment of EC.[10]

Kwock and co-authors show *in vitro* evidence of a radiation-induced decrease in EC membrane transport of 5-hydroxytryptamine (5-HT) and amino acids. The removal of 5-HT by EC requires a Na(+)-dependent transport process, which is highly sensitive to free oxygen radicals. Thus, either radiation or hyperbaric oxygen inhibit 5-HT removal, and this defect can be corrected by superoxide dismutase.

Several groups have studied the effect of radiation on prostacyclin (PGI_2) synthesis, with confliciting results. Kwock et al.[11] summarize the observations: the hydroxyl radicals generated by ionization-increase phospholipase-A_2 activity, generating various arachidonic acid metabolites including PGI_2. It appears now that within 90 minutes following radiation exposure (e.g., 2 Gy) *in vitro*, there is a rapid decrease in PGI_2 synthesis. However, this is followed by a period of increased synthesis of PGI_2 that begins 4 to 24 hours after exposure. *In vivo*, the duration of each period may be longer. The activity of angiotensin converting enzyme (ACE) is decreased by radiation *in vitro* and *in vivo*. ACE normally inactivates bradykinin, which is a vasodilator and stimulates PGI_2 production. Thus, by inhibiting ACE production, radiation not only causes (local) vasodilation, but also indirectly promotes PGI_2 production.

Endothelial cells contain transglutaminase, a Ca^{2+}-dependent enzyme involved in cell growth and differentiation whose activity is inversely proportional to cell proliferative activity. Thus, it is not surprising that this enzyme shows a dose-dependent increase following exposures of up to 20 Gy.

The authors end their chapter with an interesting discussion of apoptotic death in endothelial cells induced by radiation, and the role of the plasma membrane in such a process. They quote extensively the work of Haimovitz-Friedman and Fuks, which is described below.[9]

Irradiation of confluent EC cultures, or of blood vessels *in vivo*, at high enough doses will cause loss of EC, leaving gaps of variable size in the endothelial layer. **Rosen and Goldberg**[15] (Chapter 5) review the mechanisms that result in the reconstitution of an endothelial surface: hypertrophy of remaining EC, spreading and/or migration of neighboring EC, and proliferation of surviving cells, or of cells from outside of the irradiated field.[15]

They describe their experiments documenting some of these mechanisms, particularly hypertrophy. Bovine aortic endothelial cells (BAEC) in culture respond to the radiation-induced loss (SD of 400, 1200, 3000 cGy) by increase in volume of the remaining cells. This hypertrophy is characterized by significant increase in cell protein, mostly resulting from increased synthesis and to a lesser degree by decreased protein turnover. They demonstrate well these findings, in a predominantly diploid population of EC. As noted above, independent morphologic (light and ultrastructural) observations support the EC hypertrophic response.[10]

Radiation-induced cell loss results in discontinuity of the monolayer and stimulates cell hypertrophy. The discontinuity may be aggravated by cell retraction. In pulmonary microvascular endothelium exposed to doses of less than 100 cGy, cell retraction is associated with reorganization of F-actin microfilament stress fibers (see also Onoda et al.,[14] below). The authors speculate about the possible role of a metabolite of arachidonic acid on this cytoskeletal rearrangement.

Rosen and Goldberg also mention their unpublished studies indicating that radiation induces in EC the expression of *scatter factor*, a cytokine also called *hepatocyte growth factor*. Some of the published work quoted in their chapter (15) does indicate that scatter factor is angiogenic *in vivo*. They also quote a study showing that conditioned medium from irradiated cultures of bovine pulmonary artery EC stimulates a two- to threefold protein synthesis in irradiated fibroblasts, without an increase in DNA. Thus, hypertrophy of EC may result from the release of cytokine(s) expressed in response to radiation, in EC or other vascular cells. The denuded surface may provide an important signal for this response.

Haimovitz-Friedman and Fuks (Chapter 6) show evidence that endothelial cells appear to be more radioresistant *in vivo* than *in vitro*, by comparing the D_0 and Dq values. They point out that this effect may result from the protection provided by the matrix surrounding vessels. In addition, autologous extracellular matrix protects more than nonautologous matrix, revealing the specificity of this protection.

They comment on the fact that basic fibroblastic growth factor (bFGF) given shortly after radiation enhances significantly the survival fraction of EC at low doses (wide shoulder), while it does not show protection if added 72 hours after exposure. Their own work suggests that there are two waves of DNA breaks caused by radiation in BAEC. bFGF does not protect against the first (which appears to be a direct effect of radiation on DNA), but there is a second wave of DNA fragmentation that bFGF does prevent. This second wave ends up in apoptotic cell death of endothelial cells which occurs in interphase (radiation-induced apoptotic cell death also occurs in the thymus,[3] in some other lymphopoietic organs,[3] and in hemopoietic tissue). Therefore, the shoulder (low-dose region of the endothelial cell survival curve) may represent the dose range where apoptotic death may prevail over postmitotic mechanisms of death. However, these two mechanisms may vary with the dose and with the cell type.

The protective effect of bFGF in preventing the apoptotic death of EC caused by radiation appears to be mediated by the binding of bFGF to endothelial cell high-affinity transmembrane receptors, and by the activation of its tyrosine kinase domain. There are several pathways of activation of tyrosine kinase. One, studied by these authors, applies to BAEC: it consists of translocation of the alpha-isotype of protein kinase C (PKC) from cytosol to membrane, resulting in activation. This activation appears to be required and sufficient to protect against radiation-induced apoptosis. Although radiation alone does not activate PKC in BAEC, in certain other cell types PKC is activated by the radiation exposure itself. Such cells include a line of human adenocarcinoma of lung and some B-lymphocyte precursors. It is also interesting that PKC activation down-regulates radiation-induced apoptosis in lymphoid cells.

The signal that leads to radiation-induced apoptosis in endothelial cells is extranuclear (independent of nuclear DNA) and mediated by the hydrolysis of sphingomyelin, in the outer layer of the plasma membrane. This hydrolysis is carried out by sphingomyelinases, resulting in the formation of ceramide, which through serine/threonine kinases and other mediators, leads to apoptosis. PKC activation inhibits the sphingomyelin-ceramide cascade. Interestingly enough, tumor necrosis factor alpha (TNF) appears to produce a similar sequence of events leading to the apoptotic death of endothelial cells. bFGF not only protects EC from radiation but also protects EC from TNF-induced apoptosis. These processes are illustrated in Figure 1.1.

Haimovitz and Fuks have carried their experiments further to an *in vivo* system. Aware that endothelial cells of microvessels appear morphologically to be more radiation sensitive than those of larger vessels,[6] they studied apoptosis produced by radiation exposures of 15 to 25 Gy in the lungs of mice. In those experiments, apoptosis occurred mainly in the endothelial cells of capillaries and not in the endothelial cells of large vessels or in the epithelial cells of bronchi or in the pneumocytes. bFGF injected intravenously protected the capillaries from apoptosis and increased the survival of the mice.

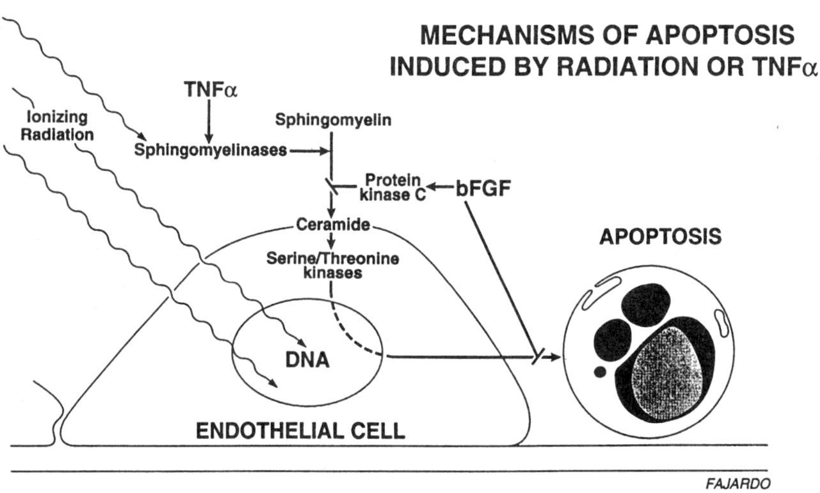

Figure 1.1. Chapter 6, by Haimovitz-Friedman and Fuks, describes various aspects of radiation-induced endothelial cell lethality. This diagram illustrates a pathway (initially independent of DNA) that leads to apoptosis via the hydrolysis of sphingomyelin in the cell membrane, with the formation of ceramide. A similar mechanism may be initiated by tumor necrosis factor-α (TNF-α), also resulting in apoptosis. Both, radiation-induced and TNF-induced apoptosis can be blocked by basic fibroblast growth factor (bFGF).

Hence, interphase apoptosis may be a prevalent mechanism of cell death at low and intermediate doses such as those used in clinical radiotherapy. Primary DNA damage with double strand breaks may be predominant in postmitotic death caused by higher doses. Probably both mechanisms play a role at any lethal dose.

Murray (Chapter 8) describes briefly some endothelial cell functions: selective barrier, vascular homeostasis, immune "accessory" function, and angiogenesis.[13] In considering the control of endothelial function, several aspects are discussed. One refers to cytokine effects, particularly the activation of EC by TNFα and Interleukin-1β (IL-1β). These cytokines modulate the expression of several effector molecules, including tissue factor (thromboplastin) and plasminogen activator inhibitor, as well as the expression of adhesion molecules such as ICAM-1, VCAM-1, and ELAM-1 (E-selectin). The up-regulation of the MHC antigens (I & II) by interferon is also mentioned.

Another aspect of endothelial activation refers to the induction of the protooncogenes c-fos and c-jun. The protein products of these protooncogenes interact with the AP-1 DNA regulatory site. The genes for E-selectin and VCAM-1 contain binding sites for the transcription factor NF-kB. Aside from TNFα, IL-1, endotoxin, and UV-light, NF-kB can be activated by reactive oxygen species (ROS).

Murray mentions some general effects of ionizing radiation on cells, emphasizing the induction of gene expression (e.g., expression of mRNA for TNFα). Probably, the activation of PKC leads to the transcription of early-response genes such as c-jun and c-fos. In some cell lines, radiation results in NF-kB activation.

In presenting specific effects of radiation of endothelial cells, Murray mentions morphologic changes resulting from >2 Gy on bovine or human EC. Also discussed are other effects. These include: decreased expression of β-actin gene and up-regulation of γ-actin and α-tubulin; release of von Willebrand Factor; changes in the synthesis and release of prostanoids; synthesis and release of growth factors (PDGF-like and FGF-like, after SD of 20 and 60 Gy); enhanced chemoattraction for leukocytes (in normal vessels); and up-regulation of the E-selectin (ELAM-1) gene. In summarizing the chapter, the author states, "... many biochemical pathways are in place which dictate that the major response of the endothelial cell to radiation is to adopt a pro-inflammatory, procoagulant phenotype...".

Rubin, Drab-Weiss, and Blazek (Chapter 4) have written an informative review of oxidant stress and the antioxidant defenses in the irradiated endothelium.[16] Their review is based mainly on *in vitro* studies and mentions important reactive oxygen species produced by low LET radiation, especially superoxide (O_2^-), and the hydroxyl radical (OH·).

The most important part of their paper is the consideration of the antioxidant defenses (Figure 1.2). The scavenger enzyme superoxide dismutase (SOD), which normally occurs in the cytoplasm and mitochondria, protects cells by converting O_2^- to H_2O_2 (which is further scavenged by catalase, forming water and molecular oxygen). In the case of endothelial cells, the radioprotection by SOD is less clear. BAEC "primed" by 5 Gy show increased activity of the MnSOD (one of the two forms of SOD) and have increased survival. However, BAEC irradiated in the presence of CuZnSOD (the other form of SOD) do not show increased survival.

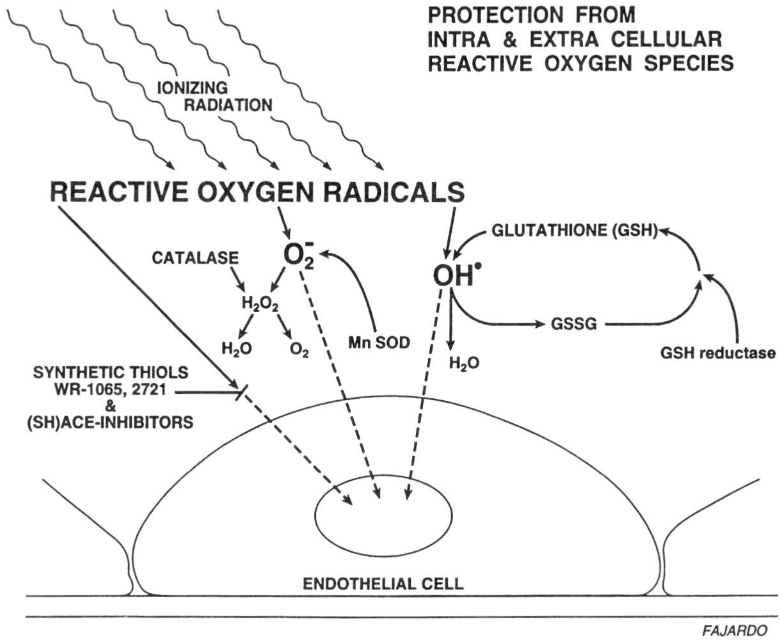

Figure 1.2. Diagram showing mechanisms of endothelial cell protection from reactive oxygen species generated by ionizing radiation. These three forms of chemoprotection are described by Rubin, Drab-Weiss, and Blazek in Chapter 4. Both the formation of free oxygen radicals and the chemoprotection, take place in and outside the cells. The pharmacologic protectors include sulfated angiotensin converting enzyme inhibitors and the synthetic aminothiols. (GSH = glutathione; GSSG = oxidized form of glutathione; Mn SOD = manganese form of superoxide dismutase [the other form, CuZnSOD, is not illustrated here]; O_2^- = superoxide; OH• = hydroxyl radical; WR-1065 and WR-2721 = synthetic aminothiols.)

Glutathione (GSH), a non-protein thiol, converts the hydroxyl radical to H_2O and a disulfide. Also, it is involved with cell proliferation: the cell content of GSH varies along the cell cycle, being lowest in G_1 and highest in S phase. This variation may explain, in part, the differences in radiation response along the cell cycle. The restoration of GSH after the conversion of hydroxyl is mediated by glutathione reductase, an enzyme that is inhibited by carmustine (BCNU, used in anti-cancer chemotherapy). EC appear to be among the first cell types damaged *in vivo* by BCNU.

Synthetic aminothiols such as WR-1065 and WR-2721 protect from radiation injury by scavenging free radicals and facilitating the repair of macromolecules. WR-1065, in particular, radioprotects BAEC by decreasing DNA injury, and promoting DNA synthesis and the progression from G_1 to S. Also, WR-1065 prevents radiation-induced morphologic alterations in BAEC. Of note, Ward and co-authors[17] also discuss how angiotensin converting enzyme (ACE) inhibitors, such as captopril, containing sulfhydryl, decrease the vascular damage produced by radiation.

Rubin et al. have observed that serum (20% in culture medium) appears to render BAEC more resistant to radiation. The authors speculate that circulating antioxidants such as vitamins C and E may be responsible for this radioprotection and compare their effects with those of the synthetic lazaroids. However, no proof for this mechanism is available. Finally, the authors mention preliminary evidence that a nitric oxide (NO) synthase inhibitor promotes DNA synthesis in irradiated endothelial cells.

A partial but excellent list of the effects of oxidant stress on endothelial cell functions is presented in their Table 4.3.

The chapter by **Graham and Peterson** (Chapter 3) deals mostly with *in vivo* studies.[8] These authors describe in detail the procedures to measure two functional parameters: vascular permeability, using various techniques, and pulmonary blood flow, using mainly ^{99}Technetium macroalbumin.

They show that in untreated rats exposed to 18 Gy, there are two peaks of increased vascular permeability in the lung, one immediate (lasting 3 days) and one delayed (at 18 days). By using a technique of left-to-right ratio of flow, they demonstrate a decrease in blood flow that begins at 3 weeks and is maintained especially after doses of 21 Gy (see also Kwock, above).[11]

The histologic studies of these authors confirm the presence of numerous mast cells in the irradiated lung, and suggest important implications related to the possible release of mast cell products. They assume that the source of most of the vasoactive agents found in irradiated lung are the mast cells (which contain more than 20 such agents). They particularly emphasize histamine.

The final portion of Chapter 3 describes experiments with 13 different drugs attempting to modify the vascular permeability and the blood flow in the irradiated lung. The authors found that the drugs most effective in correcting (decreasing) the vascular permeability were: dexamethasone, diethylcarbamazine, indomethacin, and theophylline. The most effective drugs in improving the blood flow were captopril, dexamethasone, and pentoxifylline.

Ward, Molteni, and Ts'ao (Chapter 10) review pharmacologic agents used to minimize radiation injury in the endothelium of normal tissues.[17] Most of the chapter refers to radiation-induced physiologic alterations in endothelial cells. In seven sets of tables, they compare various pharmacologic modifiers indicating the physiologic defect for which the modification has been attempted, the modifiers tested by various investigators, and the organ and species used.

The radiation-induced defects discussed are: *vascular lesions observed by histopathology* (light and electron microscopy); *endothelial cytotoxicity* (*in vitro* studies using mainly bovine aortic EC); *perfusion and hemodynamic defects; increase in vascular wall permeability* (*in vivo*); *coagulopathy* (mostly *in vivo*); *ineffective fibrinolysis* (mostly *in vivo*); *abnormal arachidonate metabolism* (alterations in eicosanoid products, mostly *in vivo*); and *abnormal endothelial enzymatic activity* (*in vivo* and *in vitro*).

The information on radiation injury in blood vessels is a very useful "dissection," mainly of sublethal physiologic damage. It should become a valuable source for radiobiologists. The data on pharmacologic modification is obviously based on experimental (mostly *in vivo*) information, many of the experiments having been performed by the authors themselves.

Although this may not have been the emphasis of the authors, one is tempted to apply these data to the modification of radiation injury in humans. How is this extensive array of information to be used clinically, i.e., to protect patients from radiation injury in their normal blood vessels? Obviously, it is impractical to prescribe one compound for each of the major potential EC lesions, and there is no single compound that would protect against all alterations of EC. Then perhaps one could use a substance that is effective in preventing most of the lesions, or the most important lesions, or the most common lesions.

The most "consistently effective" modifiers are steroids; dexamethasone is listed in the majority of the tables. It is difficult to decide which are the most important radiation-induced alterations in EC; not only do they vary from one site to another, but from one patient to the next. As for the most common (and the earliest) lesions, Ward et al. tell us that it is hyperpermeability. Of the compounds listed, a good many have not been used in humans and only a few are regularly used in the clinic, notably steroids and ACE inhibitors. Nevertheless, this chapter could be a springboard from which we may move into the practical prevention of vascular radiation injury.

The prevention of injury in the vessels of the irradiated normal tissue may raise the possibility of protecting also the vessels of the neoplasms. Fortunately, the microvessels of neoplasms are inherently abnormal in morphology (tortuosity, frequent gaps between EC, blind ends, frequent A-V anastomoses) and function (increased permeability, slow blood flow), and have poor physiologic reserve, often being unable to compensate for thrombosis, collapse, etc.[1,2] Also, tumor vessels have a much higher proportion of proliferating EC,[2] which renders them more sensitive to various forms of therapy.[7] These characteristics of neoplastic microvessels provide an important therapeutic advantage.

Chapter 9, by **Onoda, Kantak, and Diglio,** deals with the modulatory roles of the extracellular matrix (collagen I and collagen IV in particular) on the effect of low-dose radiation in mouse microvascular pulmonary endothelial cells grown in monolayers.[14] The authors describe experiments using single X-ray doses of 0.125 to 1.0 Gy. Grading systems were developed for cell detachment (phase contrast microscopy) and for depolymerization of F-actin filaments (immunofluorescence). These low doses produced reversible detachment of endothelial cells with loss of cell-to-cell contact. This retraction was mediated by concomitant, reversible reorganization of F-actin microfilaments in the EC. Such changes occurred at lower doses and persisted longer in the cells grown on a substrate containing 98% collagen type I and 2% collagen type III (Vitrogen), than in cells grown on collagen type IV.

The radiation exposure stimulated the expression, on the EC apical surface, of the $\alpha_v\beta_3$ integrin adhesion receptor (i.e., vitronectin receptor). This enhanced expression of the $\alpha_v\beta_3$ receptor increased 15 minutes postradiation and also was reversible. However, unlike the EC retraction and F-actin reorganization, the expression of the $\alpha_v\beta_3$ receptor was the same for cells exposed either to collagen I or to collagen IV. The authors emphasize the importance of the cell-to-cell contact and its modulation by extracellular matrix for the appropriate function of the endothelial vascular layer. This chapter is particularly interesting because it deals with microvascular EC and with sublethal doses of radiation as well as with reversible phenomena.

Milas and Milas discuss (in Chapter 11) the tumor bed effect (TBE).[12] The TBE is a well-documented experimental phenomenon and refers to the effect *in vivo* of previous high doses of radiation on the growth of tumor cells subsequently transplanted in the same site. The authors indicate that, as the result of the previous radiation exposure, the following occur: retardation of tumor growth; reduced neovascularization and tumor blood flow; increased tumor hypoxia and tumor necrosis; increased metastatic spread; and reduced radiocurability and chemocurability of the tumor. Several graphs are shown as examples of these effects. There are no clear explanations of these phenomena, particularly the enhancement of metastases. Perhaps they result from effects of cytokines, modifications of the extracellular matrix, and/or expression of adhesion molecules (D. Rubin, personal communication) (also see Chapters 8 and 9 by Murray[13] and Onoda et al.,[14] respectively). The information about the effect on angiogenesis is elegantly presented.

The general concept of the TBE is interesting as an experimental fact and appears to be well established. Although the authors do not dwell on this point, it would be interesting to find human counterparts to the tumor bed effect. A somewhat analogous situation could be that of prophylactic radiotherapy to some tissues that subsequently may be seeded by neoplasm; for instance, the brain prophylactically irradiated in children who have leukemia.

West and Moore (Chapter 12) discuss photodynamic therapy (PDT) in relation to endothelial cells.[18] (The general goals and the methodology of this exciting treatment modality are not included in this chapter.) From the literature, the authors have compared the sensitivity of various cell lines, human and not human, to PDT. In all cases, the endothelium (mostly BAEC, sometimes HUVEC) was the most sensitive cell type even when compared to tumor cells. This increased sensitivity may be associated with high numbers of low-density lipoprotein (LDL) receptors on the endothelial cell surface; the photosensitizers used in PDT circulate in association with macromolecules such as LDL and albumin. It appears that the phototoxic vascular injury involves activation of the complement system, degranulation of mast cells, and migration of leukocytes to the site of injury.

Within a few hours of exposure to PDT, there is a marked increase in the permeability of vessels and conformational changes, particularly segmental vasoconstrictions in arterioles. Thrombosis occurs as the result of EC detachment, with exposure of the thrombogenic subendothelial layers, followed by local platelet aggregation. Rapid, PDT-induced, release of von Willebrand Factor (vWF) contributes to the thrombosis. Vascular occlusion may appear as early as 45 seconds after PDT and persists for hours. There is decreased nitric oxide (NO) production and/or release, and this impairs vascular relaxation and promotes platelet aggregation.

Eicosanoids are released during PDT. However, the type of prostaglandin detected has varied from one experiment to another (e.g., one released $PGF_{2\alpha}$, a vasoconstrictor; another released $PGF_{1\alpha}$, a vasodilator) perhaps because of the different types of EC used. It is known that EC from different vessels produce different amounts of the various prostaglandins, such as PGI_2, PGE_2, and $PGF_{2\alpha}$.[4] In any case, the cyclooxygenase inhibitor indomethacin blocks some of the effects of PDT. Aspirin, given after exposure, can abolish PDT-induced vascular occlusion.

It is difficult to make a summary statement bridging the multiple and varied conclusions expressed by the authors of the chapters that follow. The unifying concepts are those of endothelial cells and radiation. Because of the gaps in knowledge about the radiation impact, many of the discussions refer to endothelial cell physiology in general and independently of radiation. Most chapters are concerned with *in vitro* studies, but at least four describe important *in vivo* experiments. There are relatively few data on human studies. These characteristics reflect the state of endothelial cell research today. The individual and basic metabolic functions of EC are best studied in cultured EC. The *in vivo* experiments are quite pertinent but more difficult to interpret because of the complexity of the multiple mechanisms involved. A graphic summary is presented in Figure 1.3.

The information accumulated in this book is of great importance in advancing our understanding of the complex endothelial cell functions in general and in particular on the effects of radiation on such cells. This book will become a very useful guide and source of information for researchers in and out of the discipline of radiation biology who are working in the fascinating field of endothelial cells.

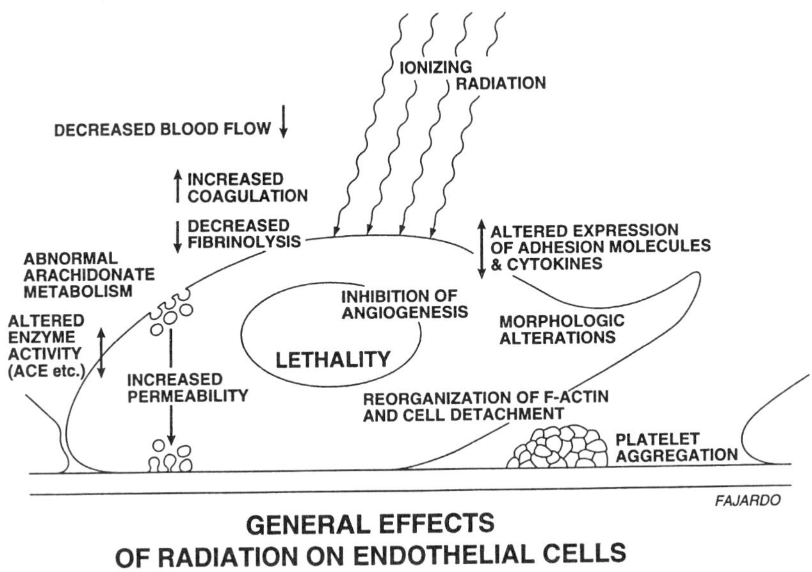

Figure 1.3. This diagram is a summary of the most important, lethal and sublethal, effects of ionizing radiation on endothelial cells. It combines *in vivo* and *in vitro* data and is based on information from several chapters in this book as well as from other publications (see references).

REFERENCES

1. Denekamp, J., Inadequate vasculature in solid tumors: consequences for cancer research. *Br. J. Radiol.*, 24 (Suppl.), 111, 1992.
2. Denekamp, J. and Hobson, B., Endothelial cell proliferation in experimental tumors. *Br. J. Cancer*, 46, 711, 1982.
3. Fajardo, L. F., *Pathology of Radiation Injury.* Masson Publishing USA, Inc., New York, 1982.
4. Fajardo, L. F., The complexity of endothelial cells. *Am. J. Clin. Pathol.*, 92, 241, 1989.
5. Fajardo, L. F., The unique physiology of endothelial cells and its implications in radiobiology. *Front. Radiat. Ther. Oncol.*, 23, 96, 1989.
6. Fajardo, L. F. and Stewart, J. R., Pathogenesis of radiation-induced myocardial fibrosis. *Lab. Invest.*, 29, 244, 1973.
7. Fajardo, L. F. and Prionas, S. D., Endothelial cells and hyperthermia. *Int. J. Hyperthermia*, 10, 347, 1994.
8. Graham, M. M. and Peterson, L. M., Functional measures of endothelial integrity and pharmacologic modification of radiation injury. Chapter 3, in *The Radiation Biology of the Vascular Endothelium*, D. B. Rubin, Ed., CRC Press LLC, Boca Raton, FL, 1998.
9. Haimovitz-Friedman, A. and Fuks, Z., Signaling in the radiation response of endothelial cells. Chapter 6, in *The Radiation Biology of the Vascular Endothelium*, D. B. Rubin, Ed., CRC Press LLC, Boca Raton, FL, 1998.
10. Johnson, L. K., Longenecker, J. P., and Fajardo, L. F., Differential radiation response of myocytes. Symposium on radiation injury (by invitation). *Anal. Quant. Cytol.*, 4, 188, 1982.
11. Kwock, L., Blackstock, A. W., and Friedman, M., Effect of ionizing radiation on endothelial cell plasma membrane processes. Chapter 7, in *The Radiation Biology of the Vascular Endothelium*, D. B. Rubin, Ed., CRC Press LLC, Boca Raton, FL, 1998.
12. Milas, L. and Milas, M., Radiation biology and the endothelial cell. Tumor bed phenomenon: effect on tumor progression and therapy. Chapter 11, in *The Radiation Biology of the Vascular Endothelium*, D. B. Rubin, Ed., CRC Press LLC, Boca Raton, FL, 1998.
13. Murray, J. C., Inflammatory cytokines, radiation and endothelial gene products: a common role for reactive oxygen intermediates. Chapter 8, in *The Radiation Biology of the Vascular Endothelium*, D. B. Rubin, Ed., CRC Press LLC, Boca Raton, FL, 1998.
14. Onoda, J. M., Kantak, S. S., and Diglio, C. A., Extracellular matrix regulates endothelial cell morphology, receptor expression, and response to low dose radiation. Chapter 9, in *The Radiation Biology of the Vascular Endothelium*, D. B. Rubin, Ed., CRC Press LLC, Boca Raton, FL, 1998.
15. Rosen, E. M. and Goldberg, I. D., Protein synthesis in irradiated endothelial cells. Chapter 5, in *The Radiation Biology of the Vascular Endothelium*, D. B. Rubin, Ed., CRC Press LLC, Boca Raton, FL, 1998.

16. Rubin, D. B., Drab, E. A., and Blazek, E. R., Antioxidant defenses and the irradiated endothelium. Chapter 4, in *The Radiation Biology of the Vascular Endothelium*, D. B. Rubin, Ed., CRC Press LLC, Boca Raton, FL, 1998.
17. Ward, W. F., Molteni, A., and Ts'ao, C.-h. Endothelial-oriented strategies to spare normal tissues. Chapter 10, in *The Radiation Biology of the Vascular Endothelium*, D. B. Rubin, Ed., CRC Press LLC, Boca Raton, FL, 1998.
18. West, D. C. and Moore, J. V., Photodynamic therapy and endothelial cells. Chapter 12, in *The Radiation Biology of the Vascular Endothelium*, D. B. Rubin, Ed., CRC Press LLC, Boca Raton, FL, 1998.

CHAPTER 2

The Histopathology of Irradiated Endothelium

David B. Rubin and Melvin L. Griem

CONTENTS

I. INTRODUCTION

Radiation toxicity in many tissues is expressed earliest by perturbations of the vascular endothelium.[13] Moreover, each stage of tissue toxicity can be related to a disturbance of vascular endothelial function and replicated *in vitro*.[4] This link can be made because the vascular endothelium regulates the interactions between the circulation and surrounding tissues.[5]

1-8493-4840-4/98/$0.00+$.50

Within 3 years after the discovery of X-rays, Gassmann studied the histology of irradiated skin and reported[6] on the primary disturbance of the endothelium. Over the next 50 years, several reports presented evidence that the response of the vascular lining to radiation exposure was pivotal to tissue damage. These investigations have examined the impact of both photons and charge particles[7] on vascular damage. Table 2.1 summarizes some of these histological observations made in the first 50 years of radiation's clinical application. Aspects of these reports are reviewed in key articles by Roscher et al.,[8] Reinhold,[9] Maisin,[1] Fonkalsrud et al.,[10] and Griem et al.[7]

Reinhold[9] criticizes many of the histological studies cited in this and other reviews because of "...the obvious differences in reaction among the various species...(and that) the doses employed in experiments are often higher than those currently in use for radiotherapeutic practice." He adds that many of these investigations are qualitative and not quantitative. In part, because of this admonition delivered in the early 1970s, investigators subsequently responded by quantifying the damage of irradiated vessels (see Table 2.3). Furthermore, these studies on the radiation response of the endothelium have contributed to much of what is known about clinical aspects of tissue injury and healing. For example, histological studies have shown that irradiation aggravates vascular damage caused by hypertension[11,12] and atherosclerosis.[13,14]

Prior review articles on the role of endothelium in radiation toxicity[7,13,15] emphasize the following general points. First, vascular endothelial damage precedes late

Table 2.1. Descriptions of Endothelial Damage Caused By Irradiation: The First Half-Century of Radiation

Year	Investigators (Reference)	Observation
1899	A. Gassmann (6)	Dermatitis in man: endothelial proliferation, swelling, bulging into vascular lumen
1909	S. Wolbach	Endothelial swelling occludes lumen of capillaries and leads to ulcers
1930	T. Takahashi (53)	Rat abdominal surgical wound: young capillaries and fibroblasts very radiosensitive; radiation at high and low doses inhibits growth of new vessels
1935	W. Scholz	Human brain: arteriolar sclerosis around areas of radionecrosis
1937	F. Windholz (54)	Human skin: intimal proliferation, partial obliteration of arteries, and complete obliteration of veins
1942	S. Warren (55)	Human skin: endothelial proliferation occludes vessels and sometimes leads to necrosis and thrombosis in deep vessels
1944	J.F. Sheehan (56)	Human pelvic arterioles: focal thickening by foam cells, fibrin deposition and subendothelial hyalin deposition; associated with perivascular fibrosis, necrosis, and lymphocytic infiltrate
1948	Pennybacker and Russell (57)	Human brain: necrosis due to microvascular damage associated with fibrosis and thrombosis
1955	Merwin and Hill (58)	Injured mouse skin: radiation inhibits new capillary growth

stages of tissue injury: atrophy, fibrosis, and necrosis. Second, functional changes of irradiated endothelium are evident well before cell destruction. Third, early- and late-responding tissues express endothelial damage differently. In early-responding tissues, such as the skin and intestine, where acute epithelial damage is prominent, there appear two phases of vascular damage: an early phase associated with a transient inflammation, and a later phase associated with endarteritis and telangiectasia. In late-responding tissues, such as the lung, the vascular damage is more gradual without a prominent early inflammation. Fourth, the radiation response of vessels depends on the caliber of the vessels, which can vary according to species or organ studied. In general, the smaller vessels that are comprised mostly of endothelium, appear more radiosensitive than larger vessels. Fifth, much of the endothelial damage appears multifocal, in which damaged vessels are juxtaposed to normal-appearing vessels. Table 2.2 summarizes the key histopathological features of radiation-induced endothelial damage. Table 2.3 outlines details from key investigations that provide evidence for the "endothelial-based" hypothesis on radiation-induced tissue toxicity. Although the histopathology is not pathognomonic of radiation toxicity,[16,17] these endothelial alterations reflect reliably the pathogenesis of tissue damage in the setting of radiation exposure.

II. EARLY AND INTERMEDIATE CHANGES

A. Structure

Within hours to weeks following irradiation, the first visible changes noted are in the capillary endothelium. Endothelial damage is a heralding event in the radiation toxicity of the liver,[18] lung,[19,20] kidney,[15] brain,[12] and heart.[21] The histological perturbations are associated with the early disturbance in vascular function such as vasodilation (Figure 2.1), altered blood flow (Figures 2.1 and 2.2), and perivascular edema (see Chapters 3 and 10 and Figures 2.1–2.4). Specifically, there appears a multifocal disruption of cell adhesion to the basement membrane (Figures 2.3 and 2.4), nuclear swelling and pyknosis, and organelle disruption (for references, see Table 2.3).

Table 2.2. Stages in the Pathogenesis of Vascular Endothelial Damage Induced by Irradiation

Early (Minutes-Weeks)	Intermediate Weeks to 6 Months	Late >6 Months
Vasodilation	Cytoplasmic swelling	Perivascular cellular degeneration and fibrosis
Sloughing of endothelial cells	Sloughing	Endothelial proliferation
Nuclear swelling and pyknosis	Thrombosis	Basement membrane thickening
Vacuole formation	Endothelial proliferation	Intraluminal collagen
Basement membrane thickening/edema	Capillary obliteration	Telangiectasia
Leukocyte margination		

Table 2.3. Descriptions of Endothelial Damage Induced by Irradiation

Reference	Species	Organ(s)	Dose of Radiation (Gy)*	Time of Observation	Summary of Key Findings	Commentary
1	Dog	Lung		1 week	Focal sloughing of endothelium and vascular rupture	Early abstract that antedates general awareness of radiation-induced endothelial damage
15 59 19	Human	 Heart Lung				Key early articles that promoted the clinical relevance of endothelial radiobiology
60	Rabbit	Brain	20–200 protons and α-particles	1–3 months	Basement membrane thickening adjacent to necrosis: increased multivesicular bodies, vacuoles Decreased endoplasmic reticulum and ribosomes	More pronounced radionecrosis in white matter, which is less vascularized than grey matter
30	Rabbit	Heart	2	24 hours 20–60 days 134 days	Transient granulocytic infiltrate increased endothelial cell proliferation Confined to capillaries, cell swelling, and pseudopod-like projections and thrombi occlude lumen Ratio of capillaries to myocytes reduced by 45%	Quantitative histological study of labeling indices of cells by autoradiography
9	Pig	Skin grafts	5–20	3 weeks	48 hours after grafting skin, decreased colony survival of endothelial cells determined in newly growing vessels by counting connecting capillary loops filled with dye	An elegant quantitative histological technique that establishes a D_o of approximately 10 Gy

61	Pig	Skin	16.5–26.2	21–70 days	Two waves of skin necrosis: 21–36 days due to loss of epithelial cells; 36–70 days due to destruction of capillaries	A representative study from investigators who have quantified the dermal endothelial damage
39	Mouse	Muscle, skin, femur	2–4	1–22 days	No increase in endothelial proliferation	Three studies that contradict findings of other studies
62	Rabbit	Heart	20	3 minutes–49 days (lasting 134 days) 6–48 hours >70 days	Myocardial capillaries: thinning of endothelial cytoplasm, cytoplasmic projections, increased lysosomes, lamellar bodies, prominence of endoplasmic reticulum; sloughing of endothelium, and nuclei swelling and obstruction of lumen; thrombi with fibrin, rupture of capillaries; pancarditis; Progressive pericardial fibrosis and effusion, and myocardial fibrosis	One of the earliest detailed descriptions of early ultrastructural changes
25	Rabbit	Ear chamber	10–20	During/ immediate	Leukocyte influx and margination	Degree of leukocyte adhesion is related to dose, dissipates 1 day after 10-Gy exposure
52	Rabbit	Ear chamber	2–20	1–5 days	Vessels <10 μm most radiosensitive, increased intercapillary distance, dilation of surviving vessels	Microangiography allows instantaneous assessment of blood flow and perivascular edema *in vivo*
31	Rat	Heart	20	2 months	Injury to capillaries first evident: cytoplasmic swelling, vacuolization, loop formation; decreased number of capillaries, perivascular edema, and collagen deposition	For all organs, 20 Gy elicits only morphological changes to the endothelial cell

Table 2.3. (Continued) Descriptions of Endothelial Damage Induced by Irradiation

Reference	Species	Organ(s)	Dose of Radiation (Gy)*	Time of Observation	Summary of Key Findings	Commentary
31 (continued)				8 months	Endothelial budding from occluded capillaries; capillaries irregularly dilated and contain more pericytes	No damage noted to muscle under 15 Gy; skeletal muscle endothelium responds similar to endothelium in the myocardium; occasional mast cells are most prominent leukocyte in muscle tissues
		Kidney		2 months	Tubule capillaries: endothelial cells displaced from basement membrane, perivascular edema	
					Threshold for lung endothelial changes is 15 Gy; after 30 Gy, epithelial cells are also disrupted and no repair of the original architecture appears possible	
		Lung		6 weeks–2 months	Capillary endothelial cells lift off the basement membrane, platelets aggregate, vessels occluded with endothelial nuclei and cell fragments; lymphocytic infiltrate	
				2–4 months 3–6 months	Mast cell infiltrate, capillaries begin to recanalize; recanalization widespread and complete	
				6–10 months	Thickened alveolar basement membrane	

40	Mouse	Mesentery	20–45	3 weeks	Increased endothelial cell labeling index	Postulates that endothelial cell depletion is explained by a very small growth fraction that contains a subpopulation of cells with a short cell cycle time
				6 weeks	Endothelial cell depletion	
63	Dog	Corneal neovascularization	4–18 β radiation	6–48 hours	Dose-related inhibition of neovascularization; SLDR evident below 3.5 Gy	Proliferating endothelial cells appear more radiosensitive than quiescent cells
1	Mouse	Lung	20	20 minutes 3 hr	Increased pinocytotic vesicles, focal areas of perivascular edema, and ultrastructural changes: dilated cisternae of endoplasmic reticulum, lysosomal bodies, condensed nuclei;	Early sustained increases in pinocytosis may account for cell swelling and basement membrane edema One of the major early reviews on the histology of the irradiated endothelium
				6 hr	Lipid droplets and cytoplasmic condensations, platelets accumulate	Defines timing of histological effects: hours = immediate, days = early, 1–7 months = intermediate, late = >8 months
				1-7 days	Endothelial loops, separation from basement membrane	
				1-2 months	Platelet aggregation, lipid inclusions, and collagen fibrils obstruct capillaries	
				3-4 months	Cytoplasmic vacuoles and osmiophilic precipitations, excess free ribosomes and disrupted endoplasmic reticulum	
				8-15 months	Thickened capillary basement membrane, lumen replaced by collagen, new capillary growth	

Table 2.3. (Continued) Descriptions of Endothelial Damage Induced by Irradiation

Reference	Species	Organ(s)	Dose of Radiation (Gy)*	Time of Observation	Summary of Key Findings	Commentary
64	Cattle	Pulmonary artery (culture)	30	4 hr	Stress fibers containing actin reorganize, gaps form between cells, increased flux of albumen, ruffled apical membrane; process reversible	Edema proposed to be secondary to early endothelial shape change
32	Sheep	Lung	15, bilat	1 week	Multifocal perivascular edema while fibroblasts and lymphocytes infiltrate, especially around veins and lymphatics	Pulmonary hypertension may evolve as a result of the early perivascular inflammation
				2 weeks	Multifocal damage to nonmuscular vessels: pyknotic nuclei, thickened basal lamina	
65	Rat	Lung	30, (5x6Gy)	2 weeks	Focal denudation of arteriolar endothelium and cells bulging into the lumen	Endothelial destruction could lead to vascular occlusion secondary to hyperplasia of either endothelial or smooth muscle cells; fractionation of radiation has a sparing effect
				4 weeks	Angiography shows beading of arterioles and loss of capillaries	
66	Rat	Carotid anastomoses	20	10 weeks	1–6 hr after anastomosis, increased platelet and fibrin deposition; 24 hours after anastomosis, greater leukocyte infiltration	Impaired surgical wound healing in irradiated tissues can be related to perturbed endothelial function expressed early after suturing vessels Effect can be remote to irradiation

67	Rat	Brain	Neutron (fluence of 4 × 10E12 n/cm^2)	3 days	Increases in endothelial fenestrations and hydration of the basal lamina Possible increase in pinocytotic activity	Changes exacerbated by boron compounds and more evident in endothelium surrounding implanted gliomas, which may imply greater sensitivity of dividing endothelial cells Boron itself has no impact
20	Rat	Lung	6.5 and 11 whole-body 30 to a hemithorax	2 days after 11 and 5 days after 6.5	Earliest injury evident in the lung alveolar endothelium: vacuolation, swelling, and intraluminal bulging blocking capillaries, followed by denudation Thrombocytopenia evident and related to microthrombosis	Fibrosis may result from early endothelial damage
				28 days-3 months	Perivascular fibrosis and recanalization of capillaries	
10	Dog	Femoral vessels	40 (4 × 10)	2 days	Scanning electron microscopy: linear convolutions flattened, intercellular bridges lost, apical surface cratered, increased platelet and erythrocyte adhesion Slight adventitial fibrosis	Large vessels demonstrate early changes to the endothelium without apparent disruption of blood flow The later destructive changes to large vessels may result from the damage to the endothelium of their vaso vasorum Vascular rupture may result from early changes to the endothelium
				1 week	Loss of nuclei from cell remnants	
				6 wks–3 months	Repopulation of endothelial cells with above irregularities, but little thrombosis	
68	Guinea pig	Skin	β (500–3000 rep)	1–18 hr	Increased perivascular edema and intravascular blood volume	Biphasic response may be related to activation of mast cells and local increase in vasoactive substances

Table 2.3. (Continued) Descriptions of Endothelial Damage Induced by Irradiation

Reference	Species	Organ(s)	Dose of Radiation (Gy)*	Time of Observation	Summary of Key Findings	Commentary
69	Rabbit	Skin	5–40	10–55 min	Increased permeability to dye	Triphasic response: hours, 2–3 days, 10–17 days Anti-protease EACA attenuates increased permeability
14	Rabbit (hypercholesterolemic)	Carotid artery	20	24 hr 3 days 1 week	Increased luminal alkaline phosphatase Increased luminal β-glucuronidase Increased foam cells, endothelial denudation, plaque formation	Radiation-induced damage to the endothelium promotes atherosclerosis
70	Chicken embryo	Whole body	11	hours	Vasodilation and leukocyte adhesion Death due to microcirculatory failure, which is prevented by a prior exposure to 7 Gy (Stearner 1970) or protease inhibition by soy bean Trypsin inhibitor	
3	Rat	Brain (choroid plexus)	17.5–25.0	1–8 weeks	Numbers of endothelial cell nuclei reduced	In brain, initial loss of cells not related to dose, but after 6 months the loss *was* related to dose
	Hamster	Cheek pouch	25	1 month 3–5 months 7+ months	Reduction in arterioles with diameter 7–16 μm Progressive reduction in capillaries, diameter <7 μm Loss of larger vessels, diameter >16 μm; only after 25 Gy	Smooth muscle cells appear more radioresistant
	Pig	Skin				In both rat brain and pig skin, foci of telangiectasia, necrosis, and a general atrophy were found

3 (continued)

						In pig skin, degeneration of smooth muscle cells in arterioles could result in reduced control of capillary pressure and induce telangiectasia
71	Rat	Aorta	2	4–15 weeks	Intact endothelium but no increase in the labeling index	Supports the notion of larger vessels being more radioresistant than smaller vessels Suggests that re-endothelialization in larger vessels occurs mainly by migration
26	Mouse	Mesenteric blood vessels Brain	17.5–30.0	3–24 weeks	The numbers of endothelial cells lining the walls of blood vessels declined rapidly in the first 3–9 weeks after irradiation with single doses of 17.5–30.0 Gy Increased proliferation of endothelial cells occludes blood vessels ACE decreases progressively and parallels the decrease in pulmonary arterial perfusion ACE decreases about 4% per Gy 3-6 months post-irradiation, the lumen of vessels may be occluded by cells (myointimal proliferation) Accumulation and perivascular infiltration of leukocytes was also seen in microcirculation	Up-regulation of ELAM-1 and ICAM-1 (endothelial leukocyte/intracellular adhesion molecule) for endothelial cells irradiated *in vitro* Increased ELAM-1 may explain the attachment of neutrophils, and ICAM-1 increases the attachment of lymphocytes

Table 2.3. (Continued) Descriptions of Endothelial Damage Induced by Irradiation

Reference	Species	Organ(s)	Dose of Radiation (Gy)*	Time of Observation	Summary of Key Findings	Commentary
72	Mouse	Mesentery vessels	20 30 45	6 weeks–18 months	Vein diameter, vascular volume, and permeability and the cellularity of the arterioles were altered within 3 months, gradually recovered by 12 months, and a second phase of damage appeared at 18 months Fibrosis was not seen before 12 months; but at 18 months, the vessel walls were fibrosed, sclerotic, and occasionally the lumen was occluded	The time course of changes in vascular permeability appeared to match the loss of endothelial cells lining the arterioles

[a] Gamma or X-rays unless otherwise indicated.

B. Radiosensitivity of Endothelium, Inflammation, and Blood Flow

Injured tissues express endothelial injury early, partly because endothelial cells regulate the physiology of surrounding cells. Also, endothelial cells may be inherently more radiosensitive than other vascular cells, regardless of how they interact with other cells or how they respond to the forces of circulation. Both *in vitro*[22] and *in vivo* (see Table 2.3), the vascular endothelial cells are more radiosensitive than vascular smooth muscle cells and fibroblasts.

Furthermore, radiation could activate leukocytes that exacerbate endothelial damage.[23] Leukocytes marginate to the vascular endothelium within minutes to hours of irradiation exposure.[24,25] As described by Hopewell and co-workers[26] and Onoda and co-authors in Chapter 9, the early expression of endothelial adhesion molecules within minutes[27] to hours following irradiation can explain this margination. Endothelial cells also express chemotactic activity when irradiated *in vitro*.[28,29] The type of leukocyte and the timing of the inflammation might be organ specific. For example, the irradiated heart demonstrates an immediate onset but transient infiltration of granulocytes,[30] whereas the irradiated lung shows a more gradual onset of inflammation that is predominantly lymphocytic.[31,32]

In addition, because rheological properties of the blood regulate endothelial function,[33] when radiation exposure alters blood flow (Figures 2.1 and 2.2), endothelial activation and expression of injury may ensue.[34,35] The early increase in blood flow following irradiation can be related to vasodilation promoted by activated endothelial cells. For example, the irradiated endothelial cell increases its vasodilatory activity by both producing more prostacyclin[29,36] and decreasing membrane-bound angiotensin converting enzyme activity (ACE)[22,37] (see Chapter 10 by Ward et al.). Lowered ACE could result in increased local concentrations of bradykinin, a vasodilator, and decreased concentrations of angiotensin II, a vasoconstrictor.

C. Capillary Occlusion

Irradiated capillaries also occlude at early stages of radiation toxicity. This obstruction to circulation can be ascribed to endothelial cell swelling, both nuclear and cytoplasmic, sloughing of the endothelial cells (Figures 2.3 and 2.4) as well as sludging of erythrocytes (Figure 2.2), leukocytes, platelets (Figure 2.5), and cellular debris. The formation of microthrombi and fibrin deposition likely result from the exposure of endothelial basement membrane[36] and the procoagulant biochemical function that is activated in the injured endothelial cell.[29,38]

D. Edema

Perivascular edema develops as a result of the lost semipermeable endothelial barrier and is discussed by Graham and Peterson in Chapter 3. The histological basis for the edema can be related to the early multifocal loss of endothelial cells (cited in a majority of references in Table 2.3), the increased pinocytosis,[1] the separation of endothelial cells from the basement membrane (Figures 2.3–2.5 and Table 2.3),

A)

B)

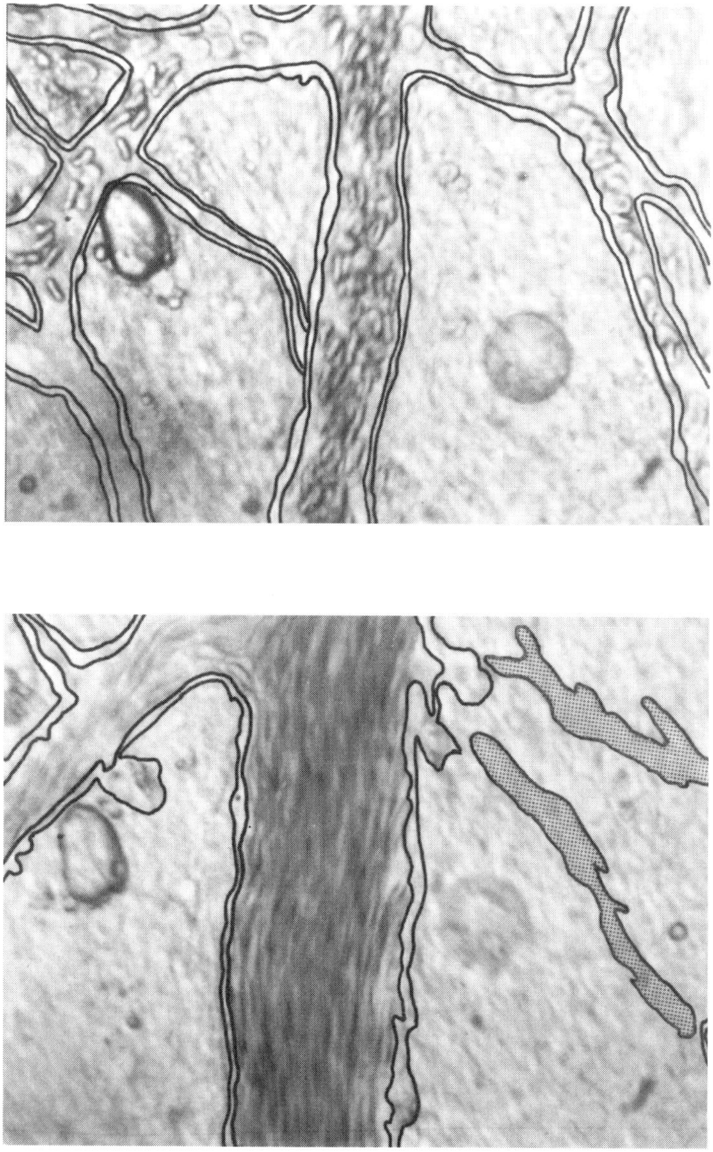

Figure 2.1. A rabbit ear chamber preparation: (A) control and (B) 8 days after 140 cGy (96×). Outlines of the vessel lumen and outside vessel wall are drawn to help demonstrate the increase in vascular diameter and the perivascular thickening (edema) observed following irradiation. Noted also are the loss of smaller capillaries and the increased blood flow. (Photographed by M.L. Griem.)

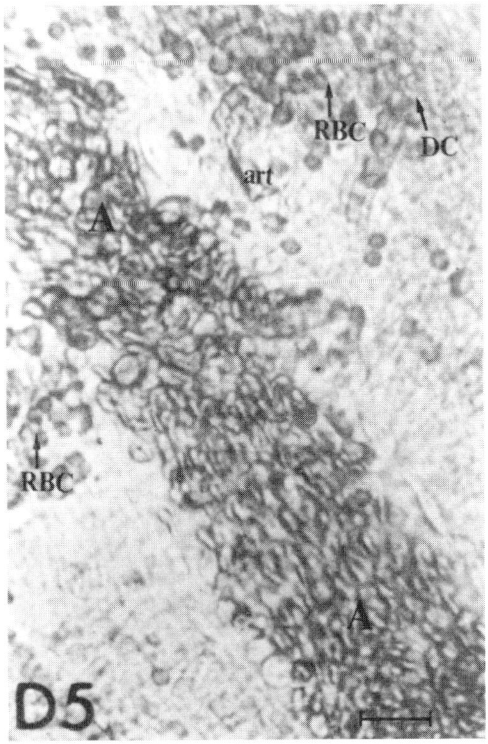

Figure 2.2. A rabbit ear chamber preparation depicted 5 days after 200 cGy (96×) demonstrates sludging of erythrocytes. Bar is 20 μm; A is arteriole lumen; DC is disrupted capillary; RBC refers to extra-vascular erythrocytes; art refers to an artifact. (From Dmitrievich, G.S. et al., *Radiat. Res.*, 99, 511–535, 1984. With permission.)

and cytoskeletal reorganization[30,39] that can contract individual cells, thereby increasing intercellular gaps.

E. Endothelial Cell Death and Proliferation

As suggested by Hirst and co-investigators,[40] it is more difficult to quantify endothelial cell turnover than it is to quantify turnover of other cell types because the circulation can wash away dying endothelial cells. Increased endothelial cell proliferation (see Figure 2.6) has been noted following irradiation and occurs before histopathology is observed in epithelial cells, fibroblasts,[20,41] and smooth muscle cells.[21,30,42] The likely explanation for increased endothelial cell proliferation is the depletion of endothelial cells.

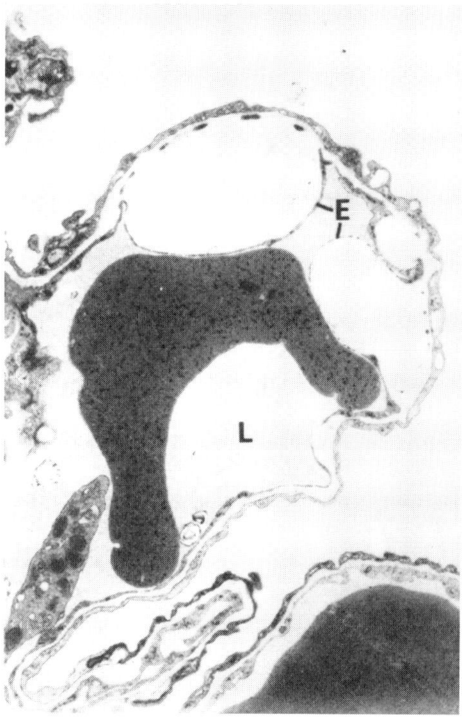

Figure 2.3. Electron micrograph of a mouse lung following 650 cGy of whole-body irradiation (3500×). After 2 weeks, endothelial cell injury and pericapillary edema are noted. The endothelial cells are designated (E) and the capillary lumen is labeled (L). Readily evident too is the filling of the vascular lumen by the detachment of endothelial cells from their basement membrane. (Photographed and kindly provided by I.Y.R. Adamson, University of Manitoba.)

Normally, endothelial cells in most vessels proliferate very slowly, with mean turnover time measured in years.[5,40] However, there appears to be a range of values for endothelial cell turnover time in uninjured vessels. For example, there is increased endothelial cell turnover in areas of high and low shear stress forces[33] that happen to be at sites prone to atherosclerosis. In their report on endothelial proliferation,[40] Hirst and colleagues reviewed prior investigations. As expected, they found that all studies showed prolonged times of turnover for the general endothelial cell population and that growth fractions were small. However, they also noted that 9 of the 17 prior studies reported that endothelial turnover in normal vessels can be as short as 2 to 10 weeks. From their results on irradiated mesenteric endothelium accompanying this review, Hirst and co-workers concluded "... that the (endothelial) cell depletion was much earlier than expected for a slowly proliferating tissue... Such early depletion is... consistent with cell death resulting from a small proportion of the cells having a short cell cycle." Hopewell and colleagues made similar conclusions from their results with the irradiated choroid plexus of rat brains.[3] They too

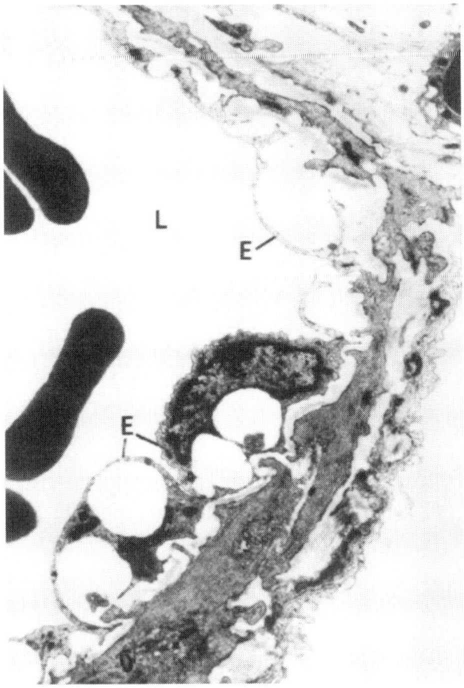

Figure 2.4. Electron micrograph of a mouse lung following 650 cGy of whole body irradiation (3500×). After 4 weeks, endothelial cell vacuolation and subendothelial edema in a venule are noted. The endothelial cells are designated (E) and the vascular lumen is labeled (L). (Original photograph kindly provided by I.Y.R. Adamson, University of Manitoba. Reprinted from Adamson, I.Y.R. and Bowden, D.H., *Am. J. Pathol.*, 112, 224–230, 1983. With permission.)

suggest that endothelial cells are lost soon after irradiation, and that this is a mitotic cell death that only occurs in a small proportion of endothelial cells having a short cell cycle time. These observations support the notion that the early depletion of endothelial cells results from a mitotic death of a rapidly dividing endothelial cell subpopulation.

Based on observations of endothelial damage first reported more than 30 years ago, we now can ascribe some of the early endothelial cell loss to apoptosis,[43] the process of programmed cell death. Haimovitz-Friedman and Fuks[44,45] discuss apoptotic endothelial cell death in Chapter 6. Key pathological features of apoptosis are observed early in the irradiated endothelium (Table 2.3). The initial disruptions to the endothelium are multifocal, are visible earliest in the plasma membrane and cytoplasm, and are associated with pyknosis. Although the consequence of multifocal apoptosis in parenchymal cells is not readily apparent, even a focal loss of endothelial cells results in a local loss of both the semipermeable barrier and circulation. As a result, the surrounding cells are disturbed.

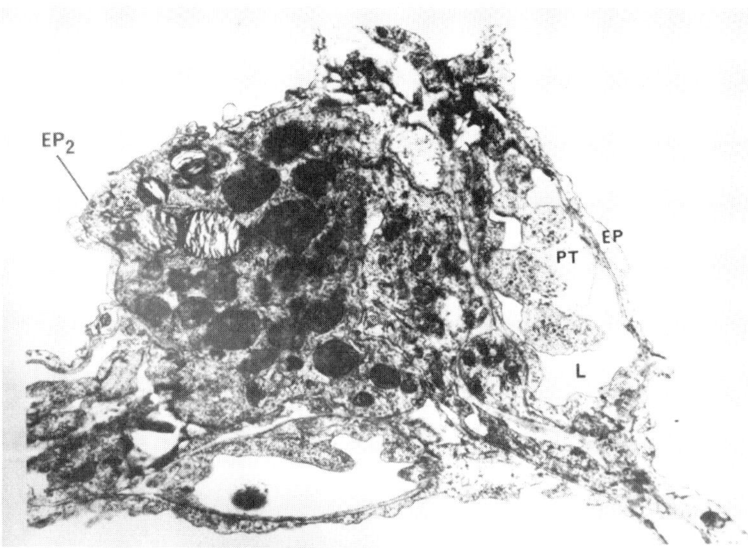

Figure 2.5 Electron micrograph of a mouse lung following 650 cGy of whole-body irradiation (3500×). Fourteen days following irradiation, a platelet thrombus (PT) is observed to be attached to the denuded basement membrane. The epithelial cells appear normal (EP, type I, and EP2, type II epithelial cell; L is vascular lumen). (Original photograph kindly provided by I.Y.R. Adamson, University of Manitoba. Reprinted from Adamson, I.Y.R. et al., *Am. J. Pathol.*, 58, 481–498, 1970. With permission.)

III. LATER CHANGES

A. Fibrosis

The development of fibrosis (see Figure 2.7) may result from the endothelial damage expressed in the earliest stages of radiation toxicity. In their review on irradiated microvasculature, Hopewell and co-authors cite evidence from various studies that atrophy or fibrosis in the rat brain and pig skin are preceded by damage to the vasculature.[3]

In a classic review on the topic,[2] Law develops the thesis that microvascular endothelial damage causes the late changes in irradiated tissues. She argues that: (1) endothelial pathology and dysfunction precede histological changes in other cell types in many irradiated tissues; and (2) fibrin deposition and perivascular edema precede collagen deposition. From these generalizations, Law proposed that fibrosis resulted from damage to the endothelial cell. Although it is not yet evident that fibrin deposition and edema cause the fibrosis, from the investigations of Franko, Ward, and co-workers,[46–48] we know that endothelial injury, including biochemical markers of endothelial function (see Chapter 10), precede and predict development of fibrosis. In their irradiated mouse lung model, they demonstrate that the tendency to develop fibrosis is genetically determined.[47,48] Furthermore, one can surmise from the investigations of Fuks, Haimovitz-Friedman, and colleagues[49,50] (see Chapter 6) that when

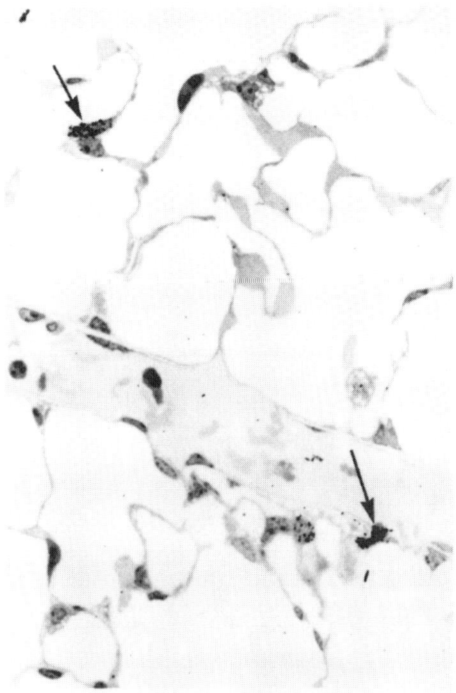

Figure 2.6 Autoradiograph of a mouse lung 2 weeks after 1000 cGy, showing ^{3}H-thymidine incorporation into endothelial cells of capillary and venule (arrows) (500×). (Original photograph kindly provided by I.Y.R. Adamson, University of Manitoba. Reprinted from Adamson, I.Y.R. and Bowden, D.H., *Am. J. Pathol.*, 112, 224–230, 1983. With permission.)

irradiated endothelial cells slough from their basement membrane, fibroblast growth factor (FGF) is activated and released from the extracellular matrix. This expression of FGF could help explain the loss of antimitogenic control that an intact endothelium exerts on the subendothelial cells.[51]

B. Telangiectasia and Sclerosis

Telangiectasia and sclerosis are the hallmark late vascular responses to irradiation and may result from loss of capillary pressure control because of smooth muscle cell loss. This ballooning of blood vessels also appears due to endothelial cell loss as reflected ultrastructurally by an onion skin-like appearance of new basement membrane laid over that of the previous occupant. Hopewell and co-investigators[3] used telangiectasia as one of four parameters that defined a tissue injury unit (TIU) in the irradiated rat brain; the other associated parameters were vessel wall thickening, enlarged endothelial cell nuclei, and hypertrophy of the adjacent astrocytes. This histopathological pattern of the disturbed vascular/glial unit is one means to quantify and predict late radiation-induced brain toxicity. For example, an increase in the TIU precedes necrosis by approximately 10 to 13 weeks in the irradiated rat fimbria.[3]

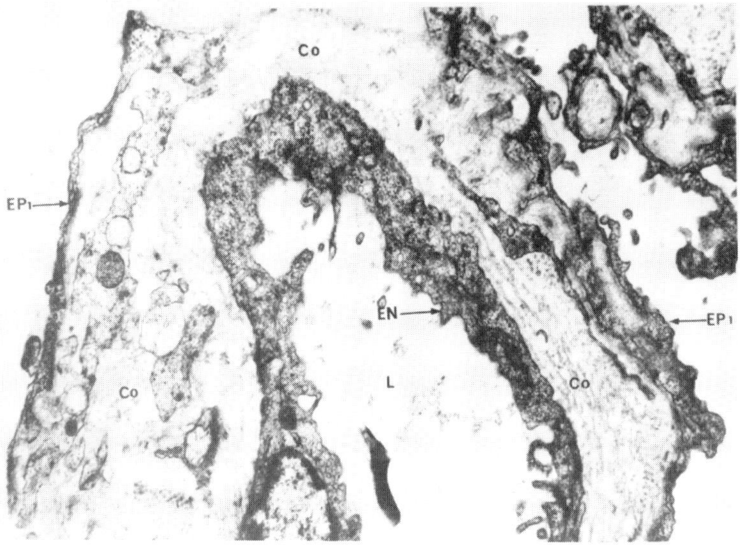

Figure 2.7. Electron micrograph of a mouse lung following 650 cGy of whole-body irradiation (15000×). After 8 weeks, there is fibrosis with excess collagen (Co) deposition in the capillary wall. Endothelial cells are labeled EN, and type I epithelial cells are labeled EP1. (Original photograph kindly provided by I.Y.R. Adamson, University of Manitoba.)

IV. CONCLUSIONS

As reflected by the work presented in subsequent chapters, the histopathology of irradiated endothelium has stimulated investigations into the mechanisms of radiation toxicity. From observations on the progressive structural changes of irradiated vessels, investigators are developing therapeutic stratagems that target the injured and activated endothelial cell. Therapeutic interventions of radiation toxicity will be based ultimately on manipulating the specific features that characterize the progression of endothelial damage. Edema could be managed more effectively by understanding the processes that stimulate endothelial cell retraction, sloughing, and pinocytosis. Control of fibrosis and vascular sclerosis may depend eventually on comprehending and regulating the adhesion of leukocytes, the loss and proliferation of endothelial cells, as well as thickening of the endothelial basement membrane. Last, the histological methods developed by radiation biologists to quantify endothelial-related pathology will provide a reliable means to assess the radiosensitization and radioprotection of normal tissues and tumors.

ACKNOWLEDGMENTS

The authors are grateful for the contribution of photographs from Dr. I.Y.R. Adamson, University of Manitoba, Canada; and for the editorial remarks of Dr. L.F. Fajardo, Stanford University, and Ms. Elizabeth Drab-Weiss, Rush Medical College.

REFERENCES

1. Maisin, J., Chapter III. Ultrastructure of the vessel wall. *Curr. Top. Radiat. Res.*, 10, 29, 1974.
2. Law, M. P., Radiation-induced vascular injury and its relation to late effects in normal tissues. *Adv. Radiat. Biol.*, 9, 37, 1981.
3. Hopewell, J. W., Calvo, W., Campling, D., Reinhold, H. S., Rezvani, M., and Yeung, T. K., Effects of radiation on the microvasculature. Implications for normal tissue damage. *Front. Radiat. Ther. Oncol.*, 23, 85, 1989.
4. Rubin, D. B., Drab, E. A., and Ward, W. F., Physiological and biochemical markers of the endothelial cell response to irradiation. *Int. J. Radiation Biology*, 60, 29, 1991.
5. Ryan, U. S., *Endothelial Cells*. CRC Press, Inc., Boca Raton, FL, 1988.
6. Gassmann, A., Zur histologie der röntgenulcera. *Fortschr. a. d. geb. d. Röntgenstrahlen*, 2, 199, 1899.
7. Griem, M. L., Robotewskyj, A., and Nagel, R. H., Potential vascular damage from radiation in the space environment. *Adv. Space. Res.*, 14, (10)555, 1994.
8. Roscher, A. A., Steele, B. C., and Woodard, J. S., Carotid artery rupture after irradiation of larynx. *Arch. Otolaryng.*, 83, 472, 1966.
9. Reinhold, H. S., The influence of radiation on blood vessels and circulation. Chapter II: Cell viability of the vessel wall. *Curr. Top. Radiat. Res.*, 10, 9, 1974.
10. Fonkalsrud, E. W., Sanchez, M., Zerubavel, R., and Mahoney, A., Serial changes in arterial structure following radiation therapy. *Surg. Gyn. Obst.*, 145, 395, 1977.
11. Asscher, A. W., Wilson, C., and Anson, S. G., Sensitisation of blood vessels to hypertensive damage by X-irradiation. *Lancet*, 580, 1961.
12. Hopewell, J. W. and Wright, E. A., The nature of latent cerebral irradiation damage and its modification by hypertension. *Br. J. Radiol.*, 43, 161, 1970.
13. Lindsay, S., Kohn, H. I., Dakin, R. L., and Jew, J., Aortic arteriosclerosis in the dog after localized aortic X-irradiation. *Circ. Res.*, 5, 51, 1962.
14. Konings, A. T. W., Sibinga, C. S., Aarnoudse, M. W., De Wit, S. S., and Lamberts, H. B., Initial events in radiation-induced atheromatosis. *Strahlentherapie*, 154, 795, 1978.
15. Rubin, P. and Casarett, G. W., Radiation effects on fine vasculature and interstitial connective tissue. In *Clinical Radiation Pathology* (Rubin, P. and Casarett, G. W. Eds.), Saunders, Philadelphia, 1968.

16. Kapanci, Y., Weibel, E. R., Kaplan, H. P., and Robinson, F. R., Pathogenesis and reversibility of pulmonary lesions of oxygen toxicity in monkeys. II. Ultrastructural and morphometric studies. *Lab. Invest.*, 20, 101, 1969.
17. Dal Nogare, A.R., Southwestern internal medicine conference: septic shock. *Am. J. Med. Sci.*, 302, 50, 1991.
18. Reed, G. B. and Cox, A. J., The human liver after radiation injury. *Am. J. Pathol.*, 48, 597, 1966.
19. Phillips, T. L., An ultrastructural study of the development of radiation injury in the lung. *Radiology*, 87, 49, 1966.
20. Adamson, I. Y. R., Bowden, D. H., and Wyatt, J. P., A pathway to pulmonary fibrosis: an ultrastructural study of mouse and rat following radiation to the whole body and hemithorax. *Am. J. Pathol.*, 58, 481, 1970.
21. Fajardo, L. F. and Stewart, R., Experimental radiation induced heart disease. *Am. J. Pathol.*, 59, 299, 1970.
22. Rubin, D. B., Drab, E. A., Ward, W. F., Smith, L. J., and Fowell, S. M., Enzymatic responses to radiation in cultured vascular endothelial and smooth muscle cells. *Radiat. Res.*, 99, 420, 1984.
23. Weiss, S. J., Young, J., LoBuglio, A. F., and Slivka, A., Role of hydrogen peroxide in neutrophil-mediated destruction of cultured endothelial cells. *J. Clin. Invest.*, 68, 714, 1981.
24. Stearner, S. P. and Christian, E. J. B., Protection against early radiation injury to the microvasculature. *Microvas. Res.*, 2, 202, 1970.
25. Klemm, K. J., Alterations of the vessel wall in the capillary bed after irradiation with fast electrons experimental study of the protective substance trihydroxyethylrutoside (THR) by means of rabbit's ear chamber. *Bibl. Anat.*, 9, 207, 1967.
26. Hopewell, J. W., Calvo, W., Jaenke, R., Reinhold, H. S., Robbins, M. E., and Whitehouse, E. M., Microvasculature and radiation damage. *Recent Results Cancer Res.*, 130, 1, 1993.
27. Mooteri, S. N., Podolski, J. L., Drab, E. A., Saclarides, T. J., Onoda, J. M., Kantak, S. S., and Rubin, D. B., WR-1065 and radioprotection of vascular endothelial cells II. Morphology. *Radiat. Res.*, 145, 217, 1996.
28. Dunn, M. M., Drab, E. A., and Rubin, D. B., The effect of irradiation on endothelial cell-polymorphonuclear leukocyte interaction. *J. Appl. Physiol.*, 60, 1932, 1986.
29. Eldor, A., Vlodavsky, I., Fuks, Z., Matzner, Y., and Rubin, Arachidonic metabolism and radiation toxicity in cultures of vascular endothelial cells. *Prostaglandins Leukotrienes and Essential Fatty Acids*, 36, 251, 1989.
30. Fajardo, L. F. and Stewart, R., Capillary injury preceding radiation-induced myocardial fibrosis. *Radiology*, 101, 429, 1971.
31. Vaeth, J. M., Green, J. P., Schroeder, A. F., McCormack, K. R., Baker, D. G., and Meurk, M. L., Frontiers of radiation therapy and oncology. *Front. Radiat. Ther. Oncol.*, 6, 21, 1972.
32. Martin, M., Lefaix, J., Pinton, P., Crechet, F., and Daburon, F., Temporal modulation of TGF-B1 and B-actin gene expression in pig skin and muscular fibrosis after ionizing radiation. *Radiat. Res.*, 134, 63, 1993.
33. Davies, P. F., Endothelial cells, hemodynamic forces, and the localization of atherosclerosis, in *Endothelial Cells: Volume II*, Ryan, U. S., Ed., CRC Press, Boca Raton, FL, 1988, 123.

34. Schilling, W. P., Mo, M., and Eskin, S. G., Effect of shear stress on cytosolic Ca^{2+} of calf pulmonary artery endothelial cells. *Exp. Cell Res.*, 198, 31, 1992.
35. Grabowski, E. F., Jaffe, E. A., and Weksler, B. B., Prostacyclin production by cultured endothelial cell monolayers exposed to step increases in shear stress. *J. Lab. Clin. Med.*, 105, 36, 1984.
36. Rubin, D. B., Drab, E. A., Ts'ao, C. H., Gardner, D., and Ward, W. F., Prostacyclin synthesis in irradiated endothelial cells cultured from bovine aorta. *J. Appl. Physiol.*, 58, 592, 1985.
37. Ward, W. F., Solliday, N. H., Molteni, A., and Port, C. D., Radiation injury in rat lung II. Angiotensin-converting enzyme activity. *Radiat. Res.*, 96, 294, 1983.
38. Ts'ao, C. H. and Ward, W. F., Plasminogen activator activity in lung and alveolar macrophages of rats exposed to graded single doses of rays to the right hemithorax. *Radiat. Res.*, 103, 393, 1985.
39. Tannock, I. F. and Hayashi, S., The proliferation of capillary endothelial cells. *Cancer Res.*, 32, 77, 1972.
40. Hirst, D. G., Denekamp, J., and Hobson, B., Proliferation studies of the endothelial and smooth muscle cells of the mouse mesentery after irradiation. *Cell. Tissue. Kinet.*, 13, 91, 1980.
41. Adamson, I. Y. R. and Bowden, D. H., Endothelial injury and repair in radiation-induced pulmonary fibrosis. *Am. J. Pathol.*, 112, 224, 1983.
42. Johnson, L. K., Longenecker, J. P., and Fajardo, L. F., Differential radiation response of cultured endothelial cells and smooth myocytes. *Anal. Quan. Cyto.*, 4, 188, 1982.
43. Carson, D. A. and Ribeiro, J. M., Apoptosis and disease. *Lancet*, 341, 1251, 1993.
44. Haimovitz-Friedman, A., Balaban, N., McLoughlin, M., Ehleiter, D., Michaeli, J., Vlodavsky, M., and Fuks, Z., Protein kinase C mediates basic fibroblast growth factor protection of endothelial cells against radiation-induced apoptosis. *Cancer Res.*, 54, 2591, 1994.
45. Fuks, Z., Persaud, R. S., Alfieri, A., McLoughlin, M., Ehleiter, D., Schwartz, J. L., Seddon, A. P., Cordon-Cardo, C., and Haimovitz-Friedman, A., Basic fibroblast growth factor protects endothelial cells against radiation-induced programmed cell death *in-vitro* and *in-vivo*. *Cancer Res.*, 54, 2582, 1994.
46. Ward, W. F., Molteni, A., Solliday, N. H., and Jones, G. E., The relationship between endothelial dysfunction and collagen accumulation in irradiated rat lung. *Int. J. Radiat. Oncol. Biol. Phys.*, 11, 1985, 1985.
47. Ward, W. F., Sharplin, J., Franko, A. J., and Hinz, J. M., Radiation-induced pulmonary endothelial dysfunction and hydroxproline accumulation in four strains of mice. *Radiat. Res.*, 120, 113, 1989.
48. Franko, A. J., Sharplin, J., Ward, W. F., and Hinz, J. M., The genetic basis of strain-dependent differences in the early phase of radiation injury in mouse lung. Radiat. Res., 126, 1, 1991.
49. Haimovitz-Friedman, A., Vlodavsky, I., Chaudhuri, A., Witte, L., and Fuks, Z., Autocrine effects of fibroblast growth factor in repair of radiation damage in endothelial cells. *Cancer Res.*, 51, 2552, 1991.
50. Witte, L., Fuks, Z., Haimovitz-Friedman, A., Vlodavsky, I., Goodman, D. S., and Eldor, A., Effects of irradiation on the release of growth factors from cultured bovine, porcine, and human endothelial cells. *Cancer Res.*, 49, 5066, 1989.

51. Fajardo, L. F., The unique physiology of endothelial cells and its implications in radiobiology. In *Frontiers in Radiation Therapy and Oncology*, 23rd ed. (Vaeth, J. M. and Meyer, J. L., Eds.), Karger, Basel, 1989, 96.
52. Dmitrievich, G. S., Fischer-Dzoga, K., and Griem, M. L., Radiosensitivity of vascular tissue. I. Differential radiosensitivity of capillaries: a quantitative *in vivo* study. *Radiat. Res.*, 99, 511, 1984.
53. Takahashi, T., The action of radium upon the formation of blood capillaries and connective tissue. *Br. J. Radiol.*, 3, 439, 1930.
54. Windholz, F. Dr., Zur kenntnis der blutgefäbmeränderungen im röntgenbestrahlten gewebe. *Strahlentherapie*, 59, 662, 1937.
55. Warren, S., Effects of radiation on normal tissues. *Arch. Pathol.*, 34, 1070, 1942.
56. Sheehan, J. F., Foam cell plaques in the intima of irradiated small arteries one hundred to five hundred microns in external diameter. *Arch. Pathol.*, 37, 297, 1944.
57. Pennybacker, J. and Russell, D. S., Necrosis of the brain due to radiation therapy: clinical and pathological observations. *J. Neurol. Neurosurg. Psychiat.*, 11, 183, 1948.
58. Merwin, R. M. and Hill, E. L., Effect of local Roentgen irradiation on the formation of new capillaries after injury. *J. Natl. Cancer Inst.*, 15, 1031, 1955.
59. Fajardo, L. F. and Stewart, J. R., Experimental radiation-induced heart disease. I. Light microscopic studies. *Am. J. Pathol.*, 59, 299, 1970.
60. McDonald, L. W. and Hayes, T. L., The role of capillaries in the pathogenesis of delayed radionecrosis of the brain. *Am. J. Pathol.*, 50, 745, 1967.
61. Archambeau, J. O., Ines, A., and Fajardo, L. F., Correlation of the dermal microvasculature morphology with the epidermal and the endothelial population changes produced by single X ray fractions of 1649, 2231 and 2619 rad in swine. *Int. J. Radiat. Oncol. Biol. Phys.*, 11, 1639, 1985.
62. Fajardo, L. F. and Stewart, J. R., Pathogenesis of radiation-induced myocardial fibrosis. *Lab. Invest.*, 2, 244, 1973.
63. Gillette, E. L., Maurer, G. D., and Severin, G. A., Endothelial repair of radiation damage following beta irradiation. *Radiology*, 116, 175, 1975.
64. Friedman, M., Ryan, U. S., Davenport, C. W., Chaney, E. L., Strickland, D. L., and Kwock, L., Reversible alterations in cultured pulmonary artery endothelial cell monolayer morphology and albumin permeability induced by ionising radiation. *J. Cell. Physiol.*, 129, 237, 1986.
65. Kwock, L., Davenport, W. C., Clark, R. L., Zarembra, J., Lingle, B., Chaney, E. L., and Friedman, M., Effects of ionizing radiation on the pulmonary vasculature of intact rats and isolated pulmonary endothelium. *Radiat. Res.*, 111, 276, 1987.
66. De Wilde, R., Boeckx, W., Van der Schueren, E., Guelinckx, P., and Gruwez, J., A scanning electron microscopic study of microvascular anastomoses on irradiated vessels: short-term effect of irradiation. *Microsurgery*, 4, 193, 1983.
67. Goodman, J. H., McGregor, J. M., Clendenon, N. R., Gahbauer, R. A., Barth, R. F., Soloway, A. H., and Fairchild, R. G., Ultrastructural microvascular response to boron neutron capture therapy in an experimental model. *Neurosurgery*, 24, 701, 1989.
68. Song, C. W., Anderson, R. S., and Tabachnick, J., Early effects of beta irradiation on dermal vascular permeability to plasma proteins. *Radiat. Res.*, 27, 604, 1966.
69. Eassa, E. H. M. and Casarett, G. W., Effect of ε- amino-*n*-caproic acid (EACA) on radiation induced increase in capillary permeability. *Radiobiology*, 106, 679, 1973.

70. Stearner, S. P. and Christian, E. J., Radiation effect on the microcirculation; relation to early mortality in the chick embryo. *Radiat. Res.*, 34, 138, 1968.
71. Fisher, J. J., Proliferation of rat aortic endothelial cells following X irradiation. *Radiat. Res.*, 92, 405, 1982.
72. Hirst, D. G., Denekamp, J., and Travis, E. L., The response of mesenteric blood vessels to irradiation. *Radiat. Res.*, 77, 259, 1979.

CHAPTER 3

Functional Measures of Endothelial Integrity and Pharmacologic Modification of Radiation Injury

Michael M. Graham and Lanell M. Peterson

CONTENTS

1-8493-4840-4/98/$0.00+$.50

I. INTRODUCTION

The capillary endothelial cell is a major target of radiation injury. Since these cells play a critical role in maintaining homeostasis in every organ, relatively minor disruption of the endothelium can cause significant problems. There are two overall mechanisms for this injury: hormonal effects, due to vasoactive agents that are released in response to radiation, and direct injury, which results in endothelial cell death some time later. Endothelial injury can be useful in attempting to destroy a tumor or can be a problem, causing edema, fibrosis, and even necrosis in normal tissues. Because of the central role endothelial cell injury plays in radiation effects, it is important to understand how radiation injury is manifested, the likely biochemical pathways responsible, and the possible approaches to ameliorate this injury.

One of the major functions of endothelial cells is control of permeability, which limits the flux of water, various solutes, and proteins from the plasma into tissue. The least permeable endothelium is in the brain where the blood-brain barrier blocks the passage of many materials from plasma into tissue. In other organs, such as the liver, the endothelium is extraordinarily permeable, offering virtually no barrier to materials in the plasma. The capillaries in the rest of the body lie between these extremes, where the endothelium, and underlying basement membrane, act as a semipermeable membrane offering significant resistance to the passage of large proteins, but allowing smaller solutes to pass freely. Because of the continuous return of interstitial protein to the circulation via lymphatics, the interstitial protein concentration is always lower than in the plasma. This results in significant osmotic pressure, which in turn controls the flow of water across the membrane. Any disruption of the endothelium results in increased permeability, increased flux of protein, decreased osmotic pressure, and finally edema because of accumulating water in the tissue.

In this chapter we describe the anatomic and functional changes of capillary endothelium that occur following radiation exposure. These changes are presented in a framework of possible biochemical pathways that might account for the observed injury and its ultimate long-term effects. The approach of investigating the importance of these pathways is described in some detail, using a number of agents to block the postulated pathways. This approach should ultimately lead to a better understanding of the complex series of events that occur following radiation injury to endothelial cells. Many of the studies described below investigated lung endothelium since this is the area where we have done most of our work.

II. CHANGES IN STRUCTURE FOLLOWING RADIATION

The anatomic changes in capillary endothelial cells following radiation and the process leading from radiation to fibrosis have been reviewed in detail in several papers[1–8] as well as by Fajardo in Chapter 1 and by Rubin and Griem in Chapter 2. Although the timing of events varies considerably from one tissue to another, there are a number of common features. Figure 3.1 is a diagrammatic summary of the pathways that may lead from radiation exposure to fibrosis. Within hours following radiation, there is an increase in capillary permeability without significant morphologic change except for edema. This is presumably due to release of vasoactive substances into the interstitial fluid of the irradiated tissue. Within a few days, the endothelial cell membrane becomes irregular, followed by swelling of the cells. There is a change in number of cytoplasmic organelles; in some tissues there is an increase, in others a decrease. The endothelial cells may swell to occlude the lumen and in some areas the cells become detached, exposing underlying basement membrane. This results in activation and adherence of platelets and the formation of microthrombi. The gaps in the endothelium allow large circulating cells (such as red cells, lymphocytes, and granulocytes) to move out of the bloodstream into the interstitium. Macrophages become prominent in the interstitium. Areas of microhemorrhage may occur. Capillaries become blocked with platelet aggregations and fibrin clots. With time, these regions of occluded capillaries become fibrotic with the growth of fibroblasts, and the remnants of the endothelial cells are removed by tissue macrophages. In many irradiated tissues, mast cells accumulate in large numbers. These cells are filled with granules containing numerous vasoactive substances, including histamine, which are presumably released into the interstitium. This release of histamine and other vasoactive agents is a particularly important pathway in the sequence of events leading from radiation to fibrosis because toward the end of the process, it has a feedback loop that leads back from delayed histamine release from mast cells to leakage of proteins, which has the potential for maintaining the injury.

The timing of this sequence of events depends to a large extent on the turnover time of endothelial cells. Hobson and Denenkamp,[9] in a study using tritiated thymidine, showed that endothelial turnover time varies from 47 days in small intestines to 23,000 days in the brain. These times seem to correlate with the time of onset of the delayed radiation injury. For instance, the delay period is several months in the brain[10] to a few weeks in the lung.[11] Because endothelial turnover times are longer in older animals, it is predictable that the delay between radiation and apparent injury will be longer in older individuals.

The anatomic changes associated with endothelial cells, usually seen within a few days, are the first evidence of radiation injury in many tissues. This observation supports the hypothesis that endothelial injury is a critical event in radiation damage. The corollary to this hypothesis is that any treatment that prevents or reduces endothelial injury should improve the radiation tolerance of an irradiated tissue. This is the basis for much of the work described below.

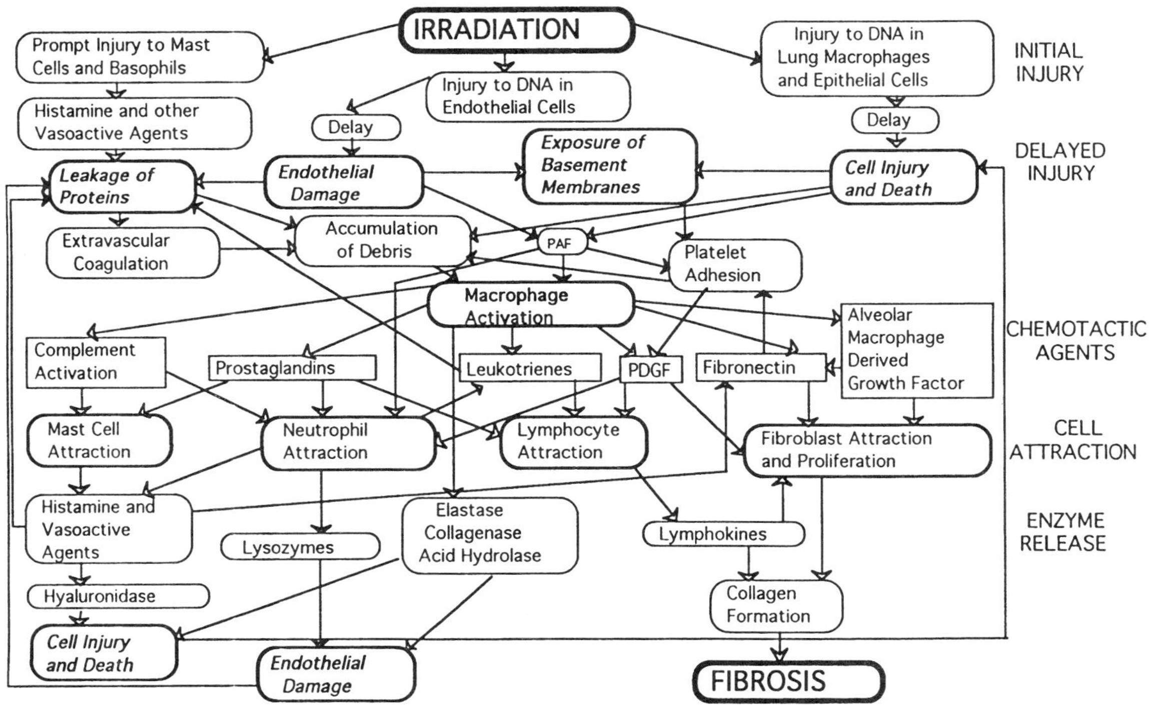

Figure 3.1. A diagram of possible pathways and events leading to fibrosis that may take place following radiation of the lung. The general hypothesized order of events is from initial injury, delayed cell death, release of chemotactic agents, attraction of mast cells, neutrophils, lymphocytes, and fibroblasts, release of proteolytic enzymes, and finally end-stage fibrosis. Note there are two positive feedback pathways postulated that may maintain or even amplify the overall process. This diagram was the basis for choosing many of the drug experiments described in Section V in an attempt to block specific pathways. (From Graham, M.M. et al., *Int. J. Radiat. Oncol. Biol. Phys.*, 19, 329, 1990. With permission.)

III. CHANGES IN FUNCTION FOLLOWING RADIATION

It is clear from numerous studies of radiation injury that endothelial cell injury is the key event in most tissues that ultimately leads to fibrosis or necrosis.[4,12–14] The sequence of endothelial injury, cell detachment, thrombosis, and fibrosis results in significant tissue injury that often limits the radiotherapist in attempting to deliver curative doses to a nearby tumor. The primary functional parameter relating to capillary endothelium is *permeability*. This relates to the leakiness of the capillary to various solutes in the plasma. In almost all studies, it has not been feasible to measure permeability alone. The measurement generally relates to the permeability-surface area product (PS). This is because the rate of efflux of a test substance through the capillary walls in a tissue depends linearly on both the permeability and the surface area available for exchange. True measurement of permeability is possible only if the surface area is known, which is usually not the case.

Secondary parameters relating to the integrity of the capillary endothelium are blood flow and perfusion pressure. Edema, as a result of increased permeability, may occur within hours of radiation. This can result in partial obstruction of the capillaries, with resultant decreased flow and increased back pressure. Later, if there is sufficient endothelial injury to result in exposure of the basement membrane, microthrombi will form, resulting in further reduction in blood flow and an increase in blood pressure. In most tissues of the body, there are sufficient alternate pathways for the blood that even total occlusion of the capillaries will not result in significant increase in blood pressure. However, if the irradiated tissue includes both lungs, then there is no alternative for the blood flow and pulmonary artery pressure will rise.[15]

There have been numerous studies of changes in capillary PS following radiation. The earliest studies used injected dyes that bind to plasma protein and accumulate more rapidly in tissues with elevated PS than in normal tissue. Most of these studies were qualitative and the tissue studied was skin. These experiments have been particularly useful in demonstrating early increases in PS. Jolles and Harrison,[16,17] using Protamine Blue, showed increased appearance of dye in the skin of a rabbit within 1 hour of irradiation with 8 Gy. Hasegawa and Wang[18] have done a similar study in mice.

A more quantitative approach to measuring capillary PS is to use radiolabeled tracers. Studies have been done with radio-iodinated and P-32-labeled albumin and with radio-iodinated fibrinogen.[19–22] Although these techniques are potentially more quantitative, a persistent problem has been how to correct for intravascular activity. A particularly convenient way to do this has been developed and used in our laboratory and is described in detail below.

Another, very different, approach to the measurement of capillary PS is to measure the flow and protein content of lymph arising from an irradiated tissue. This requires isolation and cannulation of the relevant lymphatic vessels. It has proven to be most feasible in the lung. The rationale for this approach is that increased capillary PS should result in leakage of protein into the interstitial fluid of a tissue. This results in a lowering of osmotic pressure, which is necessary to maintain water balance across the capillary wall. The increased water flux results in an increased lymph flow, and the concentration of protein in the lymph is elevated from its normal values. It is possible to calculate the mean capillary PS of a tissue from the lymph flow and the ratio of protein concentration in lymph to that in plasma. Studies such as this, in sheep, have shown lymph flow doubles at 3 hours following 15-Gy whole thorax radiation.[15]

A. Labeled Albumin Method for Measurement of Vascular Permeability

I-131 albumin is injected intravenously; 20 minutes later, I-125 albumin is injected intravenously. After another 10 minutes, the animal (usually a rat or mouse) is sacrificed and tissue and blood samples are obtained. Weighed tissue samples and pipetted plasma samples are counted in a gamma counter set up to separate the I-125 counts from I-131. The rate of leakage of albumin from the plasma into tissue is calculated using the following model. Albumin is assumed to be distributed between plasma and interstitial fluid (ISF) separated by a permeable membrane, the endothelial cell and basement membrane. It is assumed that the plasma concentration of tracer albumin is constant and that the rate of leakage out of the vascular space into the interstitial compartment is also constant and proportional to the plasma concentration. Given these assumptions, the equation describing the amount of tracer albumin in tissue is:

$$A = Cp \cdot Vp + LR \cdot Cp \cdot t \tag{1}$$

Where:
A = total activity of tissue (μCi/g)
Cp = concentration of tracer in plasma (μCi/ml)
Vp = volume of plasma in tissue (ml/g)
LR = leakage rate from plasma into tissue ($ml \cdot min^{-1}\ g\ tissue^{-1}$)
t = time elapsed since injection of tracer (min)

Because the two tracer albumins are injected at different times, the expressions for their activity will be somewhat different. The equations for I-131 albumin and I-125 albumin, injected 10 and 30 minutes prior to sacrifice, respectively, are:

$$A_{131} = Cp_{131} \cdot Vp + LR \cdot Cp_{131} \cdot 10 \tag{2}$$

$$A_{125} = Cp_{125} \cdot Vp + LR \cdot Cp_{125} \cdot 30 \tag{3}$$

A_{131}, A_{125}, Cp_{131}, and Cp_{125} can all be measured using either gamma or liquid scintillation counting.

Equations (2) and (3) are solved for the unknowns LR and Vp:

$$LR = 1/20 \cdot (A_{125}/Cp_{125} - A_{131}/Cp_{131}) \tag{4}$$

$$Vp = 1/2 \cdot (3 \cdot A_{131}/Cp_{131} - A_{125}/Cp_{125}) \tag{5}$$

Since LR is directly proportional to PS, this approach provides a measure of capillary PS. This method has been validated[23,24] and has been used for several years in the study of changes in PS following radiation.[11,23]

Figure 3.2 shows a biphasic pattern of increased PS following 18-Gy Cs-137 irradiation given to the entire thorax of Sprague-Dawley rats in a single dose. There is a prompt increase in PS immediately following radiation. PS remains elevated for 3 days, falls back to normal levels, and then increases markedly at 18 days after irradiation. In other studies,[11] this response is shown to be somewhat dose dependent but the delay to the second peak does not seem to depend on dose. This is probably because the delay is related to endothelial turnover time, which is not particularly affected by radiation injury, whereas the magnitude of the change is directly related to the severity of the injury to the endothelial cells.

B. Lung Blood Flow Measured with Macro-Aggregated Albumin

Measurement of relative lung blood flow in the rat or other animals is simple and repeatable. This makes it an ideal parameter to follow vascular injury following irradiation of one lung.[25] Tc-99m macro-aggregated albumin (MAA) is injected intravenously, and an image is acquired of the distribution of the radioactive particles in the rat's lungs with a gamma camera and computer. MAA particles are irregular and about 50 μ in diameter. They are trapped by the first capillary bed they encounter. Following intravenous injection, the first capillary bed is in the lungs. This method is essentially the technique used in routine clinical lung perfusion imaging, a common procedure in all nuclear medicine clinics. Forty μCi (1.48 MBq) of ^{99m}Tc-MAA suspended in 0.2 ml saline was injected into the tail vein of each rat. The rat was

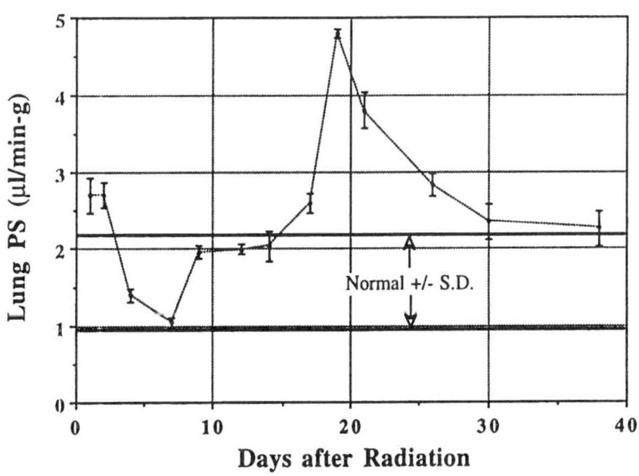

Figure 3.2. Lung permeability-surface area product (PS) at several times following 18-Gy whole-thorax irradiation of Sprague-Dawley rats. The method used is described in Section IIIA. The normal range ±1 standard deviation is shown. The error bars on the data points are ±1 standard error of the mean. (From Graham, M.M. and Evans, M.L., *Microvasc. Res.*, 42, 266, 1991. With permission.)

then imaged for 2 minutes on a gamma camera. Rectangular regions-of-interest were drawn over the lungs, and a left-to-right perfusion ratio was calculated. The MAA particles occlude only a very small fraction of the lung capillaries and break down in a few hours. This means the test can be repeated many times without injuring the animals or affecting subsequent measurements. This approach was used to follow changes in blood flow in irradiated left lungs of Sprague-Dawley rats for up to a year following hemithorax irradiation. The radiation was Cs-137 radiation given in a single dose. The results are shown in Figure 3.3. Beginning at about 3 weeks following radiation, blood flow to the irradiated lung begins to decrease. At high doses (25 Gy), the blood flow reaches a minimum of 15% of normal. There is virtually no recovery of flow at this dose, but at lower doses partial recovery occurs in a dose-dependent fashion. This method was used in the studies of the effects of different blocking agents on lung injury, described in Section V.

The decrease of blood flow beginning at 3 weeks fits in well with the observed changes in vascular permeability in the lung. The most consistent hypothesis explaining both the peak of increased permeability and the onset of decreased flow at 3 weeks after radiation is that there is a delayed endothelial injury that becomes maximal at 3 weeks. At that time there is apparently sufficient disruption of the endothelium that platelets and fibrin begin to adhere to the exposed basement membrane and blood vessels become occluded. This progresses until a large fraction of the vessels are blocked. Vascular permeability rises as the vessels become injured

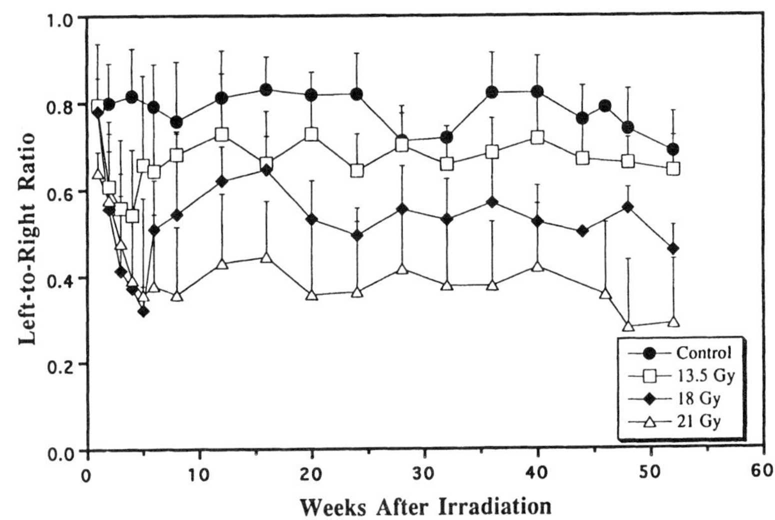

Figure 3.3. Left-to-right lung blood flow ratios at several times following left lung irradiation of Sprague-Dawley rats. The blood flow begins to decrease at 2 weeks after radiation, reaching a minimum at 4 to 5 weeks. At 21 Gy, there is no recovery of blood flow, while at lower doses there is some recovery. Error bars shown are ±1 standard error of the mean for 7 to 13 rats/dose group. (From Peterson, L.M. et al., *Radiat. Res.*, 129, 139, 1992. With permission.)

and leaky. The permeability seems to drop at the time blood flow drops because fewer vessels are open and the total surface area available for exchange drops. Note that the measurement of vascular permeability is a measure of the permeability-surface area product. Apparently the permeability rises up to 3 weeks after radiation and then the surface area available for exchange decreases. Thus, both increased permeability and decreased blood flow are useful functional measures of radiation injury to endothelial cells in the lung.

IV. HORMONAL CHANGES IMMEDIATELY FOLLOWING RADIATION AFFECTING ENDOTHELIAL CELLS

The first phase of the biphasic increase in vascular permeability following radiation is seen within minutes or hours after exposure. This time is before any anatomic changes are apparent. The only reasonable mechanism to account for this early increase in permeability is release of various vasoactive agents that cause vasodilation and increased permeability.

The source of these vasoactive agents is undoubtedly the mast cell, which is packed with numerous granules containing over 20 agents including histamine, serotonin, and bradykinin. Mast cells degranulate in response to radiation, releasing these agents into the interstitium. Arachidonic acid is also released from the phospholipid portion of the cell membrane. The arachidonic acid is rapidly metabolized to more than 90 other agents including several prostaglandins. The mechanism for mast cell degranulation in response to radiation is not well established. However, the mast cell membrane can be stabilized and degranulation can be blocked quite effectively by cromoglycate. Administration of this agent prior to radiation blocks virtually all the early effects of radiation.[26]

Histamine is certainly the most important of the various agents released by mast cells. This has been demonstrated by blocking its effects with antihistamines. Willoughby,[27] using Trypan Blue, showed that antihistamines blocked the early increased PS in rat intestine. Mount and Bruce,[28] using I-125 albumin, showed similar blocking in studies of irradiation of the ears of rabbits. Graham,[23] using the dual albumin method described above, showed complete blocking of the early increase in PS of irradiated rat intestine by cyproheptadine, a blocker of both histamine and serotonin. A recent study[26] supports the concept that histamine is the most important factor in causing the increased permeability seen shortly after radiation.

The success of antihistamines in blocking the early increase in capillary PS following radiation suggests that the other agents released by mast cells play an insignificant role in this phenomenon. Serotonin certainly is released at the time of irradiation, as evidenced by the efficacy of ondansetron, a specific 5-HT3 antagonist, in blocking postradiation nausea and vomiting.[29] Bradykinin and prostaglandins are also released into the interstitium at the time of radiation but their effects on vascular permeability are relatively minor. Thus, if the effects of histamine are blocked, there is essentially no early increase in capillary PS following radiation.

V. PHARMACEUTICALS USED TO CONTROL ENDOTHELIAL INJURY FOLLOWING RADIATION

In referring to Figure 3.1, it is logical to assume that if one or more of the pathways leading from radiation to fibrosis could be blocked, the injury caused by radiation could be reduced or eliminated. Several pharmaceuticals have been used to attempt to block the changes in vascular permeability and blood flow in order to determine which pathway or pathways play important roles in the sequence of events leading from initial radiation damage to disruption of the endothelial cells and ultimately to late fibrosis. Thirteen drugs were studied in our laboratory.

PS and blood flow studies were all done with 4- to 5-week-old Sprague-Dawley rats. Irradiation was single-dose Cs-137 radiation, usually at 18 Gy, unless stated otherwise. All the water-soluble drugs were administered in the drinking water at the doses shown in Table 3.1. Four drugs (cobra venom factor, cromolyn, vitamin D_3, and trapidil) were administered by intramuscular injection. In an attempt to maximize the effect, the dose chosen was just below toxic levels. The drugs were started 2 days before radiation and continued for the entire study period after radiation. Exceptions to this approach were: dexamethasone, which was tapered over an 8-week period starting at 4 or 8 weeks after radiation in the blood flow studies; and cobra venom factor, which was given in two doses just prior to irradiation.

Figures 3.4 and 3.5 show examples of the results of these blocking agents on postradiation changes in vascular permeability and blood flow. For more detailed descriptions of the findings, please refer to References 25 and 30.

A. Captopril

Captopril is an angiotensin converting enzyme (ACE) inhibitor that may have some anti-inflammatory effects and has been shown to inhibit vascular permeability

Table 3.1 Drug Treatments

Drug	Dose
Captopril	500 mg/l, DW
Cobra Venom Factor	150 units IM at 13 and 1 hr before irradiation
Cromolyn	400 mg/kg/day IM
Cyproheptadine	15 mg/l, DW
Vitamin D-3	2000 IU/kg/2 days, IM every other day
Dexamethasone	0.25 mg/l, DW (tapered for 4 weeks in BF studies)
Diethylcarbamazine	400 mg/l, DW
ε-Aminocaproic acid	2000 mg/l, DW
Indomethacin	4.0 mg/l, DW
Penicillamine	200 mg/l, DW, with Pyridoxine, 20 mg/l, DW
Pentoxifylline	500 mg/l, DW
Theophylline	400 mg/l, DW
Trapidil	400 mg/kg/day, IM

Note: All drugs were administered in drinking water (DW) or injected intramuscularly (IM) daily from 2 days before irradiation, unless otherwise noted.

From Graham, M.M. et al., *Int. J. Radiat. Oncol. Biol. Phys.*, 19, 329, 1990. With permission.

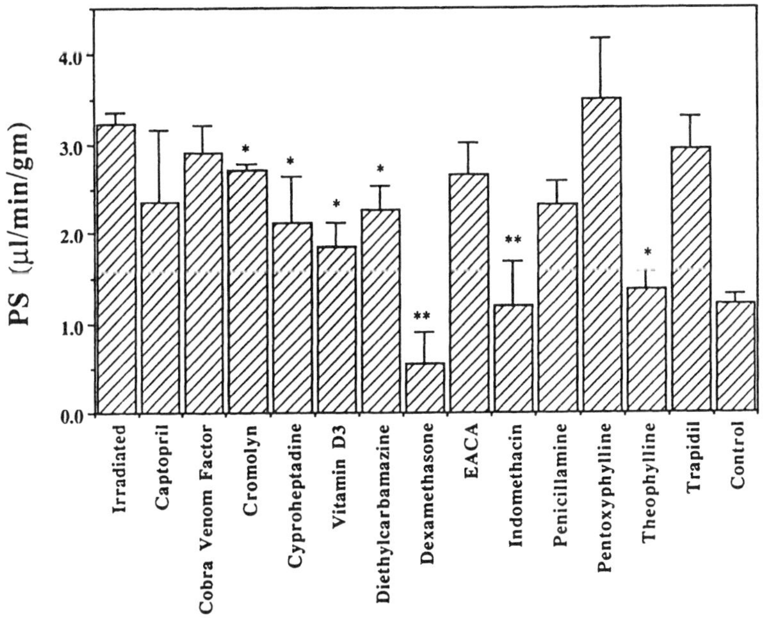

Figure 3.4. Lung permeability-surface area product at 21 days following 18-Gy irradiation of the left lung of Sprague-Dawley rats. The rats were treated with the indicated drugs at the doses shown in Table 3.1. The most effective drugs are clearly dexamethasone, indomethacin, and theophylline. A less marked but statistically significant effect was seen with cromolyn, cyproheptadine, vitamin D_3, and diethylcarbamazine. Error bars shown are ±1 standard error of the mean for 3 to 5 rats/group. ($^*p < 0.01$, $^{**}p < 0.001$.) (From Graham, M.M. et al., *Int. J. Radiat. Oncol. Biol. Phys.*, 19, 329, 1990. With permission.)

induced by injection of vasoactive mediators such as histamine, bradykinin, and serotonin.[6] It increases levels of prostaglandins[31] and has been used in studies of platelet and vascular prostaglandin metabolism.[32] Other uses of captopril include inhibiting the renin-angiotensin system to reduce blood pressure in hypertension[33] and reducing atherosclerosis in heart disease.[34] It has been shown to modify radiation-induced pulmonary endothelial dysfunction,[35] to reduce monocrotaline-induced cardiopulmonary injury,[36] to reduce collagen and mast cell accumulation in irradiated rat lung,[37] and to preserve kidney function and enhance survival in experimental radiation nephropathy.[38] Captopril did not significantly suppress the radiation-induced increase in lung PS,[30] but did delay the onset and reduce the severity of early blood flow changes.[25] Thus, captopril seems to be somewhat successful in decreasing lung injury, perhaps more by decreasing the vascular obstruction by modifying platelet behavior than by protecting endothelial cells.

B. Cobra Venom Factor (CoVF)

Cobra venom factor is used to deplete complement. Complement activation leads to chemotaxis and activation of neutrophils, increased endothelial permeability, and

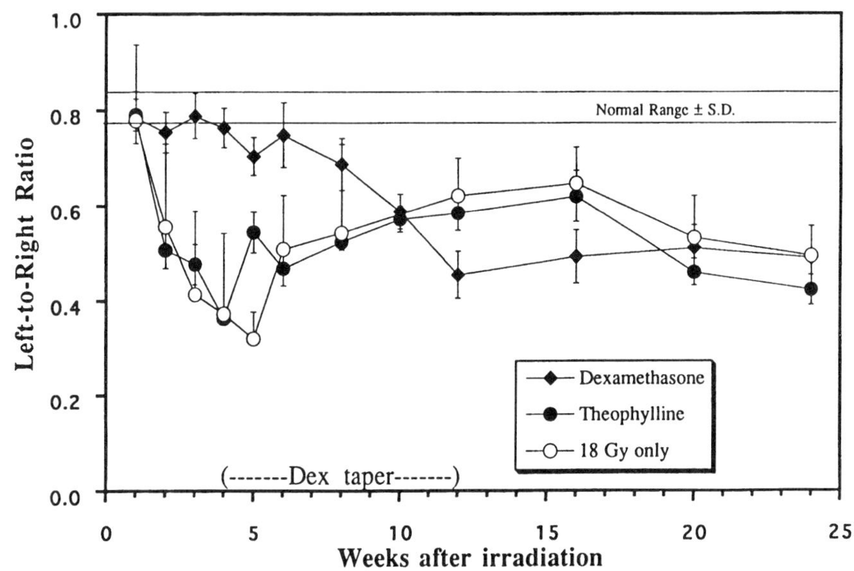

Figure 3.5. Left-to-right lung blood flow ratios at several times following 18-Gy left-lung irradiation of Sprague-Dawley rats. The untreated rats showed a nadir at 5 weeks. Theophylline did not significantly change the time course of blood flow after radiation. Dexamethasone completely blocked the decrease in blood flow but when it was tapered, the blood flow dropped well below untreated levels until 20 weeks, when all levels were approximately the same. Error bars shown are ±1 standard error of the mean for 4 to 6 rats/group. (Adapted from Peterson, M.L. et al., *Radiat. Res.*, 129, 139, 1992.)

damage to endothelial cell membranes that ultimately lead to cell death.[39] CoVF effectively suppresses radiation-induced vascular permeability in the skin and reduces the degree of fibrosis.[40] The success of CoVF seems to be neutrophil dependent. It has been shown to inhibit immunologic lesions that are neutrophil dependent,[41] and has recently been used in conjunction with an antibody to human IL-8 to inhibit inflammatory reactions in rat lungs.[42] An anti-human P-selectin antibody reacts with a pulmonary vascular antigen in rats after infusion of CoVF,[43] suggesting that selectin is also an important agent in the inflammatory damage pathway.[43,44] P-selectin is an agent stored in intracellular granules in endothelial cells and platelets, which when translocated to the cell membrane promotes adhesive interaction with neutrophils.[43] After injection of CoVF, studies of complement receptor type 1 activation showed reduced increases in lung vascular permeability by 67% when compared to glycogen-injected rats.[45] CoVF injected at the time of irradiation was not effective in reducing lung PS at 3 weeks after irradiation.[30] This suggests that complement alone does not induce an inflammatory response, but must be accompanied by other factors,[39,44] or that complement levels recovered sufficiently at 3 weeks that the initial suppression had little effect.

C. Cromolyn

Cromolyn inhibits allergen-induced histamine release and degranulation of mast cells.[46] Mast cells are found in the lung during radiation pneumonitis[4,7,47] and have also been noted during idiopathic pulmonary fibrosis.[48] Phagocytosis of mast cell granules by fibroblasts may lead to increased production of collagen.[49] Cromolyn decreased vascular permeability following skin injury,[50] and slightly but significantly reduced the delayed radiation-induced increase in lung PS, providing additional evidence that histamine release and mast cells continue to play an important role in the development of lung damage,[30] as well as being responsible for most of the early effects seen within hours following radiation exposure.

D. Cyproheptadine

Cyproheptadine is a histamine and serotonin antagonist.[51] Histamine released from mast cells, basophils, and platelets may lead to the increase in vascular permeability immediately following radiation exposure,[52] and antihistamines have been proven to be effective blocking agents.[23,53] Cyproheptadine significantly lowered vascular permeability in rat intestines[23,26] immediately after radiation and had a modest effect on lung PS at 21 days.[30] It had no significant effect on blood flow,[25] implying endothelial injury, and subsequent exposure of basement membrane, was not affected. This suggests that increased permeability is initiated by histamine and/or serotonin, but that endothelial injury is related to other pathways.[26]

E. Vitamin D_3

Vitamin D_3 inhibits platelet-derived growth factor (PDGF). PDGF is released by platelets and macrophages and is a chemo-attractant for neutrophils, monocytes, and fibroblasts.[54–56] It also stimulates fibroblast production,[57–59] and has been suggested as therapy for patients with radiation-induced arterial lesions.[60] Platelet extracts have also demonstrated an ability to increase vascular permeability in rabbit skin.[61] Use of vitamin D_3 resulted in a significant reduction in the degree of increased PS at 3 weeks after radiation exposure to the lung.[30] PDGF may be involved in the development of radiation fibrosis because of its influence on neutrophils and fibroblasts and its involvement with interstitial pulmonary fibrosis.

F. Dexamethasone

Dexamethasone is a glucocorticosteroid that acts as a potent anti-inflammatory agent. Several hypotheses have been suggested for the effectiveness of this pharmaceutical. The anti-inflammatory effects of corticosteroids include inhibition of edema, fibrin deposition, fibroblast proliferation, collagen deposition, and cicatrization.[62] It blocks prostaglandin synthesis by inhibiting cyclooxygenase, which then blocks the oxidation of arachidonic acid to cyclic endoperoxides and prostaglandins.[63] It stabilizes cell walls by acting as a phospholipase protein inhibitor, which protects membrane phospholipids from interacting with phospholipase.[63] Dexamethasone inhibits production of plasminogen activator from polymorphonuclear leukocytes.[64] It acts as a radioprotector at the time of radiation[65] by delaying cell division induced by radiation, thus allowing additional time for cell repair.[66] This is also seen as less histological damage postradiation.[67] Dexamethasone has been shown to reduce mortality if withdrawn after the pneumonitis period and enhance pulmonary mechanics in mice. This is probably due to improved lung compliance at the alveolar fluid surface by increasing surfactant phospholipid production.[68–70] Dexamethasone inhibits the enzyme phospholipase A_2, which reduces the formation of prostaglandins, leukotrienes, and O_2 radicals. It inhibits the release of histamine and lysosomal constituents, which affects the margination of white cells in the circulation, and inhibits tumor necrosis factor.[71] Recent studies indicate that dexamethasone can be used to suppress myointimal hyperplasia in rabbits,[72] inhibit gene expression of JE mRNA, and inhibit ischemia-induced elevation (JE is one of the most abundant genes induced by platelet-derived growth factor). It encodes a low molecular weight glycoprotein that functions, in part, as a monocyte chemo-attractant,[73] and inhibits thrombin-stimulated release of mitogenic activity.[74]

Early studies of acute radiation reactions show that dexamethasone reduced the incidence of the reactions in the lung within 50 days postirradiation.[75] In studies of vascular permeability, dexamethasone has been extremely successful in reducing vascular permeability in whole, hemithorax, and fractionally irradiated rat lungs.[30,76,77] It also significantly reduced the severity of the decline in lung blood flow.[25,77] It is obvious that dexamethasone blocks several of the pathways leading from radiation to fibrosis, and it is reasonable to assume that the effectiveness of this steroid can be attributed to blocking more than one pathway. Even though it is extremely effective, because dexamethasone acts in so many ways, its effects do not identify a primary pathway responsible for radiation lung injury. Significant drawbacks of using dexamethasone exclusively is that it is not well tolerated for long periods of time, weight gain is slow, and withdrawal still allows delayed radiation injury to occur.[25,30,69,70,77,78]

G. Diethylcarbamazine

Diethylcarbamazine inhibits leukotriene synthesis by specific inhibition of the 5-lipoxygenase pathway of arachidonic acid metabolism to prostaglandins. Leukotrienes attract neutrophils and eosinophils to the site of inflammation where they probably release vasoactive and proteolytic agents, resulting in constriction of blood vessels.[79–81] Leukotrienes have been shown to increase vascular permeability.[81–83] Studies of histamine release, slow-reacting substance of anaphylaxis, and prostaglandins in the guinea-pig lung suggest that diethylcarbamazine reactions are similar to isoprenaline and methylxanthines, and that the inhibition of leukotriene synthesis may depend, in part, on elevated cAMP.[80] Leukotriene inhibitors also reduce hydrogen peroxide-induced lung injury in the rat.[84] Arachidonic acid metabolites are important mediators in local and systemic response to skeletal muscle ischemia in dogs. Diethylcarbamazine reduced both white cell activation and infarct size in the dog gracilis muscle.[85] Diethylcarbamazine was effective in reducing vascular permeability at 3 weeks after radiation exposure,[30] and mediating Interleukin-2-induced lung permeability.[86] This provides further evidence of the importance of leukotrienes and prostaglandins in the endothelial injury seen after radiation.

H. ε-Aminocaproic Acid (EACA)

Epsilon amino caproic acid (EACA) acts as an antiproteolytic agent that inhibits agents released from macrophages and neutrophils in response to injury.[87] One result is to inhibit fibrinolysis by inhibiting plasminogen activation.[88–90] EACA has been shown to reduce radiation-induced vascular permeability in rabbit and guinea-pig skin.[16,91,92] Although PS also appeared to be reduced in the irradiated rat lung after EACA treatment, the effect was not significant.[30] While proteolytic enzymes are certainly involved in the radiation reaction, alternative pathways appear to compensate for reduced protease activity when EACA is given.

I. Indomethacin

Indomethacin is an anti-inflammatory drug that inhibits neutrophil aggregation[93] and the cyclooxygenase enzyme pathway for conversion of arachidonic acid to prostaglandins. The anticancer effect of indomethacin may by related to its antagonism to a specific family of peptides and/or its agonistic activity to thromboxane A_2 and melatonin.[94] While the importance of neutrophils has been questioned,[50,95] prostaglandins most likely play a major role in radiation response. Prostaglandins are known to cause vasodilation.[80,81,96] Indomethacin enhances immunological release of the slow-reacting substance of anaphylaxis[80] and blocks the depressor effects of captopril in patients with hypertension.[97] It has been shown to reduce leukocyte infiltration in radiation-induced gastrointestinal injury in rats, but did not affect morphologic changes.[98] It has been used to study the inhibitory effects on the *in vivo* synthesis of thromboxane and prostacyclin in humans,[99] and has been found to block platelet activating factor action, thromboxane and 6-keto-PGF (the stable metabolite of prostacyclin) in pigs.[100] Levels of the prostaglandin PGI_2, a vasodilator, increase steadily in the rat lung between 2 and 6 months after irradiation and may be a compensatory response to lung hypoperfusion.[101] Indomethacin decreases vascular permeability in irradiated guinea pig skin[92] and the rat lung.[30] The effects of indomethacin treatment seem to be surpassed only by dexamethasone, but without the side effects of dexamethasone. This indicates that the prostaglandin pathway is a major mechanism for delayed lung injury and blocking it may be useful clinically.

J. Penicillamine

Penicillamine inhibits collagen crosslinking and maturation.[102] It has been used mostly to ameliorate late effects of radiation where it reduces pulmonary fibrosis, perfusion defects, collagen accumulation, and endothelial enzyme dysfunction.[103–105] Penicillamine was ineffective against radiation pneumonitis and the radiation-induced increase in vascular permeability.[30,106] This result is not surprising in that it seems unlikely that collagen synthesis plays a major role in the early inflammatory process.

K. Pentoxifylline

Pentoxifylline is a xanthine derivative that may have the ability to reduce radiation- and chemotherapy-induced toxicity.[107–110] Its major use is to treat peripheral vascular disease.[111] It enhances microvascular blood flow, decreases platelet aggregation,[108,112] and stimulates prostacyclin production and fibrinolysis.[108,113] Pentoxifylline has been studied for its ability to treat Adult Respiratory Disease Syndrome and has been shown to decrease neutrophil migration.[114] It reduces pulmonary leucostasis, improves capillary patency, and reduces endothelial and epithelial damage associated with peritonitis in rabbits.[115] It has not had a significant effect in preventing or reducing the acute radiation effects after large doses of radiation,[108,116,117] but it appears to improve late radiation complications.[108,109,116] Recent studies, however, show that following radiation doses of 15 to 18 Gy with a dose of 1385 mg/m^2/day (295 mg/kg/day) (comparable to a dose well tolerated in patients), pentoxifylline significantly improves blood flow during the acute injury phase in rat lung, but also seems to slightly impair recovery of blood flow at later times (Stelzer, unpublished results). Pentoxifylline should be an effective drug in controlling postradiation permeability changes, but further studies are needed to confirm its efficacy.

L. Theophylline

Theophylline blocks platelet aggregation by inhibiting arachidonic acid metabolism.[118,119] It blocks histamine release from mast cells[120,121] and is a bronchodilator.[122] Theophylline has also been shown to increase the release of alveolar surfactant[122] and to reduce endotoxin-induced microvascular injury in pig lungs.[123] It reduces the migration of endothelial cells, which are important in the initiation and progression of atherosclerosis.[124] Theophylline was also effective in preventing the radiation-induced increase in vascular permeability at 3 weeks after exposure in the rat lung.[30] Since arachidonic acid metabolism leads to prostaglandin production, these results further support the importance of the pathways that lead from histamine release to prostaglandin production as a major mechanism behind delayed radiation injury to endothelial cells in the lung.

M. Trapidil

Trapidil is an antagonist to platelet-derived growth factor.[125] The potential role for PDGF in lung radiation injury includes stimulation of both vascular smooth muscle cells and fibroblasts.[126] Trapidil has been effective against fibroblast proliferation *in vitro* and in rats.[126] Although vitamin D_3, another PDGF inhibitor, was effective in reducing PS after radiation, trapidil was not effective, possibly because the dose was not high enough, or because fibroblast production is a relatively late effect of radiation damage.

One pathway that was not studied is the possibility that significant intra- or extravascular coagulation occurs, with resultant vascular obstruction and/or macrophage activation. There is evidence[127] that heparin, an anticoagulant, is radioprotective in the liver. This suggests this pathway may be important and should certainly be investigated in future work.

VI. CONCLUSIONS

A growing body of evidence points to the capillary endothelial cells as the most sensitive target for radiation injury in many organs. Because endothelial cells, and the associated basement membranes, are the main barrier controlling flux of material from the bloodstream into tissue, it is not surprising that one of the first signs of radiation injury is edema. Edema can occur within minutes or hours due to release of histamine and other vasoactive agents from tissue mast cells. Later, largely as a result of endothelial cell death, a complex sequence of events occurs, often ending with loss of vasculature and fibrosis.

Several biochemical pathways have been postulated to explain this sequence of events. Attempts to define these pathways with various blocking agents, as is so often the case, are not completely definitive. However, these studies indicate that mast cells, which are seen in large numbers following radiation, are an important aspect of the sequence of events leading to late radiation injury. Prostaglandins and other derivatives of arachidonic acid, probably derived from mast cells, are also important. The results of the above blocking studies allowed us to somewhat simplify the complex scheme of events. Figure 3.6 shows the simplified scheme. It differs from Figure 3.1 in that the pathways of neutrophil activation, complement activation, and proteolytic enzymes have been removed.

Mast cells may well be very important, but why are there so many mast cells present in irradiated tissue? There must be a number of chemo-attractants that derive from other cells already present. A likely candidate for the source of these chemo-attractants is the tissue macrophage. This cell seems to become activated in radiation-injured tissue and is the source for a wide variety of vasoactive and chemo-attractant agents.

In summary, then, there seem to be three separate cell types involved with initiation, amplification, and maintenance of delayed radiation injury. The endothelial cells are the primary cells that maintain homeostasis in the tissue and when they are injured, permeability rises and capillaries become obstructed. Both macrophages and mast cells release a wide variety of biologically active agents that tend to maintain or exacerbate the injury.

This process is of particular interest in the context of radiation oncology. Normal tissue injury is usually the limiting factor in defining a given radiation dose in attempting to treat most tumors. Irradiation of the lung is a particular problem, since late pulmonary fibrosis results in nonfunctional lung and considerable associated morbidity. Studies such as those described here should help define new strategies

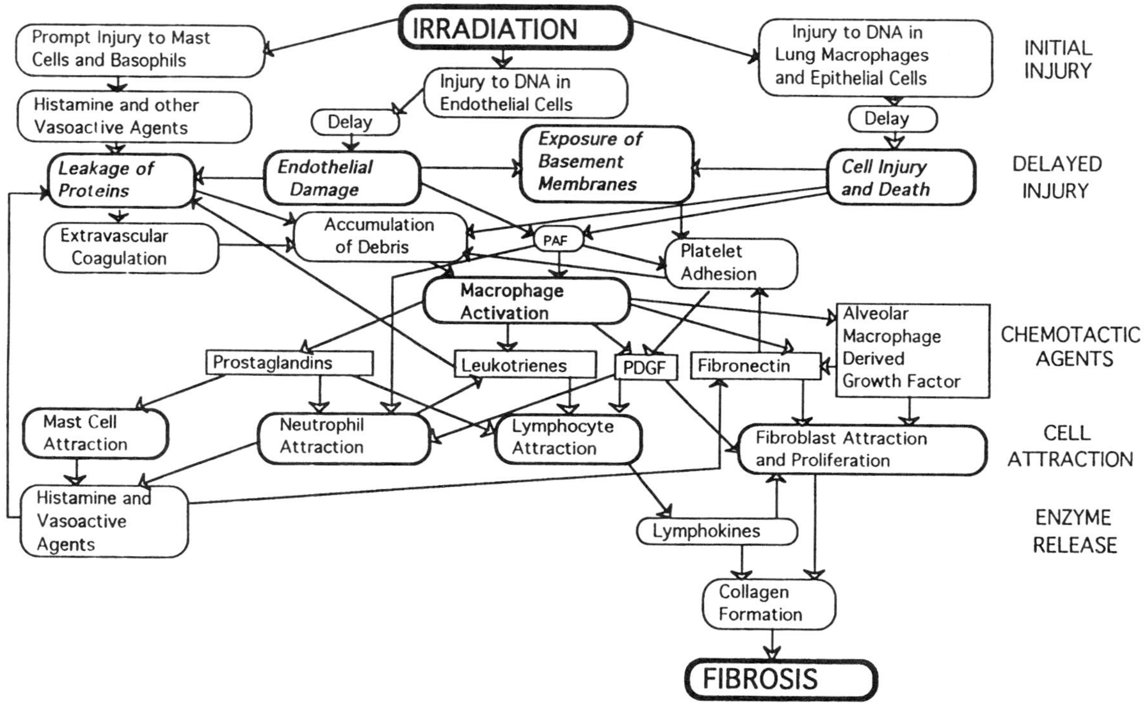

Figure 3.6. Modified diagram of possible pathways and events that may take place following radiation of the lung. The diagram is similar to Figure 3.1, except pathways involving complement activation and proteolytic enzymes have been removed, based on the experimental results described. The process is still likely to be quite complex, involving multiple pathways. (From Graham, M.M. et al., *Int. J. Radiat. Oncol. Biol. Phys.*, 19, 329, 1990. With permission.)

that may prevent or decrease these adverse postradiation effects. It is apparent that potent corticosteroids, such as dexamethasone, can completely suppress the radiation injury effects, but unfortunately those effects often become manifest when the steroid is withdrawn. There is a wide range of possibilities in terms of different doses and tapering schedules that need to be explored to try to optimize the use of steroids. Similarly, there are a number of possible synergisms to explore using other agents such as cromolyn, indomethacin, theophylline, or pentoxifylline. It is quite likely that a combination of one or more of these agents with a corticosteroid will result in an effective treatment to decrease the endothelial injury of radiation as well as the additional late effects.

REFERENCES

1. Gross, N. J., Pulmonary effects of radiation therapy, *Ann. Int. Med.,* 86, 81, 1977.
2. Gross, N. J., The pathogenesis of radiation-induced lung damage, *Lung*, 159, 115, 1981.
3. Phillips, T. L., An ultrastructural study of the development of radiation injury in the lung, *Radiology,* 87, 49, 1966.
4. Phillips, T. L., Benak, S., and Ross, G., Ultrastructural and cellular effects of ionizing radiation, *Front. Radiat. Ther. Oncol.,* 6, 21, 1972.
5. Hunninghake, G. W., Garrett, K. C., Richerson, H. B., Fantone, J. C., Ward, P. A., Rennard, S. I., Bitterman, P. B., and Crystal, R. G., Pathogenesis of the granulomatous lung diseases, *Am. Rev. Respir. Dis.,* 130, 484, 1984.
6. Fantone, J. C., Schrier, D., and Weingarten, B., Inhibition of vascular permeability changes in rats by captopril, *J. Clin. Invest.,* 69, 1207, 1982.
7. Travis, E. L., The sequence of histological changes in mouse lungs after single doses of X-rays, *Int. J. Radiat. Oncol. Biol. Phys.*, 6, 345, 1980.
8. Cotran R. S., Kumar V., and Robbins, S. L., Inflammation and Repair, in *Robbins Pathologic Basic of Disease*, 4th ed., Cotran R. S., Kumar V., and Robbins, S. L., Eds., W.B. Saunders Co., Philadelphia, 1989, 39.
9. Hobson, B. and Denekamp, J., Endothelial proliferation in tumors and normal tissues: continuous labelling studies, *Br. J. Cancer,* 49(4), 405, 1984.
10. Remler, M. P., Marcussen, W. H., and Tiller-Borsich, J., The late effects of radiation on the blood brain barrier, *Int. J. Radiat. Oncol. Biol. Phys.*, 12(11), 1965, 1986.
11. Evans, M. L., Graham, M. M., Mahler, P. A., and Rasey, J. S., Changes in vascular permeability following thorax irradiation in the rat, *Radiat. Res.,* 107, 262, 1986.
12. Law, M. P., Radiation-induced vascular injury and its relation to late effects in normal tissue, In *Advances in Radiation Biology,* Vol. 9, Lett, J. T. and Adler H., Eds., Academic Press, New York, 1981, 37.
13. Law, M. P. and Ahier R. G., Vascular and epithelial damage in the lung of the mouse after X rays or neutrons, *Radiat. Res.,* 117, 128, 1989.
14. Travis, E. L., Harley, R. A., Fenn, J. O., Klobukowski, C. J., and Hargrove, H. B., Pathologic changes in the lung following single and multi-fraction irradiation, *Int. J. Radiat. Oncol. Biol. Phys.*, 2, 475, 1977.
15. Loyd, J. E., Bolds, J. M., Sheller, J. R., Duke, S. S., Gillette, A. W., Malcolm, A. W., Meyrick, B. O., and Brigham, K. L., Acute effects of thoracic irradiation on lung function and structure in awake sheep, *J. Appl. Physiol.*, 62, 208, 1987.

16. Jolles, B. and Harrison, R. G., Enzymic processes and vascular changes in the skin radiation reaction, *Br. J. Radiol.*, 39, 12, 1966.
17. Jolles, B. and Harrison, R. G., Enzymic processes in vascular permeability and fragility changes in the skin radiation reaction, *Bibl. Anat.*, 9, 482, 1967.
18. Hasegawa, A. T. and Wang, R. I. H., Attenuation of increased vascular permeability and toxicity of a radioprotective chemical mixture by chlorpromazine in mice, *Radiat. Res.*, 45, 364, 1971.
19. Song, C. W., Anderson, R. S., and Tabachnick, J., Early effects of beta irradiation on dermal vascular permeability to plasma proteins, *Radiat. Res.*, 27, 604, 1966.
20. Baumgarten, A., Melrose, G. J. H., and Vagg, W. J., Continuous quantitative recording of changes in vascular permeability, *Experientia*, 23, 884, 1967.
21. Shaber, G. S. and Miller L. L., Studies on fibrinogen turnover before and after whole body X-irradiation in the rat, *Proc. Soc. Exp. Biol. Med.*, 113, 346, 1963.
22. Studer, R. and Potchen, J., The radioisotopic assessment of regional microvascular permeability to macromolecules, *Microvas. Res.*, 3, 35, 1971.
23. Graham, M.M., The measurement of capillary permeability changes in the irradiated rat using a double isotope technique, *Armed Forces Radiob. Res. Inst.*, SR71-12, 1, 1971.
24. Graham, M. M. and Evans, M. L., A simple, dual tracer method for the measurement of transvascular flux of albumin into the lung, *Microvas. Res.*, 42, 266, 1991.
25. Peterson, L. M., Evans, M. L., Graham, M. M., Eary, J. F., and Dahlen, D. D., Vascular response to radiation injury in the lung, *Radiat. Res.*, 129, 139, 1992.
26. Donlon, M. A., Role of mast cell mediators in radiation injury and protection, *Pharmacol. Ther.*, 39, 373, 1988.
27. Willoughby, D. A., Pharmacological aspects of the vascular permeability changes in the rat's intestine following abdominal radiation, *Br. J. Radiol.*, 33, 515, 1960.
28. Mount, D. and Bruce, W. R., Local plasma volume and vascular permeability of rabbit skin after irradiation, *Radiat. Res.*, 23, 471, 1964.
29. Tyers, M. B., Bunce, K. T., and Humphrey, P. P., Pharmacological and anti-emetic properties of ondansetron, *Eur. J. Cancer. Clin. Oncol.*, 25(S1), S15, 1989.
30. Graham, M. M., Evans, M. L., Dahlen, D. D., Mahler, P. A., and Rasey, J. S., Pharmacological alteration of the lung vascular response to radiation, *Int. J. Radiat. Oncol. Biol. Phys.*, 19, 329, 1990.
31. Swartz, S. L., Williams, G. H., Hollenberg, N. K., Levine, L., Dluhy, R. G., and Moore, T. J., Captopril-induced changes in prostaglandin production: relationship to vascular response in normal man, *J. Clin. Invest.*, 65, 1257, 1980.
32. Wagner, B., Eichler, H.-G., Schneider, B., Blöchl-Daum, B., Brenner, B., Gasic, S., Lechner, L., and Kyrle, P. A., Studies on the effect of two angiotensin-converting enzyme inhibitors, captopril and cilazapril, on platelet and vascular prostaglandin metabolism *in vivo*, *Naunyn-Schmiedeberg's Arch. Pharmacol.*, 346, 453, 1992.
33. Laffan, R. J., Goldberg, M. E., High, J. P., Schaetter, R. R., Waugh, M. H., and Rubin, B., Antihypertensive activity in rats of SQ14225, an orally active inhibitor of angiotensin I-converting enzyme, *J. Pharmacol. Exp. Ther.*, 204, 281, 1978.
34. Ambrosioni, E., Bacchelli, S., Esposti, D. D., and Borghi, C., ACE-inhibitors and atherosclerosis, *Eur. J. Epidemiol.*, S1, 129, 1992.
35. Ward, W. F., Kim, Y. T., Molteni, A., and Solliday, N. H., Radiation-induced pulmonary endothelial dysfunction in rats: modification by an inhibitor of angiotensin converting enzyme. *Int. J. Radiat. Oncol. Biol. Phys.*, 15, 135, 1988.
36. Molteni, A., Ward, W. F., Ts'ao, C., Solliday, N. H., and Dunnes, M., Monocrotaline-induced pulmonary fibrosis in rats: amelioration by captopril and penicillamine, *Proc. Soc. Exp. Biol. Med.*, 180, 112, 1985.

37. Ward, W. F., Molteni, A., Ts'ao, C.-h., and Hinz, J. M., Captopril reduces collagen and mast cell accumulation in irradiated rat lung, *Int. J. Radiat. Oncol. Biol. Phys.*, 19, 1405, 1990.
38. Cohen, E. P., Fish, B. L., and Moulder, J. E., Treatment of radiation nephropathy with captopril, *Radiat. Res.*, 132(3), 346, 1992.
39. Ryan, U. S. and Ryan, J. W., Endothelial cells and inflammation, *Clin. Lab. Med.*, 3, 577, 1983.
40. Ullrich R. L. and Casarett, G. W., Interrelationship between the early inflammatory response and subsequent fibrosis after radiation exposure, *Radiat. Res.*, 72, 107, 1977.
41. Cochrane, C. G., Muller-Eberhard, H. J., and Aikin, B. S., Depletion of plasma complement *in vivo* by a protein of cobra venom: its effects on various immunologic reactions, *J. Immunol.*, 105, 55, 1970.
42. Mulligan, M. S., Jones, M. L., Bolanowski, M. A., Baganoff, M. P., Deppeler, C. L., Meyers, D. M., Ryan, U. S., and Ward, P.A., Inhibition of lung inflammatory reactions in rats by an anti-human IL-8 antibody, *J. Immunol.*, 150, 5585, 1993.
43. Mulligan, M. S., Polley, M. J., Bayer, R. J., Nunn, M. F., Paulson, J. C., and Ward, P. A., Neutrophil-dependent acute lung injury, *J. Clin. Invest.*, 90, 1600, 1992.
44. Mulligan, M. S., Watson, S. R., Fennie, C., and Ward, P. A., Protective effects of selectin chimeras in neutrophil-mediated lung injury, *J. Immunol.*, 151, 6410, 1993.
45. Mulligan, M. S., Yeh, C. G., Rudolph, A. R., and Ward, P. A., Protective effects of soluble CR1 in complement- and neutrophil-mediated tissue injury, *J. Immunol.*, 148, 1479, 1992.
46. Marquardt, D. L., Walker, L. L., and Wasserman, S. I., Cromolyn inhibition of mediator release in mast cells derived from mouse bone marrow, *Am. Rev. Respir. Dis.*, 133, 1105, 1986.
47. Travis, E. L., Down, J. D., Holmes, S. J., and Hobson, B., Radiation pneumonitis and fibrosis in mouse lung assayed by respiratory frequency and histology, *Radiat. Res.*, 84, 133, 1980.
48. Lykke, A. W. J., Schonell, M. E., and Stewart, B. W., Atypical mast cell degranulation and focal hydropic degeneration of venular endothelium in diffuse fibrosing alveolitis, *Experientia,* 35, 1492, 1979.
49. Atkins, F. M., Friedman, M. M., Subba Rao P. V., and Metcalfe, D. D., Interactions between mast cells, fibroblasts and connective tissue components, *Int. Archs. Allergy Appl. Immunol.*, 77, 96, 1985.
50. Ward, P. A. and Till, G. O., Pathophysiologic events related to thermal injury of skin, *J. Trauma,* 30(12S), S75, 1990.
51. Stone, C. A., Wagner, H. C., Ludden, C. T., Stavorski, J. M., and Ross, C. A., Antiserotonin-antihistaminic properties of cyproheptadine, *J. Pharmacol. Exptl. Ther.*, 131, 73, 1961.
52. Movat, H. Z., The vascular phenomena of acute inflammation, *The Inflammatory Reaction.* Elsevier Science Publishers, Elsevier, NY, 1985, 43.
53. Slauson, D. O., Hahn, F. F., Benjamin, S. A., Chiffelle, T. L., and Jones, R. K., Inflammatory sequences in acute pulmonary radiation injury, *Am. J. Pathol.*, 82, 549, 1976.
54. Martinet, Y., Ron, W. N. G., Grotendorst, R., Martin, G. R., and Crystal, R. G., Exaggerated spontaneous release of platelet-derived growth factor by alveolar macrophages from patients with idiopathic pulmonary fibrosis, *N. Engl. J. Med.*, 317, 202, 1987.
55. Deuel, T. F., Senior, R. M., San Huang, J., and Griffin, G. L., Chemotaxis of monocytes and neutrophils to platelet-derived growth factor, *J. Clin. Invest.*, 69, 1046, 1982.

56. Seppä, H., Grotendorst, G., Seppä, S., Schiffman, E., and Martin, G. R., Platelet-derived growth factor is chemotactic for fibroblasts, *J. Cell. Biol.,* 92, 584, 1982.
57. Dosquet-Bernard, C., Wilheim, F., Lomri, N., Tobelem, G., and Caen, J. P., Dihydroxy-vitamin D_3 modulates the growth of 3T3 cells and human skin fibroblasts stimulated by platelet-derived growth factor, *Cell. Biol. Int. Rep.,* 10, 931, 1986.
58. Kohler, N. and Lipton, A., Platelets as a source of fibroblast growth-promoting activity, *Exptl. Cell. Res.*, 87, 297, 1974.
59. Rutherford, R. B. and Ross, R., Platelet factors stimulate fibroblasts and smooth muscle cells quiescent in plasma serum to proliferate, *J. Cell. Biol.,* 69, 196, 1976.
60. Sinzinger, H. and Firbas, W., Irradiation depresses prostacyclin generation upon stimulation with the platelet-derived growth factor, *Br. J. Radiol.*, 58, 1023, 1985.
61. Nachman, R. L., Weksler, B., and Ferris, B., Characterization of human platelet vascular permeability-enhancing activity, *J. Clin. Invest.,* 51, 549, 1972.
62. Haynes, R. C., Adrenocorticotropic hormone; adrenocortical steroids and their synthetic analogs; inhibitors of the synthesis and actions of adrenocortical hormones, in *The Pharmacological Basis of Therapeutics*, 8th ed., Gilman, A. G., Rall, T. W., Nies, A. S., and Taylor, P., Eds., Pergamon Press, New York, 1990, 1431.
63. Di Rosa, M., Role of inflammation of glucocorticoid-induced phospholipase inhibitory proteins, *Prog. Biochem. Pharmacol.*, 20, 55, 1985.
64. Granelli-Piperno, A., Vassalli, J.-D., and Reich, E., Secretion of plasminogen activator by human polymorphonuclear leukocytes, *J. Exp. Med.,* 146, 1693, 1977.
65. Phillips, T. L., Wharam, M. D., and Margolis, L. W., Modification of radiation injury to normal tissues by chemotherapeutic agents, *Cancer*, 35, 1678, 1975.
66. Van Den Brenk, H. A. S., Radiation effects on the pulmonary system. In *Pathology of Irradiation,* Berdjis, C., Ed., Williams and Wilkins, Baltimore, MD, 1971, 569.
67. Berdjis, C. C. and Brown, R. F., Histopathology of the effect of cortisone on the irradiated rat lung, *Dis. Chest.*, 32, 481, 1957.
68. Gross, N. J., Radiation pneumonitis in mice. Some effects of corticosteroids on mortality and pulmonary physiology, *J. Clin. Invest.,* 66, 504, 1980.
69. Gross, N. J. and Narine, K. R., Experimental radiation pneumonitis: corticosteroids increase the replicative activity of alveolar type 2 cells, *Radiat. Res.*, 115, 543, 1988.
70. Gross, N. J., Narine, K. R., and Wade, R., Protective effect of corticosteroids on radiation pneumonitis in mice, *Radiat. Res.*, 113, 112, 1988.
71. Rugstad, H. E., Antiinflammatory and immunoregulatory effects of glucocorticoids: mode of action, *Scand. J. Rheumatol.*, S76, 257, 1988.
72. Colburn, M. D., Moore, W. S., Gelabert, H. A., and Quiñones-Baldrich, W. J., Dose responsive suppression of myointimal hyperplasia by dexamethasone, *J. Vasc. Surg.*, 15, 510, 1992.
73. Poon, M., Megyesi, J., Green, R. S., Zhang, H., Rollins, B. J., Safirstein, R., and Taubman, M. B., *In vivo* and *in vitro* inhibition of JE gene expression by glucocorticoids, *J. Biol. Chem.*, 266(33), 22375, 1991.
74. Nakano, T., Raines, E. W., Abraham, J. A., Wenzel, F. G., Higashiyama, S., Klagsbrun, M., and Ross, R., Glucocorticoid inhibits thrombin-induced expression of platelet-derived growth factor A-chain and heparin-binding epidermal growth factor-like growth factor in human aortic smooth muscle cells, *J. Biol. Chem.*, 268(30), 22941, 1993.
75. Moss, W. T., Haddy, F. J., and Sweany, S. K., Some factors altering the severity of acute radiation pneumonitis: variation with cortisone, heparin, and antibiotics, *Radiology,* 75, 50, 1960.

76. Evans, M. L., Graham, M. M., Mahler, P. A., and Rasey, J. S., Use of steroids to suppress vascular response to radiation, *Int. J. Radiat. Oncol. Biol. Phys.*, 13, 563, 1987.
77. Peterson, L. M., Evans, M. L., Thomas, K. L., and Graham, M. M., Vascular response to fractionated irradiation in the rat lung, *Radiat. Res.*, 131, 224, 1992.
78. Castellino, R. A., Glatstein, E., Turbow, M. M., Rosenberg, S., and Kaplan, H. S., Latent radiation injury of lungs or heart activated by steroid withdrawal, *Ann. Int. Med.*, 80, 593, 1974.
79. Matthews, W. R. and Murphy, R. C., Inhibition of leukotriene biosynthesis in mastocytoma cells by diethylcarbamazine, *Biochem. Pharmacol.*, 31, 2129, 1982.
80. Piper, P. J. and Temple, D. M., The effect of lipoxygenase inhibitors and diethylcarbamazine on the immunological release of slow reacting substance of anaphylaxis (SRS-A) from guinea-pig chopped lung, *J. Pharm. Pharmacol.*, 33, 384, 1981.
81. Marx, J. L., The leukotrienes in allergy and inflammation, *Science*, 215, 1380, 1982.
82. Dahlen, S., Bjork, J., Hedqvist, P., Arfors, K., Hammarstrom, S., Lindgren, J., and Samuelsson, B., Leukotrienes promote plasma leakage and leukocyte adhesion in postcapillary venules: *in vivo* effects with relevance to the acute inflammatory response, *Proc. Natl. Aced. Sci.*, 78, 3887, 1981.
83. Persson, C. G. A., Erjefalt, I., and Andersson, P., Leakage of macromolecules from guinea-pig tracheobronchial microcirculation.Effects of allergen, leukotrienes, tachykinins, and anti-asthma drugs, *Acta. Physiol. Scand.*, 127, 95, 1986.
84. Burghuber, O. C., Strife, R. J., Zirrolli, J., Henson, P. M., Henson, J. E., Mathias, M. M., Reeves, J. T., Murphy, R. C., and Voelkel, N. F., Leukotriene inhibitors attenuate rat lung injury induced by hydrogen peroxide, *Am. Rev. Respir. Dis.*, 131, 778, 1985.
85. Cambria, R. A., Anderson, R. J., Dikdan, G., Lysz, T. W., and Hobson, R. W., The influence of arachidonic acid metabolites on leukocyte activation and skeletal muscle injury after ischemia and reperfusion, *J. Vasc. Surg.*, 14, 549, 1991.
86. Klausner, J. M., Goldman, G., Skornick, Y., Valeri, R., Inbar, M., Shepro, D., and Hechtman, H. B., Interleukin-2-induced lung permeability is mediated by leukotriene B_4, *Cancer*, 66, 2357, 1990.
87. Janoff, A. and Carp, H., Proteases, antiproteases, and oxidants: pathways of tissue injury during inflammation, *Monogr. Pathol.*, 23, 62, 1982.
88. Alkjaersig, N., Fletcher, A. P., and Sherry, S., Aminocaproic acid: an inhibitor of plasminogen activation, *J. Biol. Chem.*, 234, 832, 1959.
89. Ambrus, C. M., Ambrus, J. L., Lassman, H. B., and Mink, I. B., Studies on the mechanism of action of inhibitors of the fibrinolysin system, *Ann. NY Acad. Sci.*, 146, 430, 1968.
90. Ts'ao, C., Ward, W. F., and Port, C. D., Radiation injury in rat lung. III. Plasminogen activator and fibrinolytic inhibitor activities, *Radiat. Res.*, 96, 301, 1983.
91. Eassa, E. M. and Casarett, G. W., Effect of epsilon-amino-*n*-caproic acid (EACA) on radiation-induced increase in capillary permeability, *Radiology*, 106, 679, 1973.
92. Song, C. W., Drescher, J. J., and Tabachnick, J., Effects of anti-inflammatory compounds on beta-irradiation-induced increase in vascular permeability, *Radiat. Res.*, 34, 616, 1968.
93. Abrahamson, S., Edelson, H., Kaplan, H., Ludewig, R., and Weissmann, G., Inhibition of neutophil-activation by nonsteroidal anti-inflammatory drugs, *Am. J. Med.*, 4B, 3, 1984.
94. Szkudlinski, M. W., A proposed mechanism of anticancer activity of indomethacin, *Med. Hypoth.*, 39, 265, 1992.

95. Rinaldo, J. E. and Rogers, R. M., Adult respiratory distress syndrome, *N. Engl. J. Med.,* 315, 578, 1986.
96. Young, M. R. and Knies, S., Prostaglandin E production by Lewis lung carcinoma: mechanism for tumor establishment *in vivo*, *JNCI,* 72, 919, 1984.
97. Seto, S., Aoi, W., Iwami, K., Yamaguchi, T., Ashizawa, N., Kusano, S., Baba, K., Doi, Y., Kuramochi, M., and Hashiba, K., Effect of propranolol and indomethacin on the depressor action of captopril in patients with essential hypertension, *Clin. Exper. Theory and Pract.*, A9(2&3), 623, 1987.
98. Rose, P. G., Halter, S. A., and Su, C. M., The effect of indomethacin on acute radiation induced gastrointestinal injury: a morphologic study, *J. Surg. Oncol.*, 49, 231, 1992.
99. Drvota, V., Vesterqvist, O., and Green, K., Effects of non-steroidal anti-inflammatory drugs on the *in vivo* synthesis of thromboxane and prostacyclin in humans, *Adv. Prost. Thromb. Leuko. Res.*, 21, 153, 1990.
100. Goldstein, R. E., Feuerstein, G. Z., Bradley, L. M., Stambouly, J. J., Laurindo, F. R. M., and Davenport, N. J., Cardiovascular effects of platelet-activating factor, *Lipids*, 26, 1250, 1991.
101. Ts'ao, C., Ward, W. F., and Port, C. D., Radiation injury in rat lung. I. Prostacyclin production, arterial perfusion, and ultrastructure, *Radiat. Res.*, 96, 284, 1983.
102. Deshmukh, K. and Nimni, M., A defect in the intramolecular and intermolecular cross-linking of collagen caused by penicillamine, *J. Biol. Chem.,* 244, 1787, 1969.
103. Ward, W. F., Shih-Hoellwarth, A., Port, C. D., and Kim, Y. T., Modification of radiation-induced pulmonary fibrosis in rats, *Radiology,* 131, 751, 1979.
104. Ward, W. F., Radiation-induced pulmonary arterial perfusion defects: modification by D-penicillamine, *Radiology,* 139, 201, 1981.
105. Ward, W. F., Molteni, A., Ts'ao, C., and Solliday, N. H., Radiation injury in rat lung. IV. Modification by D-penicillamine, *Radiat. Res.*, 98, 397, 1984.
106. Port, C. D. and Ward, W. F., The ultrastructure of radiation injury in rat lung: modification by D-penicillamine, *Radiat. Res.,* 92, 61, 1982.
107. Bianco, J. A., Appelbaum, F. R., Nemunaitis, J., Almgren, J., Andrews, F., Kettner, P., Shields, A., and Singer, J. W., Phase I-II trial of pentoxiphylline for the prevention of transplant related toxicities following bone marrow transplantation, *Blood*, 78, 1205, 1991.
108. Dion, M. W., Hussey, D. H., and Osborne, J. S., The effect of pentoxiphylline on early and late radiation injury following fractionated irradiation in C3H mice, *Int. J. Radiat. Oncol. Biol. Phys.*, 17, 101, 1989.
109. Dion, M. W., Hussey, D. H., Doornbos, J. F., Bigliotti, A. P., Wen, B., and Anderson, B., Preliminary results of a pilot study of pentoxiphylline in the treatment of late radiation soft tissue necrosis, *Int. J. Radiat. Oncol. Biol. Phys.*, 19, 401, 1990.
110. Werner-Wasik, M. and Madoc-Jones, H., Trental (pentoxiphylline) relieves pain from postradiation fibrosis (letter), *Int. J. Radiat. Oncol. Biol. Phys.*, 25, 757, 1993.
111. Rall, T. W., Drugs used in the treatment of asthma: the methylxanthines, cromolyn sodium, and other agents, in *The Pharmacologic Basis of Therapeutics*, 8th ed., Gilman, A. G., Rall, T. W., Nies, A. S., and Taylor, P., Eds., Pergamon Press, New York, 1990, 618.
112. Song, C. W., Hasegawa, T., Kwon, H. C., Lyons, J. C., and Levitt, S. H., Increase in tumor oxygenation and radiosensitivity caused by pentoxiphylline, *Radiat. Res.*, 130, 205, 1992.
113. Matzky, R., Darius, H., and Schror, K., The release of prostacyclin (PGI2) by pentoxiphylline from human vascular tissue, *Arzeneimittelforschung*, 32(10), 1315, 1982.

114. Mandell, G. L., ARDS, neutrophils, and pentoxiphylline, *Am. Rev. Respir. Dis.*, 138, 1103, 1988.
115. Tighe, D., Moss, R., Heath, M. F., Hynd, J., and Bennett, E. D., Pentoxiphylline reduces pulmonary leucostasis and improves capillary patency in a rabbit peritonitis model, *Circ. Shock*, 28, 159, 1989.
116. Koh, W.-J., Stelzer, K. J., Peterson, L. M., Staker, B. L., Ward, W. F., Russell, K. J., and Griffin, T. W., Effect of pentoxiphylline on radiation-induced lung and skin toxicity in rats, *Int. J. Radiat. Oncol. Biol. Physc.*, 31, 71, 1995.
117. Ward, W. F., Kim, Y. T., Molteni, A., Ts'ao, C., and Hinz, J. M., Pentoxiphylline does not spare acute radiation reactions in rat lung and skin, *Radiat. Res.*, 129, 107, 1992.
118. Fuster, V., Adams, P. C., Badimon, J. J., and Cheseboro, J. H., Platelet-inhibitor drugs' role in coronary artery disease, *Prog. Cardiovas. Dis.*, 29, 325, 1987.
119. Soslau, G. and Parker, J., Modulation of platelet function by extracellular adenosine triphosphate, *Blood*, 74(3), 984, 1989.
120. Pearce, F. L., Flint, K. C., Leung, K. B. P., Hudspith, B. N., Seager, K., Hammond, M. D., Brostoff, J., Geraint-James, D., and Johnson, N. M., Some studies on human pulmonary mast cells obtained by bronchoalveolar lavage and by enzymic dissociation of whole lung tissue, *Int. Archs. Allergy Appl. Immunol.*, 82, 507, 1987.
121. Kurosawa, M., Mori, H., Nagai, H. and Koda, A., Change in the activity of the cyclic AMP-dependent protein kinase in antigen-stimulated sensitized mast cells and effect of drugs inhibiting allergic mediator release, *Jpn. J. Pharmacol.*, 43, 454, 1987.
122. Persson, C. G. A., Overview of effects of theophylline, *J. Allergy Clin. Immunol.*, 78, 780, 1986.
123. Urbain, B., Gustin, P., and Ansay, M., Endotoxin-induced microvascular injury in isolated and perfused pig lungs, *Vet. Res. Commun.*, 16(6), 453, 1992.
124. Naito, M., Hayashi, T., Kuzuya, M., Funaki, C., Asai, K., and Kuzuya, F., Vascular endothelial cell migration *in vitro*: roles of cyclic nucleotides, calcium ion, and cytoskeletal system, *Artery*, 17(1), 21, 1989.
125. Ohnishi, H., Yamaguchi, K., Shimada, S., Suzuki, Y., and Kumagai, A., A new approach to the treatment of atherosclerosis and trapidil as an antagonist to platelet-derived growth factor, *Life Sci.*, 28, 1641, 1981.
126. Tiell, M. L., Sussman, I. I., Gordon, P. B., and Saunders, R. N., Suppression of fibroblast proliferation *in vitro* and of myointimal hyperplasia *in vivo* by the trazolopyrimidine, trapidil, *Artery*, 12, 33, 1983.
127. Kinzie, J., Studer Blanca Perez, R. K., and Potchen, E. J., Noncytokinetic radiation injury: anticoagulants as radioprotective agents in experimental radiation hepatitis, Science, 175(29), 1481, 1972.

CHAPTER 4

Antioxidant Defenses and the Irradiated Endothelium

David B. Rubin, Elizabeth A. Drab-Weiss, and Ed Robert Blazek

CONTENTS

I. INTRODUCTION

Endothelial cells are important targets of injury because they line blood vessels and regulate the interaction between the blood and surrounding tissue.[1] Among the most common causes of endothelial injury is oxidative stress from various sources, including oxidized lipids,[2] inflammation,[3] therapy with oxygen,[4] anti-cancer drugs,[5,6] and radiation.[7] Two-thirds of the cytotoxicity created by low linear energy transfer (LET) radiation, the type used in radiation therapy, is likely caused by the generation of reactive oxygen species (ROS).[8–10] In cells, low LET radiation deposits energy primarily to water, creating ROS including superoxide, O_2^-, and the hydroxyl radical,

1-8493-4840-4/98/$0.00+$.50

OH·. Because OH· is relatively long-lived (10^{-5} seconds), it is regarded as the primary mediator of the oxidant stress created by exposure to low LET radiation.[10]

As discussed in Chapter 2 on histopathology, the endothelial cell frequently appears to be the first cell perturbed after irradiation and other types of oxidative stresses.[7,11–13] It can be argued that an oxidant stress that perturbs the endothelium is one of the most common events in the pathogenesis of tissue toxicity. Experimentally, strengthening antioxidant defenses attenuates vascular endothelial damage, while weakening these defenses exacerbates the damage.[14–21] Hypothetically, if endothelial antioxidant defenses were bolstered, cell survival and the capacity to restore a damaged endothelium could be promoted in order to reduce tissue toxicity and enhance healing.

Although the importance of antioxidant defenses to irradiated endothelial cells is assumed, compared to other types of oxidant stress, there are relatively few studies on the topic. Much of what is known about the endothelial antioxidant defenses is derived from investigations on injury caused by exposure to H_2O_2[3,14,22–38] and hyperoxia.[4,39–45] It is understood, however, that the endothelial cell response to an oxidant stress depends on the agent creating the ROS. For example, endothelial cells surviving exposure to hyperoxia, paraquat, or O_2^- generation, manifest different profiles of antioxidant enzyme activities.[39] Also in endothelial cells, although OH· mediates the cytotoxicity of exposure to either radiation or H_2O_2, radiation creates both single- and double-strand DNA breaks, while H_2O_2 generates only single-strand breaks.[46] Finally, when a particular antioxidant defense is supplemented in cultured endothelial cells, cytoprotection depends on the source of the ROS. Nevertheless, at the end of the chapter, endothelial toxicity induced by H_2O_2 will be further discussed as a means to fill gaps in the knowledge of the role of oxidants and the irradiated endothelial cell.

II. SUPEROXIDE DISMUTASE (SOD)

SOD is an antioxidant enzyme discovered in prokaryotes and later found in eukaryotes.[47] SOD scavenges O_2^- and converts it to H_2O_2. SOD comes in two forms, CuZnSOD located in the cytoplasm and MnSOD found in the mitochondria.[48,49] CuZnSOD is available for administration *in vivo* and *in vitro* and has a longer half-life when conjugated to polyethylene glycol (PEG-CuZnSOD).[50,51] Treatment with CuZnSOD protects rodents against radiation toxicity.[52–56] Also, irradiation can induce MnSOD.[57] The induction of MnSOD by a low-level oxidant stress may protect tissues against toxicity created by a subsequent high-level oxidant stress. This priming of antioxidant defenses is well documented in rats exposed to hyperoxia[58–60] and also reported for cultured endothelial cells exposed to endotoxin.[16] The radioprotection of rats rendered by pre-exposure to hyperoxia[61] or endotoxin[57] has been associated with elevated levels of SOD activity. Although baseline levels of SOD activity do not correlate with the radiosensitivity of human cancer cell lines,[49] inducible SOD activity could be important in the development of radiation resistance.

In cultured endothelial cells exposed to hyperoxia, both an elevated level of SOD activity[43] and treatment with PEG-CuZnSOD[51] relate to improved cell survival following oxidant stress. In contrast, the role of SOD in the endothelial cytotoxicity caused by irradiation is less clear. Irradiated bovine aortic endothelial cells (BAEC) that remain adhered 1 day after exposure to 5 Gy have elevated MnSOD activity compared to nonirradiated cells: +70% when corrected for DNA content per culture.[62] Also, bovine aortic smooth muscle cells, which are more radioresistant than endothelial cells, have 2 to 3 times more MnSOD activity than found in the BAEC.[62] Although the elevated SOD activity is associated with endothelial cell survival, when the BAEC were irradiated in the presence of either unconjugated CuZnSOD or PEG-CuZnSOD, the loss of adhered cells was not attenuated despite an increase in total cellular SOD activity.[63]

SOD activity may regulate cell division.[64] The cytokine, interleukin-3 inhibits the division of mouse bone marrow cells through the activity of CuZnSOD. Because the state of cell proliferation determines the response to irradiation, understanding the role of SOD in endothelial cell proliferation could be important. In cultured BAEC, however, the level of SOD activity, when corrected for protein content, does not relate to the proliferative state of the endothelial cells (Figure 4.1), which is unlike the proliferation-related changes of glutathione (GSH) content that are described below.

III. GLUTATHIONE (GSH)

After SOD catalyzes the dismutation of O_2^- to form H_2O_2, catalase promotes the detoxification of H_2O_2 to form H_2O and O_2. If H_2O_2 is not removed, it is reduced to 2 OH· with the help of metal cations (the Fenton reaction). OH· is highly reactive with sulfhydryl groups and is scavenged by both protein thiols and non-protein thiols (NPSH). GSH, the tripeptide γ-glutamyl-cysteinyl-glycine, is the most abundant NPSH in the cell and when its sulfhydryl group donates a H· to OH·, H_2O and the disulfide of GSH, GSSG, are formed. This reaction is catalyzed by glutathione peroxidase (GSH-Px). The rate-limiting step of GSH synthesis is γ-glutamyl-cysteinyl synthetase,[65] which is specifically inhibited by the methionine analog, buthionine sulfoximine (BSO).[66]

The content of GSH has been related to both cell proliferation[67–72] and resistance to irradiation.[73–76] As depicted in Figure 4.1, GSH content increases as cell cultures make the transition from lag phase to logarithmic growth. As found in other cell types (e.g., in stimulated lymphocytes,[77]) GSH content is also related to cell cycle progression and appears lowest in early G_1 phase and peaks in S phase (Figure 4.2). This relationship of GSH content to cell cycle distribution may partly explain why BAEC enriched in S phase are more radioresistant than EC in G_1 phase.[78]

Lowering GSH content, however, is not clearly associated with inhibited endothelial cell growth. On one hand, decreasing GSH content in BAEC with various agents, including BSO, lowers colony formation in both control and irradiated cells (Table 4.1). Also, on occasion, in asynchronously dividing BAEC treated with BSO, one documents retarded cell cycle progression from both G_1 to S phase and at G_2M (Figure 4.3). We have been unable to demonstrate consistently, however, that lowering levels of GSH with BSO to 10% of control values slows BAEC proliferation or alters distribution in the cell cycle (data not shown). On the other hand, supplementing culture medium with extra cysteine (2 m*M*) prior to exposure to 5 Gy,

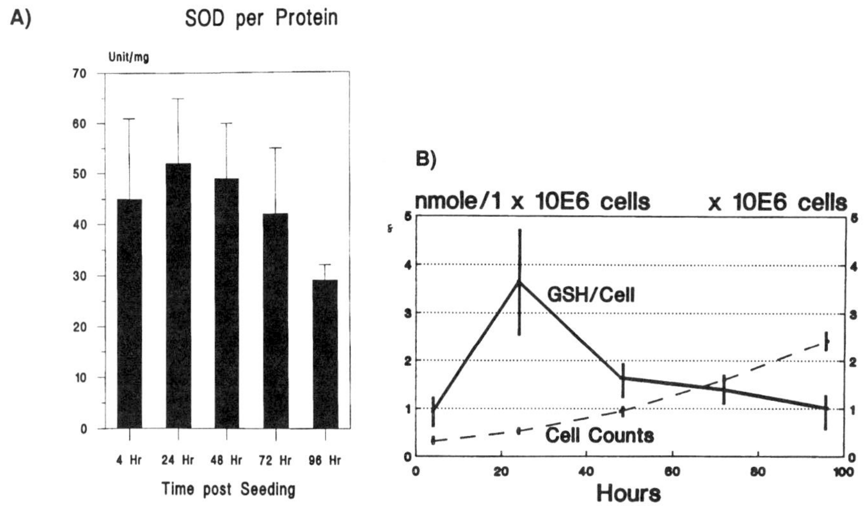

Figure 4.1. Superoxide dismutase (SOD) activity and glutathione (GSH) content as related to cell proliferation. Bovine aortic endothelial cells (BAEC) were seeded in culture. During lag phase and logarithmic growth, A: total SOD activity and B: GSH content were measured as reported.[64,82] Cell counts are shown in B. Values are mean ± SEM, n = 6.

promotes both DNA synthesis (% ^{3}H-Tdr incorporation/10^6 cells: Control=13%, 5 Gy=3%, Cys+5Gy= 6%; mean of n=3) and an increased S phase distribution (Figure 4.4).

The entry of various NPSH into cells promotes both the uptake of cysteine and GSH synthesis.[79–81] In a similar fashion, the radioprotective aminothiol, *N*-(2-mercaptoethyl)-1,3-diaminopropane (WR-1065), elevates GSH content in BAEC.[82] This elevation of GSH content is prevented by BSO exposure, implying that WR-1065 uptake provides increased substrate for GSH synthesis, possibly by promoting cysteine uptake. Regardless of the GSH content, however, WR-1065 radioprotects BAEC with equal effectiveness, which indicates that GSH does not mediate the radioprotection. This observation is similar to that made in other cell types.[83]

An oxidative stress created by irradiation should transiently increase GSSG but this has not been documented in endothelial cells or other cell types. Twenty-four hours after BAEC are exposed to radiation, there are small increases of both the total cellular GSH content and the fluorescence of a thiol probe, monochlorobimane[84,85] (Figure 4.5). Although these measurements indicate that there is an increase in thiol content per cell, these changes appear to be related to an increase in cell volume and total protein, and not an increased thiol concentration.

The increased formation of protein-(NPSH) mixed disulfides, however, provides indirect evidence that thiols are oxidized in irradiated BAEC.[82] Formation of protein-(NPSH) mixed disulfides may be an important post-translational modification that alters protein function.[86,87] Because GSH is the most abundant NPSH, protein-GSH is likely the most common type of protein mixed disulfide.[86] Protein-GSH mixed disulfides are created when GSH thiolates a thiyl radical (RS·). Protein-(NPSH)

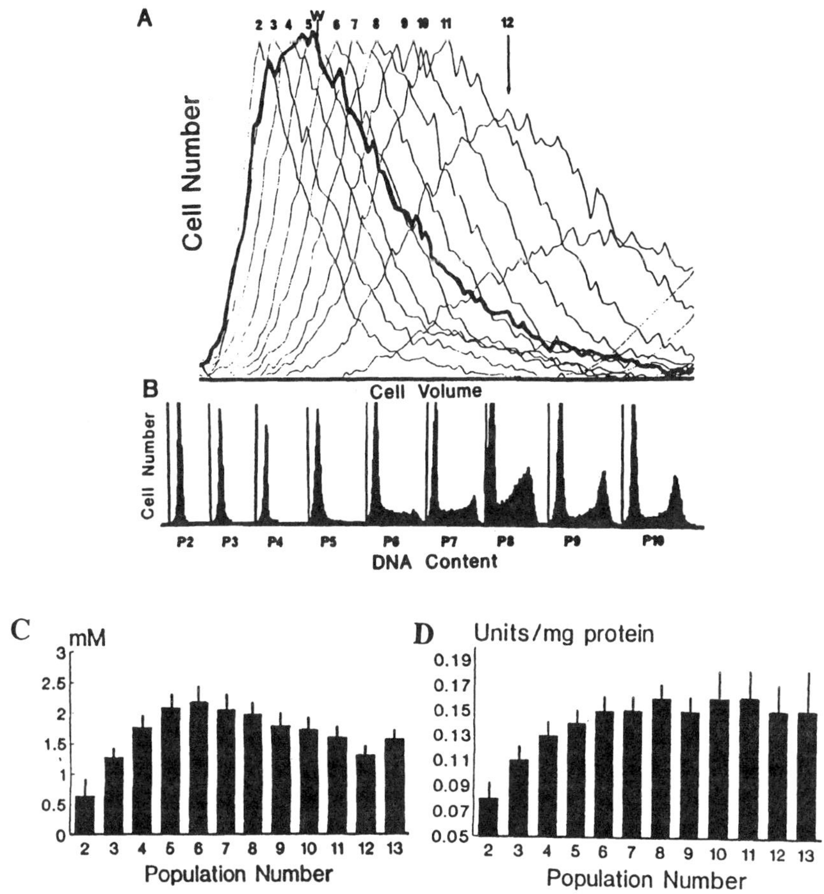

Figure 4.2. The relationship between cell cycle and GSH content and glutathione reductase (GSSG-R) activity. BAEC were separated by centrifugal elutriation as reported.[78] A: An example of volume distribution, volume on the x-axis and cell number on the y-axis. Before separation, bold line designated "W" for whole population; and in the separated EC populations, thin lines that are numbered 2 through 12, the sequence of collection. Population 1 is discarded because there were too few cells to analyze. B: DNA cytohistograms are shown that correspond to the depicted volume distributions in populations 2 through 10 (P2-P10). C: GSH was measured in the separated EC populations and values were corrected for cell number and volume to derive concentration. D: GSSG-R was measured in the separated EC populations and the values were corrected for total cell protein. Values depicted are mean ± SEM, n = 8.

mixed disulfides turnover when a thiol:disulfide exchange occurs between the mixed disulfide and GSSG or a disulfide of another NPSH. The creation or turnover of protein-NPSH mixed disulfides appear to be important to the enzymes, carbonic anhydrase, glyceraldehyde-3-phosphate dehydrogenase, xanthine dehydrogenase conversion to xanthine oxidase, phosphorylase b, and actin.[86–91]

Table 4.1 Glutathione, Proliferation, and Radiation Sensitivity in Bovine Aortic Endothelial Cells

Drug Endpoint	Control	OTZ	BSO	DEM	ETOH .1% (DEM Control)
GSH content per cell (nmol/10[a])	1.8 ±0.2	2.9[a] ±0.1	0.3[b] ±0.1	0.02[c] ±0.02	2.3 ±0.2
Colony number without irradiation	165 ±13	135 ±15	127 ±48	38[c] ±13	162 ±8
Percent survival after 5 Gy	3.6 ±0.4	3.9 ±1.1	2.6[a] ±0.4	3.2[c] ±1.4	5.5[a] ±0.6

Note: Endothelial cells were cultured from bovine aortas and exposed to 2-oxo-4-thoazolidine (OTZ) 10 m*M* for 2 hr, buthionine sulfoxine (BSO) 50 μ*M* for 20 hr, diethyl maleate (DEM) 1 m*M* for 1 hr, and ethanol 1% for 1 hr. DEM was dissolved in ETOH, and the other drugs were dissolved in PBS. After drug exposures, cultures were exposed to 5 GY (^{137}Cs source) and colony formation was assessed after 1 week. Percent survival after 5 Gy is calculated from a ratio of irradiated colonies, with a correction made for plating efficiency. Results represent mean ±SEM, n = 5.

[a] In comparison to control conditions without ethanol, $p < 0.05$.
[b] In comparison to control conditions without ethanol, $p < 0.01$.
[c] In comparison to ethanol control for DEM, $p < 0.01$.

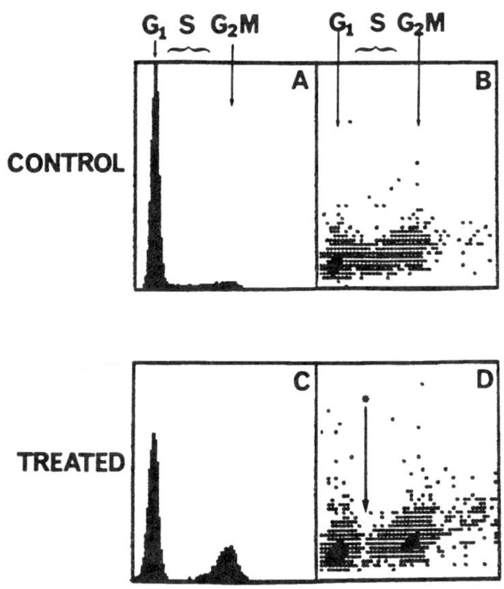

Figure 4.3. Cell cycle phase distribution in bovine aortic endothelial cells after BSO treatment (50 μ*M* for 24 hr) which depleted GSH. A and C are two-dimensional DNA cytohistograms: x-axis is DNA content based on fluorescence of propidium iodide; y-axis is cell number. B and D are the corresponding three-dimensional histograms with DNA content on the x-axis, forward angle light scatter on the y-axis, and cell number on the z-axis. A and B are the controls, and C and D are the result of BSO exposure. The arrow in D indicates the block between G_1 and S after GSH is depleted. Also noted is the increased distribution in late S and G_2M phases.

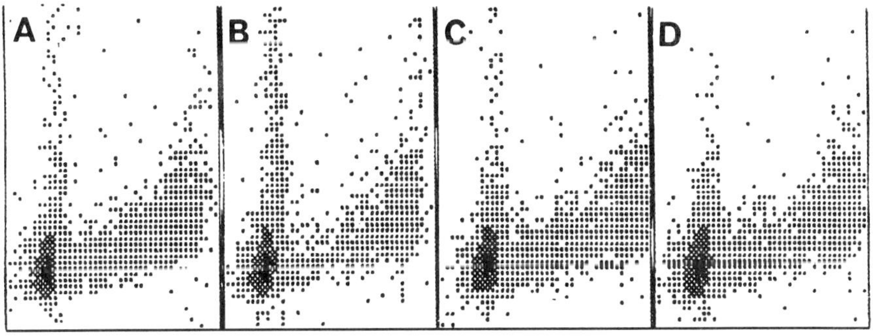

Figure 4.4. Cell cycle phase distribution following irradiation and cysteine treatment. Cysteine 2 m*M* was incubated with the BAEC for 2 hr prior to exposure to 5 Gy (^{137}Cs) and 24 hr after irradiation, cell cycle phase distribution was measured.[78] A description of the axes is in the legend of Figure 4.3. A: control, B: 5 Gy, C: cysteine 2 m*M*, and D: cysteine prior to 5 Gy. As suggested by the observations of DNA synthesis (see text), the cysteine treatment increased S phase distribution in the irradiated cells.

Results from experiments on irradiated BAEC indicate that within minutes of radiation exposure, actin-GSH is one of the five prominent protein-NPSH formed.[91] The increased turnover of actin-GSH preceded the increased expression of apical membrane integrin $\alpha_5\beta_1$, the reorganization of stress fibers that contain actin, the increased F-actin and the cell hypertrophy. The increased F-actin, implies that the cytoskeletal reorganization was not necessarily caused by actin depolymerization. In noncellular solutions consisting of purified actin and GSH, formation of actin-GSH creates fragile actin filaments.[92] If formation of actin-GSH inside a cell also forms a fragile microfilament tending to fragment, the result could contribute to the morphological adaptations of endothelial cells following irradiation.

IV. CATALASE (CAT) AND GLUTATHIONE PEROXIDASE (GSH-PX)

In endothelial cells that survive hyperoxia, endogenous CAT activity decreases while GSH-Px activity increases.[43] Also, selenium treatments boost endothelial cell GSH-Px activity while attenuating cytotoxicity induced by hyperoxia.[15] However, endogenous levels of the two enzyme activities have not been examined in irradiated endothelial cells. Addition of unconjugated or PEG-conjugated CAT to cultures of BAEC, did not protect against loss of adhered cells following irradiation.[63]

V. GLUTATHIONE REDUCTASE (GSSG-R)

After GSSG is formed, cells restore GSH levels via the activity of GSSG-R. 1,3-Bis[2-chloroethyl]-1-nitrosourea (BCNU or Carmustine) inhibits GSSG-R activity in a relatively selective manner[93] and has been the most commonly applied inhibitor of this enzyme in experiments with endothelial cells. The inhibition of

GSSG-R activity aggravates tissue toxicity caused by an oxidant stress.[94] When BCNU inhibits GSSG-R activity in cultured endothelial cells, the cells have a greater tendency to lyse following exposure to H_2O_2.[14,17,22]

BCNU and radiation therapy are combined in the treatment of certain cancers, particularly malignancies of the brain.[95] As observed in tissues exposed to radiation, endothelial cells appear to be among the first cell types damaged by BCNU exposure.[5,96] Because these agents perturb endothelial cells in the early stages of tissue toxicity, and because GSSG-R activity is inhibited by BCNU, one might conjecture that a portion of the endothelial cytotoxicity results when the inhibition impairs the restoration of GSH and the cells are less able to resist the oxidant stress of radiation.

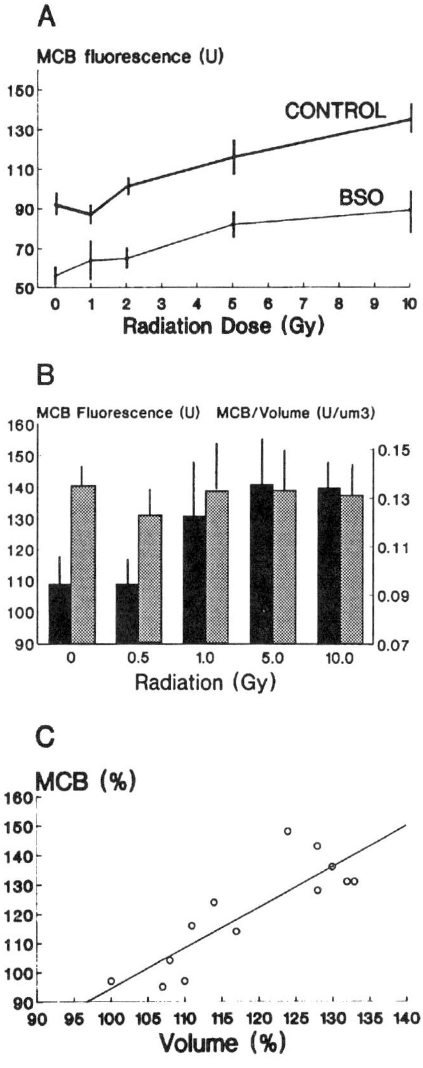

Figure 4.5. The relationship between irradiation and thiol content as reflected by fluorescence of the thiol probe, monochlorobimane (MCB). Twenty-four hours following irradiation (0, 1, 2, 5, and 10 Gy), cells were stained with MCB for 1 hr under ambient conditions, (500 μM, Molecular Probes, Eugene, OR). Cells were counted and volumes were measured as reported.[78] A: MCB fluorescence versus radiation dose in BAEC treated with or without buthionine sulfoximine (BSO) 50 μM for 24 hr prior to irradiation in order to inhibit GSH synthesis and lower GSH content to 10% of control values. Observed is that BSO lowers but does not eliminate MCB fluorescence and that radiation exposure increases the fluorescent signal per cell. Values are n = 6, mean ± SEM. Co-measured GSH content showed that mean control values were 3.5 nmole/10^6 cells versus 4.1 nmole/10^6 cells after 5 Gy. B: When correcting the MCB fluorescence per cell for cell volume, it is evident that the radiation-induced increase is related to cell enlargement. C: When compared to control values, the change in MCB fluorescence correlates (r = 0.89) to the change in cell volume.

This premise, however, is difficult to prove because of the other effects of BCNU,[97] especially the alkylation of DNA.

In experiments designed to determine the role of GSSG-R in the endothelial cytotoxicity caused by BCNU and radiation exposure, BAEC were treated with BCNU at 10 μ*M* for 30 minutes prior to irradiation. This BCNU treatment most selectively inhibited GSSG-R activity (by 50 to 90%) compared to SOD activity and the activity of other enzymes (data not shown). As demonstrated in Figure 4.6, this selective inhibiting dose of BCNU lowers GSSG-R but has little impact on cell proliferation, protein content, and GSH content. Higher doses of BCNU, however, that less selectively inhibit GSSG-R, inhibit BAEC proliferation, elevate protein content, and lower GSH content.

When BCNU inhibits GSSG-R in irradiated BAEC, there is a greater loss of both adhered cells and colony formation (Table 4.2); however, DNA damage is also more aggravated (Figure 4.7), likely because of alkylation. Without BCNU exposure, when culture medium is supplemented with GSH (2 m*M*), cellular GSH content

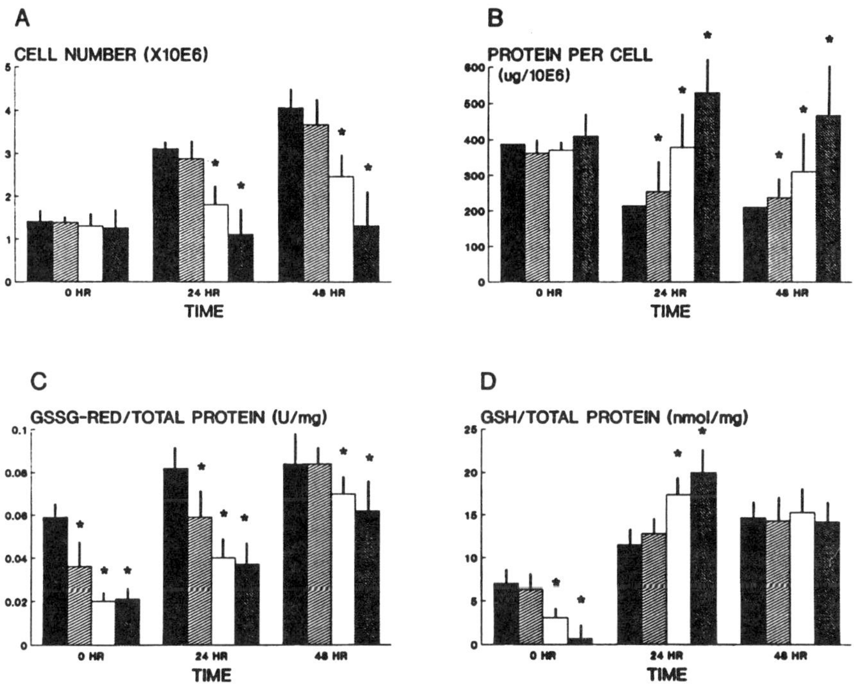

Figure 4.6 The impact of 1,3-bis[2-chloroethyl]-1-nitrosourea (BCNU) on cell proliferation and glutathione reductase (GSSG-R) activity and GSH content was examined. BCNU was added to the endothelial cells for 45 min: control (■), 10 μ*M* (▨), 50 μ*M* (□), 100 μ*M* (▩). Soon after BCNU exposure, and after 24 and 48 hr, we recorded (A) adhered cell number, (B) protein content per cell, a hypertrophy index, (C) GSSG-R per total protein content, and (D) GSH corrected for total protein content. Data are means ± SEM, n = 7; (*) the difference from control ($p < 0.05$).

Table 4.2 **Adhered Cell Number and Colony Formation in Endothelial Cells Exposed to BCNU 10 μM and Radiation**

	Radiation Dose (cGy)							
	0	**50**	**100**	**200**	**300**	**400**	**500**	**1000**
Cell counts ($\times 10^6$)								
Control	2.15	2.11	2.00	1.78	1.74	1.65	1.57	1.45
	±0.5	±0.7	±0.7	±0.7	±0.7	±0.7	±0.7	±0.8
BCNU	1.95[a]	1.95[a]	1.85[a]	1.77	1.68	1.61	1.56	1.45
	±0.6	±0.7	±0.7	±0.7	±0.7	±0.7	±0.7	±0.7
Colonies[a]								
Control	430	392	352	162	107	—	31	—
	±239	±268	±221	±108	±72	—	±22	—
BCNU[b]	238	226	169[c]	55	25	—	16	—
	±215	±228	±150	±46	±29	—	±13	—

Note: Cultures of confluent bovine aortic endothelial cells were exposed to BCNU and gamma-radiation. Adhered cells were counted 24 hr after exposure. Cells were seeded for colony assay immediately after irradiation. Data are mean ± S.D. n = 7. For adhered cell number: [a] $p < 0.01$ for BCNU versus control at same radiation dose. For colony assay, [b] $p < 0.001$ for all BCNU versus all control; [c] $p < 0.05$ for BCNU versus control at same radiation dose. Manova testing indicated that lowered cell counts and colonies were related to radiation dose.

increases and the DNA strand breakage attenuates slightly (Figure 4.7), but the cell loss is unchanged (data not shown). In contrast, if GSH is added prior to BCNU exposure, DNA damage appears to be greater because the latter portion of the elution profile is steeper. We speculate that GSH supplementation increases the rate of DNA elution because DNA/GSH crosslinks elute at normal rates and compete with the more slowly eluting DNA/DNA or DNA/protein crosslinks induced by BCNU.

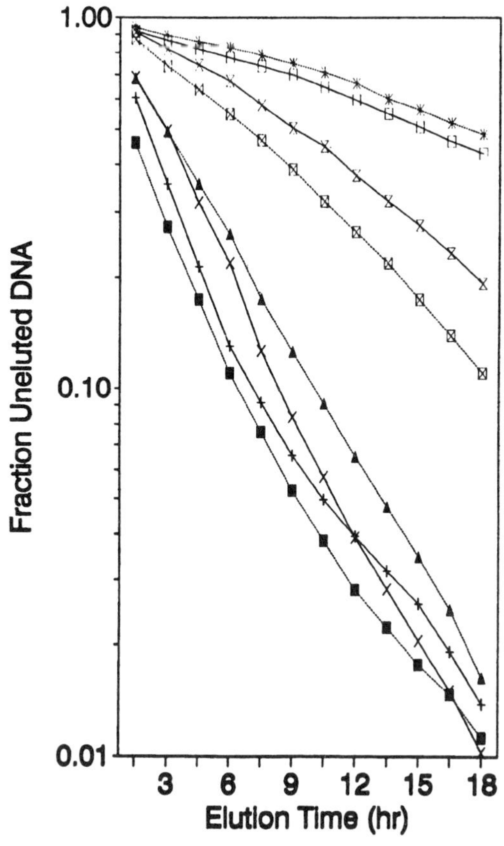

Figure 4.7. Alkaline (pH 12.1) filter elution study of DNA damage produced in endothelial cells by BCNU exposure and its modulation by previous exposure of the cells to GSH. Symbols represent means of duplicate channels in one experiment; three separate experiments confirmed these results. Solid lines indicate no pre-exposure of cells to GSH; dashed lines indicate the GSH pre-exposure. The symbols indicate the following treatments: (□) untreated cells; (*) GSH (2 m*M*, 2 hr); (X) BCNU (50 μ*M*, 45 min); (⊠) GSH before BCNU; (X̲) 3 Gy gamma-rays; (▲) GSH before 3 Gy; (+) BCNU before 3 Gy; (■) GSH before BCNU before 3 Gy.

VI. RADIOPROTECTION WITH THE AMINOTHIOL WR-1065

The aminothiols may radioprotect by scavenging free radicals and facilitating repair of damaged macromolecules.[98–100] Although WR-1065 and other aminothiols were developed as radioprotectors of normal tissues, there are few reports on the radioprotection of vascular endothelium. WR-2721 promotes the response of irradiated rabbit aortic rings to thromboxane mimetics.[101] In irradiated rat lungs, however, WR-2721 does not modify angiotensin converting enzyme activity,[102] an indicator of endothelial cell dysfunction following irradiation.[62,103] Incidentally, Ward and co-workers have demonstrated that sulfhydryl containing ACE inhibitors, such as captopril, attenuate the vascular damage induced by irradiation.[104–106]

WR-1065 radioprotects BAEC against cytotoxicity.[82,91] Specifically, WR-1065 attenuates DNA damage, promotes DNA synthesis and G1 to S phase progression, and diminishes the extent to which both adhered cells and colonies are lost.[82] WR-1065 also radioprotects BAEC against morphological alterations,[91] implying that WR-1065 targets molecules other than DNA. In irradiated BAEC, WR-1065 prevents cell enlargement, promotes the retention of stress fibers that contain F-actin, and down-regulates the apical membrane surface expression of an integrin receptor for fibronectin, $\alpha_5\beta_1$.

VII. RADIOPROTECTION WITH SERUM, LAZAROIDS, AND VITAMIN C

The endothelium *in vivo* is bathed in both antioxidant and pro-oxidant agents. The specific role of circulating antioxidant agents in reducing the radiation-induced damage to the endothelium is not known. We and others have observed (not shown) that irradiated BAEC are more resistant when exposed in culture medium containing 20% (v/v) serum compared to that containing 2% serum. Some of these circulating antioxidants include Vitamins E and C. Lazaroids are synthetic sterol compounds (21 amino steroids) that resemble Vitamin E and, like the vitamin, insert themselves into membranes. Lazaroids protect coronary endothelial cells from H_2O_2-induced toxicity and also have been reported to radioprotect rat gut. In pilot experiments, we found that exposure to lazaroids elevates GSH content in BAEC (twice that of control for U78517F and threefold greater for U74389). The lazaroids, however, had no impact on either cell proliferation, DNA synthesis, or colony formation following irradiation (not shown).

Vitamin C treatment may attenuate radiation-induced damage to BAEC. Although Vitamin C has no affect on loss of cultured endothelial cells, it promotes the recovery of prostacyclin production,[108] a common marker of endothelial function that is perturbed following irradiation.[109,110]

VIII. NONCYTOTOXIC ENDPOINTS: OXIDATIVE STRESS AND ENDOTHELIAL DYSFUNCTION

Most of the results thus far reviewed focus on the relationship between endothelial cytotoxicity and antioxidant properties. The irradiated endothelial cell, however, does more than just die — it becomes dysfunctional. The result of this endothelial dysfunction is inflammation, edema, abnormal vasomotor control, and perhaps the loss of an antimitogenic control exerted on subendothelial cells.[111–113] Table 4.3 is presented as a partial list of how a more thoroughly studied oxidant stress, exposure to H_2O_2, has been related to vascular endothelial functions. From these examples, one can speculate how the oxidative stress of radiation exposure impairs endothelial function.

Future investigations of antioxidant functions in the endothelial response to irradiation will focus on the role that oxidoreduction of molecules plays in signal transduction, including the modification of adhesion molecules,[114] kinases,[115] and transcription factors.[116] Based on the observations[82,91] with WR-1065 and on the formation of thiol mixed disulfides, one can conjecture that antioxidant radioprotectors do more than attenuate DNA damage. The other impact of radioprotective thiols may be exerted on reactive sulfhydryl groups that are located on key molecules of signal transduction and cell cycle regulation pathways. Examples of these molecules and their reactive sulfhydryls are presented in Table 4.4. Beyond preventing fixation with oxygen, antioxidant thiols also might instigate post-translational modifications that regulate protein function. One of the best-known examples of this regulation is the activation of the transcription factor NfkB,[117,118] which mediates the impact of TNF and other cytokines. NfkB activity is diminished by the addition of thiol reducing agents, including pyrrilodine dithiocarbamate and N-acetylcysteine (NAC).[119,120] NAC is an NPSH used therapeutically for lung disease (Mucomyst®) and acetaminophen poisoning. Clinical studies suggest that NAC ameliorates acute lung injury that results from damage to the vascular endothelium.[121]

Also, as suggested by the examples in Table 4.3, future investigations that examine radiation-induced oxidant stress to the endothelium will likely include the study of nitric oxide (NO). NO, a derivative of arginine metabolism, was originally detected as endothelial-derived relaxing factor.[122] We speculate that NO-thiol moieties will prove to be important in moderating both vasomotor activity and endothelial cell proliferation. For example, we have preliminary evidence showing that an NO synthase inhibitor promotes DNA synthesis in irradiated BAEC (presented at *Radiation Research Society, 44th Meeting*, April 1996).

IX. CONCLUSIONS

This review presents evidence that antioxidant functions ameliorate and perhaps mediate endothelial cell damage caused by radiation exposure. But most of this evidence is on cultured endothelial cells and not *in vivo*. Also, while antioxidant mechanisms are of presumed importance in the irradiated endothelium, relative to other types of oxidative stresses, much less is known. Intensification of cancer

Table 4.3 Endothelial Response to H_2O_2

Endothelial Function	Endothelial Source	Observation (Ref.)
Vasomotor regulation	Feline cerebral arteries	Catalase, not SOD, abolishes acetylcholine-induced vasodilation; H_2O_2 induces vasodilation (123)
	Rabbit aortic rings	H_2O_2 induces endothelial-dependent vasorelaxation, which is mediated by NO generation (124)
	Calf pulmonary arteries	Lactate relaxes vessels precontracted with K^+, NADH oxidase is activated to make O_2^- which increases H_2O_2 that activates guanyl cyclase and increases cGMP independent of NO (26)
	Canine coronary arteries	When NOS is activated in vessels that are depleted of tetrahydrobiopterin, H_2O_2 is generated (125)
	Rat pulmonary arteries	Both H_2O_2 and NO maintain similarly the low pressures during normoxia (125)
Interaction with other cells	Canine carotid arteries	H_2O_2 increases leukocyte adhesion to the endothelium (126)
	Rat mesenteric venules	NOS inhibition induces intravascular H_2O_2 production mostly through the adhesion and activation of leukocytes (127)
	Rat pulmonary endothelium	Adhesion of metastatic cells to the endothelium produces locally damaging H_2O_2 (128)
	Human umbilical vein endothelial cells	H_2O_2 promotes PMN adhesion via increased expression of ICAM-1 (33)
Coagulation	Human umbilical vein endothelial cells	O_2^- but not H_2O_2 increases cytosolic free Ca^{2+}, which stimulates the release of von Willebrand Factor (129)
Permeability	Bovine aortic endothelial cells	μM of H_2O_2 redistributes actin containing cytoskeleton without increased permeability, but mM dose causes dense peripheral bands to disappear and increase permeability (130); H_2O_2 stimulates endocytosis (131)
	Pig pulmonary artery endothelial cells	Transendothelial solute flux is increased by H_2O_2, which is inhibited by elevating cAMP (132)
	Rabbit lung endothelium	Edema caused by H_2O_2 is attenuated by elevating cAMP (133)
	Rat lung endothelium	Edema caused by H_2O_2 is attenuated by the radioprotector WR-2721 (134)
Signal transduction and vascular repair	Bovine glomerular endothelial cells	H_2O_2 induces mRNA of Mn-SOD (135)

Table 4.3 (continued) Endothelial Response to H_2O_2

Endothelial Function	Endothelial Source	Observation (Ref.)
	Bovine aortic endothelial cells	H_2O_2 lyses cells, which is partly prevented by exposure to one NOS inhibitor but not to another (136); H_2O_2 activates protein kinase C, which increases phospholipase A_2 and arachidonate production (137); μM dose of H_2O_2 induces apoptosis and both creates multilobed nuclei and delays division in the survivors (138); H_2O_2 increases cytosolic free Ca^{2+} to stimulate NO synthesis by promoting the formation of L-citrulline from L-arginine (139)
	Bovine pulmonary artery endothelial cells	H_2O_2: Induces mRNA of platelet-derived growth factor (27); Activates phospholipase D to produce phosphatidic acid and diacylglycerol (140); Decreases xanthine oxidase activity through NO production (141); May increase the release of cytosolic free Ca^{2+} and attenuates this release due to bradykinin exposure (142); Stimulates Na^+/K^+ pump activity (143); Induces mRNA of heat shock protein 70 while increasing the activities of SOD, catalase, and glutathione peroxidase (144); Translocates cytosolic PKC to the membrane and increases diacylglycerol (145)
	Human umbilical vein endothelial cells	H_2O_2: Redistributes ICAM-1, ICAM-2, and the β1 and β3 integrins (146); Induces the S-thiolation of glyceraldehyde-3-dehydrogenase (90) and other proteins (147); Depletes ATP and promotes efflux of adenine metabolites, effects reduced by cysteine and a sulfhydryl containing ACE inhibitor (148); Increases cytosolic Ca^{2+}, an effect prevented by catalase and N-acetylcysteine (149)
	Rabbit lung endothelium	Perfusion with H_2O_2 increases the release of fibronectin (150)
	Pig coronary artery	A lazaroid inhibits the elevation of cytosolic Ca^{2+} induced by H_2O_2 (29)

TABLE 4.4 Potentially Reactive Cysteinyl Sites on Proteins that Regulate Signal Transduction and Cell Cycle Progression

Molecule	Species	Number of Cys	Reactive Cys (CK,CR,CH)	"Hyper" reactive Cys
Beta-1 (integrin)	Human	47	19	574,CKCR; 618,CKC; 691,HCKCK
Beta-1 precursor	Frog	57	14	300,KCH; 414,RKC; 434,HKC; 489,CRC; 516,CRR; 538,CKKR; 574,CKCR; 618,RCKC; 691,HCK
Actin	Urchin	5	3	375,KC-GSH disulfide is fragile F-actin
pp125FAK	Frog	13	5	
PKC	Mouse	18	5+	PKCα,β,Γ- 2 CK in catalytic domain
MEKK	Mouse	14	5	336,KCK 280,KCR
MAPK	Mouse	4	1	
Cyclin A	Mouse	4	2	183,KCK
Cyclin A	Hamster	4	3	181,KCK
Cyclin B	*C. elegans*	7	4	
Cyclin E	Mouse	13	2	
cdk1/cdc2	Cow	1	1	119, CHSRRVLHR
cdk1/cdc2	Human	1	1	119, CHSRRVLHR
cdk2-α	Rat	3	2	
cdk2-β	Rat	6	4	203, HHAKCC
cdk4	Frog	7	1	
p53	Mouse	11	3	172, RRCPHHERC
Tyr PTPase*	Human		Catalytic site	HCSAGIGRSG
cdc25-Tyr/Thr PTPase for CDK	Human		Catalytic site	HCEFSSERGP

* All tyrosine phosphatases except calcineurin.

therapy that combines accelerated radiation and chemotherapies requires that normal tissue toxicity be reduced. Further, cytokine therapies and gene therapies directed at treating malignancies will circulate and have an early and perhaps critical impact on the normal tissue vascular endothelium. Many of these new modalities will likely create and may exacerbate oxidative stress to both the tumor and normal tissue endothelium. These more aggressive attempts at curing malignancies will require better understanding of how antioxidant functions protect the endothelium and perhaps promote normal function and healing of the damaged endothelium.

REFERENCES

1. Ryan, U. S., *Endothelial Cells.* CRC Press, Inc. Boca Raton, FL, 1988.
2. Kuzuya, M., Naito, M., Funaki, C., Hayashi, T., Asai, K., and Kuzuya, F., Protective role of intracellular glutathione against oxidized low density lipoprotein in cultured endothelial cells, *Biochem. Biophys. Res. Commun.,* 163, 1466, 1989.
3. Vercellotti, G. M., Dobson, M., Schorer, A. E., and Moldow, C. F., Endothelial cell heterogeneity: antioxidant profiles determine vulnerability to oxidant injury, *Proc. Soc. Exp. Biol. Med.,* 187, 181, 1988.

4. Kapanci, Y., Weibel, E. R., Kaplan, H. P., and Robinson, F. R., Pathogenesis and reversibility of pulmonary lesions of oxygen toxicity in monkeys. II. Ultrastructural and morphometric studies, *Lab. Invest.* 20, 101, 1969.
5. Schuller, H. M., Smith, A. C., Gregg, M., and Boyd, M. R., Sequential pathological changes induced in rats with the anti-cancer drug 1,3-bis (2-chloroethyl)-1-nitrosourea (BCNU). *Exp. Lung. Res.,* 9, 327, 1985.
6. Meyers, C. E., McGuire, W. P., Kiss, R. H., Ifrim, I., Grotzinger, K., and Young, R., Adriamycin. The role of lipid peroxidation in cardiac toxicity and tumor response. *Science,* 197, 165, 1977.
7. Hopewell, J. W., The importance of vascular damage in development of late radiation effects in normal tissues. In *Radiation Biology in Cancer Research* (R.E. Meyn and H.R. Withers, Eds.), Raven Press, New York, 1980, 449.
8. Gerschman, R., GIlbert, D. L., Nye, S. W., Dwyer, P., and Fenn, W. O., Oxygen poisoning and x-irradiation: a mechanism in common, *Science,* 119, 623, 1954.
9. Hall, E. J., *Radiobiology for the Radiologist,* 3rd ed., J.B. Lippincott, Philadelphia, 1988.
10. Von Sonntag, C., Radiation chemistry in the 1990s: pressing questions relating to the areas of radiation biology and environmental research, *Int. J. Radiat. Biol.,* 65, 19, 1994.
11. Law, M. P., Radiation-induced vascular injury and its relation to late effects in normal tissues, *Adv. Radiat. Biol.,* 9, 37, 1981.
12. Adamson, I. Y. R. and Bowden, D. H., Endothelial injury and repair in radiation-induced pulmonary fibrosis, *Am. J. Pathol.,* 112, 224, 1983.
13. Adamson, I. Y. R., Bowden, D. H., and Wyatt, J. P., A pathway to pulmonary fibrosis: an ultrastructural study of mouse and rat following radiation to the whole body and hemithorax, *Am. J. Pathol.,* 58, 481, 1970.
14. Harlan, J. M., Levine, J. D., Callahan, K. S., Schwartz, B. R., and Harker, L. A., Glutathione redox cycle protects cultured endothelial cells against lysis by extracellularly generated hydrogen peroxide, *J. Clin. Invest.,* 73, 706, 1984.
15. Housset, B., Ody, C., Rubin, D. B., Elemer, G., and Junod, A. F., Oxygen toxicity in cultured aortic endothelium: selenium-induced partial protective effect, *J. Appl. Physiol.,* 55(2), 343, 1983.
16. Visner, G. A., Block, E. R., Burr, I. M., and Nick, H. S., Regulation of manganese superoxide dismutase in porcine pulmonary artery endothelial cells, *Am. J. Physiol.,* 260, L444, 1991.
17. Suttorp, N., Toepfer, W., and Roka, L., Oxidant defense mechanisms of endothelial cells: glutathione redox cycle versus catalase, *Am. J. Physiol.,* 251, 671, 1986.
18. Das, S. K., White, A. C., and Fanburg, B. L., Modulation of transforming growth factor β1 antiproliferative effects on endothelial cells by cysteine, cystine and N-acetylcysteine, *J. Clin. Invest.,* 90, 1649, 1992.
19. Turrens, J. F., Crapo, J. D., and Freeman, B. A., Protection against oxygen toxicity by intravenous injection of liposome-entrapped catalase and superoxide dismutase. *J. Clin. Invest.,* 73, 87, 1984.
20. White, A. C., Das, S. K., and Fanburg, B. L., Reduction of glutathione is associated with growth restriction and enlargement of bovine pulmonary artery endothelial cells produced by transforming growth factor-β1, *Am. J. Respir. Cell. Mol. Biol.,* 6, 364, 1992.
21. Freeman, B. A., Turrens, J. F., Mirza, Z., Crapo, J. D., and Young, S. L., Modulation of oxidant lung injury by using liposome-entrapped superoxide dismutase and catalase, *FASEB.,* 44, 2591, 1985.

22. Andreoli, S. P., Mallett, C. P., and Bergstein, J. M., Role of glutathione in protecting endothelial cells against hydrogen peroxide oxidant injury, *J. Lab. Clin. Med.*, 108, 190, 1986.
23. Ishii, Y., Partridge, C. A., Del Vecchio, P. J., and Malik, A. B., Tumor necrosis factor-alpha-mediated decrease in glutathione increases the sensitivity of pulmonary vascular endothelial cells to H_2O_2, *J. Clin. Invest.*, 89, 794, 1992.
24. Kinnula, V. L., Whorton, A. R., Chang, L. Y., and Crapo, J. D., Regulation of hydrogen peroxide generation in cultured endothelial cells, *Am. J. Respir. Cell. Mol. Biol.*, 6, 175, 1992.
25. Weiss, S. J., Young, J., LoBuglio, A. F., and Slivka, A., Role of hydrogen peroxide in neutrophil-mediated destruction of cultured endothelial cells, *J. Clin. Invest.*, 68, 714, 1981.
26. Omar, H. A., Mohazzab, K. M. H., Nortelliti, M. P., and Wolin, M., O_2 dependent modulation of calf pulmonary artery tone by lactate potential role of H_2O_2 and cGMP, *Am. J. Physiol.*, 264, 1141, 1993.
27. Montisano, D. F., Mann, T., and Spragg, R. G., H_2O_2 increases expression of pulmonary artery endothelial cell platelet-derived growth factor mRNA, *Am. J. Physiol.*, 73, 2255, 1992.
28. Birnboim, A. S., Goligorsky, M. S., Del Vecchio, P. J., and Malik, A. B., Activation of protein kinase c pathway contributes to hydrogen peroxide induced increase in endothelial permeability, *Lab. Invest.*, 67, 24, 1992.
29. Kimura, M., Maeda, K., and Hayashi, S., Cytosolic calcium increase in coronary endothelial cells after H_2O_2 exposure and the inhibitory effect of U78217F, *Br. J. Pharmacol.*, 107, 488, 1992.
30. Katusic, Z. S., Schugel, J., Cosentino, F., and Vanhoutte, P. M., Endothelium-dependent contractions to oxygen-derived free radicals in the canine basilar artery, *Am. J. Physiol.*, 264, H859, 1993.
31. Yang, W. and DeBono, D., Myoglobin protects against endothelial cell membrane damage associated with hydrogen peroxide or xanthine/xanthine oxidase, *FEBS*, 319, 145, 1993.
32. Bhat, G. B., Tinsley, S. B., Tolson, K. J., Patel, J. M., and Block, E. R., Hypoxia increases the susceptibility of pulmonary artery endothelial cells to hydrogen peroxide injury, *J. Cell. Physiol.*, 151, 2228, 1992.
33. Lo, S. K., Janakidevi, K., Lai, L., and Malik, A. B., Hydrogen peroxide-induced increase in endothelial adhesiveness is dependent on ICAM-1 activation, *Am. J. Physiol.*, 264, L406, 1993.
34. Aalto, T. K. and Raivio, K. O., Mechanisms of adenine nucleotide depletion from endothelial cells exposed to reactive oxygen metabolites, *Free Radic. Biol. Med.*, 14, 177, 1993.
35. Vercellotti, G. M., Severson, S. P., Duane, P., and Moldow, C. F., Hydrogen peroxide alters signal transduction in human endothelial cells, *J. Lab. Clin. Med.*, 117, 15, 1991.
36. Cronstein, B. N., Oxidative insults: sublethal injury to the endothelium by H_2O_2. *J. Lab. Clin. Med.*, 117, 6, 1991.
37. Thomas, J. P., Geiger, P. G., and Girotti, A. W., Lethal damage to endothelial cells by oxidized low density lipoprotein: role of selenoperoxidases in cytoprotection against lipid hydroperoxide and iron-mediated reactions, *J. Lipid. Res.*, 34, 479, 1993.
38. Hertig, I. A., Hassoun, P. M., Zulueta, J. J., Thannickal, V. J., and Fanburg, B. L., Mechanism of basal and transforming growth factor β-1 stimulated H_2O_2 release by endothelial cells, *Trans. Assoc. Am. Physicians*, 106, 179, 1993.

39. Ody, C. and Junod, A. F., Direct toxic effects of paraquat and oxygen on cultured endothelial cells, *Lab. Invest.*, 52, 77, 1985.
40. Martin, Jr., W. J. and Kachel, D. L., Oxygen-mediated impairment of human pulmonary endothelial cell growth: evidence for a specific threshold of toxicity, *J. Lab. Clin. Med.*, 113, 413, 1989.
41. Rubin, D. B., Housset, B., Martin, Y., and Junod, A. F., Effects of hyperoxia on biochemical indices of pig aortic endothelial function, *In Vitro*, 19, 625, 1983.
42. Jackson, R. M., Chandler, D. B., and Fulmer, J. D., Production of arachidonic acid metabolites by endothelial cells in hyperoxia, *J. Appl. Phys.*, 61, 584, 1986.
43. Jornot, L. and Junod, A. F., Response of human endothelial cell antioxidant enzymes to hyperoxia, *Am. J. Respir. Cell. Mol. Biol.*, 6, 107, 1992.
44. Thies, R. L. and Autor, A. P., Reactive oxygen injury to cultured pulmonary artery endothelial cells: mediation by poly (ADP-ribose) polymerase activation causing NAD depletion and altered energy balance, *Arch. of. Biochem. Biophys.*, 2, 353, 1991.
45. Brieland, J. K., Clarke, S. J., Karmiol, S., Phan, S. H., and Fantone, J. C., Transferrin: a potential source of iron for oxygen free radical-mediated endothelial cell injury, *Arch. Biochem. Biophys.*, 294, 265, 1992.
46. Lesko, S. A., Lorentzen, R. J., and Ts'o, P. O., Role of superoxide in deoxyribonucleic acid strand scission, *Biochemistry*, 19, 3023, 1980.
47. Fridovich, I., The biology of oxygen radicals, *Science*, 201, 875, 1978.
48. Westman, N. G. and Marklund, S. L., Copper and zinc containing superoxide dismutase and manganese containing superoxide dismutase in human tissues and human malignant tumors, *Cancer. Res.*, 41, 2962, 1981.
49. Marklund, S. L., Westman, N. G., Roos, G., and Carlsson, J., Radiation resistance and the CuZn superoxide dismutase, Mn superoxide dismutase, catalase, and glutathione peroxidase activities of seven human cell lines, *Radiat. Res.*, 100, 115, 1984.
50. Mcdonald, R. J., Berger, E. M., White, C. W., White, J. G., Freeman, B. A., and Repine, J. E., Effect of superoxide dismutase encapsulated in liposomes or conjugated with polyethylene glycol on neutrophil bactericidal activity *in vitro* and bacterial clearance *in vivo*, *Am. Rev. Respir. Dis.*, 131, 633, 1985.
51. Beckman, J. S., Minor, R. L., White, C. W., Repine, J. E., Rosen, G. M., and Freeman, B. A., Superoxide dismutase and catalase conjugated to polyethylene glycol increases endothelial enzyme activity and oxidant resistance, *J. Biol. Chem.*, 263, 6884, 1988.
52. Petkau, A., Chelack, W. S., and Pleskach, S. D., Protection by superoxide dismutase of white blood cells in x-irradiated mice, *Life Sci.*, 22, 867, 1978.
53. Petkau, A., Kelly, K., Chelack, W. S., and Barefoot, C., Protective effect of superoxide dismutase on erythrocytes of X-irradiated mice, *Biochem. Biophys. Res. Commun.*, 70, 452, 1976.
54. Abe, M., Nishidai, T., Yukawa, Y., Takahashi, M., Ono, K., Hiraoka, M., and Ri, N., Studies on the radioprotective effects of superoxide dismutase in mice, *Int. J. Radiat. Oncol. Biol. Phys.*, 7, 205, 1981.
55. Gray, B. H. and Stull, R. W., Radioprotection by polyethyleneglycol-protein complexes in mice, *Radiat. Res.*, 93, 581, 1983.
56. Petkau, A. and Chelack, W. S., Radioprotection by superoxide dismutase on macrophage progenitor cells from mouse bone marrow, *Biochem. Biophys. Res. Commun.*, 119, 1089, 1984.
57. Weiss, J. F. and Kumar, K. S., Antioxidant mechanisms in radiation injury and radioprotection, in *Cellular Antioxidant Defense Mechanisms*, Chow, C.K., Ed., CRC Press, Boca Raton, FL, 1988, 163.

58. Crapo, J. D. and Tierney, D. F., Superoxide dismutase and pulmonary oxygen toxicity, *Am. J. Physiol.*, 226, 1401, 1974.
59. Stevens, J. B. and Autor, A. P., Induction of superoxide dismutase by oxygen in neonatal rat lung, *J. Biol. Chem.*, 252, 3509, 1977.
60. Krizala, J., Stoklasova, A., Kovarova, W., and Ledvina, M., The effect of gamma-irradiation and cystamine on superoxide dismutase activity in the bone marrow and erythrocytes of rats, *Radiat. Res.*, 91, 507, 1982.
61. Misra, H. P. and Fridovich, I., Superoxide dismutase and the oxygen enhancement of radiation lethality, *Arch. Biochem. Biophys.*, 176, 577, 1976.
62. Rubin, D. B., Drab, E. A., Ward, W. F., Smith, L. J., and Fowell, S. M., Enzymatic responses to radiation in cultured vascular endothelial and smooth muscle cells, *Radiat. Res.*, 99, 420, 1984.
63. Drab, E. A., Hansra, I., and Rubin D., The effect of superoxide dismutase (SOD) and catalase (CAT) addition to cultures of irradiated vascular endothelial cells, (Abstract) *Radiat. Res.*, Abstract supplement-#129, 1990.
64. Pluthero, P. G., Shreeve, M., Eskinazi, D., van der Gaag, H., Huang, K. S., Hulmes, J. D., Blum, M., and Axelrad, A. A., Purification of an inhibitor of erythroid progenitor cell cycling and antagonist to Interleukin-3 from mouse marrow cell supernatants and its identification as cytosolic superoxide dismutase, *J. Cell. Biol.*, 111, 1217, 1990.
65. Meister, A. and Anderson, M. E., Glutathione, *Ann. Rev. Biochem.*, 52, 711, 1983.
66. Griffith, O. W. and Meister, A., Potent and specific inhibition of glutathione synthesis by BSO, *J. Biol. Chem.*, 254 (16), 7558, 1979.
67. Harris, J. W. and Teng, S. S., Sulfhydryl groups during the S-phase: comparison of cells from G_1, plateau-phase G_1, G_0, *J. Cell. Physiol.*, 81, 91, 1972.
68. Mauro, A. Grasso, L.J. Tolmach, Variations of cellular sulfhydryl disulfide and protein content during synchronous and asynchronous growth of HeLa cells, *Biophys. J.*, 9, 1377, 1969.
69. Harris, J. W. and Patt, H. M., Non-protein sulfhydryl content and cell-cycle dynamics of Ehrlich ascites tumor, *Exp. Cell. Res.*, 56, 134, 1969.
70. Lee, F. Y. F., Siemann, D. W., Allalunis-Turner, M. J., and Keng, P. C., Glutathione contents in human and rodent tumor cells in various phases of the cell cycle, *Cancer Res.*, 48, 3661, 1988.
71. Mallery, S. R., Laufman, H. B., Solt, C. W., and Stephens, R. E., Association of cellular thiol redox status with mitogen-induced calcium mobilization and cell cycle progression in human fibroblasts, *Cell. Biochem.*, 45, 82, 1991.
72. Mallery, S. R., Lantry, L. E., Laufman, H. B., Stephens, R. E., and Brierley, G. P., Modulation of human microvascular endothelial cell bioenergetic status and glutathione levels during proliferative and differentiated growth, *J. Cell. Biochem.*, 53, 360, 1993.
73. Revesz, L., Bergstrand, H., and Modig, H., Intrinsic non-protein sulphydryl levels and cellular radiosensitivity, *Nature*, 198, 1275, 1963.
74. Harris, J. W., Painter, R. B., and Hahn, G. M., Endogenous non-protein sulfhydryl and cellular radiosensitivity, *Int. J. Radiat. Biol.*, 15, 289, 1969.
75. Ohara, H. and Terasima, T., Variations of cellular sulfhydryl content during cell cycle of HeLa cells and its correlation to cyclic change of X-ray sensitivity, *Exp. Cell. Res.*, 58, 182, 1969.
76. Biaglow, J. E., Varnes, M. E., Clark, E. P., and Epp, E. R., The role of thiols in cellular response to radiation and drugs, *Radiat. Res.*, 95, 437, 1983.
77. Fidelus, R. K. and Tsan, M. F., Enhancement of intracellular glutathione promotes lymphocyte activation by mitogen, *Cell. Immunol.*, 97, 155, 1986.

78. Rubin, D. B., Drab, E. A., and Bauer, K. D., Endothelial cell subpopulations in vitro: cell volume, cell cycle, and radiosensitivity, *J. Appl. Physiol.*, 67, 1585, 1989.
79. Deneke, S. M. and Fanburg, B. L., Regulation of cellular glutathione, *Am. J. Physiol.*, 257, L163, 1989.
80. Phelps, D. T., Deneke, S. M., Daley, D. L., and Fanburg, B. L., Elevation of glutathione levels in bovine pulmonary artery endothelial cells by N-acetylcysteine, *Am. J. Respir. Cell. Mol. Biol.*, 7, 293, 1992.
81. Issels, R. D. and Nagele, A., Promotion of cystine uptake, increase of glutathione biosynthesis, and modulation of glutathione status by *S*-2-(3-aminopropylamino)ethyl phosphorothioic acid (WR-2721) in Chinese hamster cells, *Cancer Res.*, 49, 2082, 1989.
82. Rubin, D. B., Drab, E. A., Kang, H. J., Baumann, F. E., and Blazek, E. R., WR-1065 and radioprotection of vascular endothelial cells. I. Cell proliferation, DNA synthesis and damage, *Radiat. Res.*, 145, 210, 1996.
83. Grdina, D. J., Shigematsu, N., Dale, P., Newton, G. L., Aguilera, J. A., and Fahey, R. C., Thiol and disulfide metabolites of the radiation protector and potential chemopreventive agent WR-2721 are linked to both its anti-cytotoxic and anti-mutagenic mechanisms of action, *Carcinogenesis*, 16, 767, 1995.
84. Rice, G. C., Bump, E. A., Shrieve, D. C., Lee, W., and Kovacs, M., Quantitative analysis of cellular glutathione by flow cytometry utilizing monochlorobimane: some applications to radiation and drug resistance *in vitro* and *in vivo*, *Cancer Res.*, 46, 6105, 1986.
85. Kosower, N. S., Kosower, E. S., Newton, G. L., and Ranney, H. M., Bimane fluorescent labels: labeling of normal human red cells under physiological conditions, *Proc. Natl. Acad. Sci. U.S.A.*, 76, 3382, 1979.
86. Thomas, J. A. and Sies, H., Protein S-thiolation and dethiolation, in *The Post-Translational Modification of Proteins. Roles in Molecular and Cellular Biology*, E.S. Tuboi, N. Taniguchi and N. Katunuma, Eds., Japan Scientific Societies Press/CRC Press, Tokyo and Boca Raton, FL, 1992, 35.
87. Ravichandran, V., Seres, T., Moriguchi, T., Thomas, J. A., Johnston, R. B., S-thiolation of glyceraldehyde-3-phosphate dehydrogenase induced by the phagocytosis associated respiratory burst in blood monocytes, *J. Biol. Chem.*, 269, 25010, 1991.
88. Rokutan, K., Thomas, J. A., and Johnston, R. B., Phagocytosis and stimulation of the respiratory burst by phorbol diester initiate S-thiolation of specific proteins in macrophages, *J. Immunol.*, 147, 260, 1991.
89. Keire, D. A., Strauss, E., Guo, W., Noszal, B., and Rabenstein, D. L., Kinetics and equilibria of thiol/disulfide interchange reactions of selected biological thiols and related molecules with oxidized glutathione, *J. Org. Chem.*, 57, 123, 1992.
90. Schuppe-Koistinen, I., Moldeus, P., Bergman, T., and Cotgreave, I. A., S-thiolation of human endothelial cell glyceraldehyde-3–phosphate dehydrogenase after hydrogen peroxide treatment, *Eur. J. Biochem.*, 221, 1033, 1994.
91. Mooteri, S. N., Podolski, J. L., Drab, E. A., Saclarides, T. J., Onoda, J. M., Kantak, S. S., and Rubin, D. B., WR-1065 and radioprotection of vascular endothelial cells. II. Morphology. *Radiat. Res.*, 145, 217, 1996.
92. Drewes, G. and Faulstich, H., The enhanced ATPase activity of glutathione-substituted actin provides a quantitative approach to filament stabilization, *J. Biol. Chem.*, 265, 3017, 1990.
93. Frischer, H. and Ahmad, T., Severe generalized glutathione reductase deficiency after antitumor chemotherapy with BCNU [1,3–bis(chloroethyl)-1-nitrosourea], *J. Lab. Clin. Med.*, 89, 1080, 1977.

94. Frischer, H., Chu, L. K., Ahmad, T., Justice, P., et al., Superoxide dismutase and glutathione peroxidase abnormalities in erythrocytes and lymphoid cells in Downs syndrome, in *The Red Cell: Fifth Ann Arbor Conference. Progress in Clinical and Biological Research.* Vol. 55 (G.J. Brewer, Ed.), Alan R. Liss, New York, 1981, 269.
95. Wilson, C. B., Boldrey, E. B., and Enot, K. J., 1,3-Bis(2–chloroethyl)1–nitrosourea (NSC-409962) in the treatment of brain tumors, *Can. Chem. Rep.,* 54, 273, 1970.
96. Szczech, J. and Godlewski, A., Appraisal of blood vessels in the rat brain following administration of nitrosourea derivatives, *Neuropat. Pol.*, 24, 573, 1986.
97. Siemann, D. W., Interactions between nitrosoureas and X-irradiation, in *Antitumor Drug-Radiation Interactions* (B.T. Hill and A.S. Bellamy, Eds.), CRC Press, Boca Raton, FL, 1990, 141.
98. Brown, P. E., Mechanism of action of aminothiol radioprotectors, *Nature*, 211, 363, 1967.
99. Purdie, J. W., Inhaber, E. R., Schneider, H., and Labelle, J. L., Interaction of cultured mammalian cells with WR-2721 and its thiol, WR-1065: implications for mechanisms of radioprotection, *Int. J. Radiat. Biol.*, 43, 517–527, 1983.
100. Schuchter, L. M. and Glick, J. H., The current status of WR-2721 (Amifostine): a chemotherapy and radiation therapy protector. In *Biologic Therapy of Cancer*, 3rd ed., DeVita, V. T., Hellmanand, S., and Rosenberg, S. A., Eds., J.B. Lippincott, New York, 1993, 1.
101. Warfield, M. E., Schneidkraut, M. J., Ramwell, P. W., and Kot, P. A., WR2721 ameliorates the radiation-induced depression in reactivity of rat abdominal aorta to U46619, *Radiat. Res.*, 121, 63, 1990.
102. Travis, E. L., Meistrich, M. L., Finch-Neimeuer, M. V., Watkins, T. L., and Kiss, I., Late functions and biochemical changes in mouse lung after irradiation: differential effects of WR-2721, *Radiat. Res.*, 103, 219, 1985.
103. Ward, W. F., Solliday, N. H., Molteni, A., and Port, C. D., Radiation injury in rat lung. II. Angiotensin-converting enzyme activity, *Radiat. Res.*, 96, 294, 1983.
104. Ward, W. F., Lin, P. J. P., Wong, P. S., Behnia, R., and Jalali, N., Radiation pneumonitis in rats and its modification by the angiotensin-Converting enzyme inhibitor captopril evaluated by high-resolution computed tomography, *Radiat. Res.*, 135, 81, 1993.
105. Ward, W. F., Molteni, A., Ts'ao, C. H., and Hinz, J. M., Captopril reduces collagen and mast cell accumulation in irradiated rat lung, *Int. J. Radiat. Oncol. Biol. Phys.*, 19, 1405, 1990.
106. Ward, W. F., Molteni, A., Ts'ao, C., Kim, Y. T., and Hinz, J. M., Radiation pneumotoxicity in rats: modification by inhibitors of angiotensin converting enzyme, *Int. J. Radiat. Oncol. Biol. Phys.*, 22, 623, 1992.
107. Delaney, J. P., Bonsack, M. E., and Felemovicius, I., Luminal route for intestinal radioprotection, *Am. J. Surg.*, 166, 492, 1993.
108. Eldor, A., Vlodavsky, I., Riklis, E., and Fuks, Z., Recovery of prostacyclin capacity of irradiated endothelial cells and the protective effect of Vitamin C, *Prostaglandins*, 34, 241, 1987.
109. Eldor, A., Vlodavsky, I., Fuks, Z., Matzner, Y., and Rubin, D. B., Arachidonic Metabolism and Radiation Toxicity in cultures of vascular endothelial cells, *Prostaglandins, Leukotrienes and Essential Fatty Acids*, 36, 251, 1989.
110. Rubin, D. B., Drab, E. A., Ts'ao, C. H., Gardner, D., and Ward, W. F., Prostacyclin synthesis in irradiated endothelial cells cultured from bovine aorta, *J. Appl. Physiol.*, 58, 592, 1985.

111. Haimovitz-Friedman, A., Vlodavsky, I., Chaudhuri, A., Witte, L., and Fuks, Z., Autocrine effects of fibroblast growth factor in repair of radiation damage in endothelial cells, *Cancer Res.*, 51, 2552, 1991.
112. Witte, L., Fuks, Z., Haimovitz-Friedman, A., Vlodavsky, I., Goodman, D. S., and Eldor, A., Effects of irradiation on the release of growth factors from cultured bovine, porcine, and human endothelial cells, *Cancer Res.*, 49, 5066, 1989.
113. Rogelj, S., Klagsbrun, M., Atzmon, R., Kurokawa, M., Haimovitz, A., Fuks, Z., and Vlodavsky, I., Basic fibroblast growth factor is an extracellular matrix component required for supporting the proliferation of vascular endothelial cells and the differentiation of PC12 cells, *J. Cell. Biol.*, 109, 823, 1989.
114. Edwards, B. S., Curry, M. S., Southon, E. A., Chong, A. S., and Graf, L. H., Evidence for a dithiol-activated signaling pathway in natural killer cell avidity regulation of leukocyte function antigen-1: structural requirements and relationship to phorbol ester- and CD16–triggered pathways, *Blood*, 86, 2288, 1995.
115. Suzuki, M., Capparelli, A. W., Ok, J. D., and Yanagawa, N., Thiol redox and phosphate transport in renal brush-border membrane. Effect of nicotinamide, *Biochem. Biophys. Acta*, 1021, 85, 1990.
116. Abate, C., Patel, L., Rauscher, III, F. J., and Curran, T., Redox regulation of Fos and Jun DNA-binding activity *in vitro*, *Science*, 249, 1157, 1990.
117. Lenardo, M. J. and Baltimore, D., NF-kB: a pleitropic mediator of inducible and tissue specific gene control, *Cell*, 58, 227, 1989.
118. Ghosh, S. and Baltimore, D., Activation *in vitro* of NF-kB by phosphorylation of its inhibitor IkB, *Nature*, 344, 678, 1990.
119. Ferran, C., Millan, M. T., Csizmadia, V., Cooper, J. T., Brostjan, C., Bach, F. H., and Winkler, H., Inhibition of NF-kappa B by pyrrolidine dithiocarbamate blocks endothelial cell activation, *Biochem. Biophys. Res. Commun.*, 214, 212, 1995.
120. Roebuck, K. A., Rahman, A., Lakshiminarayanan, V., Janakidevi, K., and Malik, A. B., H_2O_2 and tumor necrosis factor-alpha activate intercellular adhesion molecule 1 (ICAM-1) gene transcription through distinct *cis*-regulatory elements within the ICAM-1 promoter, *J. Biol. Chem.*, 270, 18966, 1995.
121. Bernard, G. R., N-acetylcysteine in experimental and clinical acute lung injury, *Am. J. Med.*, 91, 54, 1991.
122. Ignarro, L. J., Buga, G. M., Wood, K. S., Byrns, R. E., and Chauduri, G., Endothelium-derived relaxing factor produced and released from artery and vein is nitric oxide, *Proc. Natl. Acad. Sci.*, 84, 9265, 1987.
123. Fraile, M. L., Conde, M. V., Sanz, L., Moreno, M. J., Marco, E. J., and Lopez-de-Pablo, A. L., Different influence of superoxide anions and hydrogen peroxide on endothelial function of isolated cat cerebral and pulmonary arteries, *Gen. Pharmacol.*, 25, 1197, 1994.
124. Zembowicz, A., Hatchett, R. J., Jakubowski, A. M., and Gryglewski, R. J., Involvement of nitric oxide in the endothelium-dependent relaxation induced by hydrogen peroxide in the rabbit aorta, *Br. J. Pharmacol.*, 110, 151, 1993.
125. Monaco, J. A. and Burke-Wolin, T., NO and H_2O_2 mechanism of guanylate cyclase activation in oxygen-dependent responses of rat pulmonary circulation, *Am. J. Physiol.*, 268, L546, 1995.
126. Hughes, H., Sands, M. A., McGuire, G. M., and Taylor, A. A., PAF formation by H_2O_2-stimulated perfused canine carotid arteries, *Prostaglandins Leukotrienes Essential Fatty Acids*, 51, 323, 1994.

127. Kurose, I., Wolf, R., Grisham, M. B., Aw, T. Y., Specian, R. D., and Granger, D. N., Microvascular responses to inhibition of nitric oxide production. Role of active oxidants, *Circ. Res.*, 76, 30, 1995.
128. Soares, F. A., Shaughnessy, S. G., MacLarkey, W. R., and Orr, F. W., Quantification and morphologic demonstration of reactive oxygen species produced by Walker 256 tumor cells *in vitro* and during metastasis *in vivo*, *Lab. Invest.*, 71, 480, 1994.
129. Vischer, U. M., Jornot, L., Wollheim, C. B., and Theler, J. M., Reactive oxygen intermediates induce regulated secretion of von Willebrand factor from culture human vascular endothelial cells, *Blood*, 85, 3164, 1995.
130. Liu, S. M. and Sundqvist, T., Effects of hydrogen peroxide and phorbol myristate acetate on endothelial transport and F-actin distribution, *Exp. Cell. Res.*, 217, 1, 1995.
131. Sundqvist, T. and Liu, S. M., Hydrogen peroxide stimulates endocytosis in cultured bovine aortic endothelial cells, *Acta. Physiol. Scand.*, 149, 127, 1993.
132. Suttorp, N., Weber, U., Welsch, T., and Schudt, C., Role of phosphodiesterases in the regulation of endothelial permeability *in vitro*, *J. Clin. Invest.*, 91, 1421, 1993.
133. Seeger, W., Hansen, T., Rossig, R., Schmehl, T., Schutte, H., and Suttorp, N., Hydrogen peroxide-induced increase in lung endothelial and epithelial permeability — effect of adenylate cyclase stimulation and phosphodiesterase inhibition, *Microvasc. Res.*, 50, 1, 1995.
134. Jolliet, P., Polla, B., Donath, A., and Slosman, D., Early hydrogen peroxide-induced pulmonary endothelial cell dysfunction: detection and prevention, *Crit. Care. Med.*, 22, 157, 1994.
135. Yoshioka, T., Homma, T., Meyrick, B., Takeda, M., Moore-Jarrett, T., Kon, V., and Ichikawa, I., Oxidants induce transcriptional activation of manganese superoxide dismutase in glomerular cells, *Kidney Int.*, 46, 405, 1994.
136. Shimizu, S., Nomoto, M., Yamamoto, T., and Momose, K., Reduction by NG-nitro-L-arginine of H_2O_2-induced endothelial cell injury, *Br. J. Pharmacol.*, 113, 564, 1994.
137. Boyer, C. S., Bannenberg, G. L., Neve, E. P., Ryrfeldt, A., and Moldeus, P., Evidence for the activation of the signal-responsive phospholipase A2 by exogenous hydrogen peroxide, *Biochem. Pharmacol.*, 50, 753, 1995.
138. de Bono, D. P. and Yang, W. D., Exposure to low concentrations of hydrogen peroxide causes delayed endothelial cell death and inhibits proliferation of surviving cells, *Atherosclerosis*, 114, 235, 1995.
139. Shimizu, S., Saitoh, Y., Yamamoto, T., and Momose, K., Stimulation by hydrogen peroxide of L-arginine metabolism to L-citrulline couple with nitric oxide synthesis in cultured endothelial cells, *Res. Commun. Chem. Pathol. Pharmacol.*, 84, 315, 1994.
140. Natarajan, V., Taher, M. M., Roehm, B., Parinandi, N. L., Schmid, H. H., Kiss, Z., and Garcia, J. G., Activation of endothelial cell phospholipase D by hydrogen peroxide and fatty acid hydroperoxide, *J. Biol. Chem.*, 268, 930, 1993.
141. Hassoun, P. M., Yu, F.S., Zulueta, J. J., White, A. C., and Lanzillo, J. J., Effect of nitric oxide and cell redox status on the regulation of endothelial cell xanthine dehydrogenase, *Am. J. Physiol.*, 268, L809, 1995.
142. Wesson, D. E. and Elliot, S. J., Xanthine oxidase inhibits transmembrane signal transduction in vascular endothelial cells, *J. Pharmacol. Exp. Ther.*, 270, 1197, 1994.
143. Meharg, J. V., McGowan-Jordan, J., Charles, A., Parmelee, J. T., Cutaia, M. V., and Rounds, S., Hydrogen peroxide stimulates sodium-potassium pump activity in cultured pulmonary arterial endothelial cells, *Am. J. Physiol.*, 265, 613, 1993.
144. Lu, D., Maulik, N., Moraru, I. I., Kreutzer, D. L., and Das, D. K., Molecular adaptation of vascular endothelial cells to oxidative stress, *Am. J. Physiol.*, 264, C715, 1993.

145. Taher, M. M., Garcia, J. G., and Natarajan, V., Hydroperoxide-induced diacylglycerol formation and protein kinase C activation in vascular endothelial cells, *Arch. Biochem. Biophys.*, 303, 260, 1993.
146. Bradley, J. R., Thiru, S., and Pober, J. S., Hydrogen peroxide-induced endothelial retraction is accompanied by a loss of the normal spatial organization of endothelial adhesion molecules, *Am. J. Pathol.*, 147, 627, 1995.
147. Schuppe, I., Moldeus, P., and Cotgreave, A. A., Protein specific S-thiolation in human endothelial cells during oxidative stress, *Biochem. Pharmacol.*, 44, 1757, 1992.
148. Andreoli, S. P., Captopril scavenges hydrogen peroxide and reduces, but does not eliminate, oxidant-induced cell injury, *Am. J. Physiol.*, 264, F120, 1993.
149. David, M., Horvath, G., Schimke, I., Nagy, I., and Mueller, M. M., Comparative drug influence on peroxide mediated increase of cytosolic calcium in human endothelial cells, *Clin. Chim. Acta*, 223, 1, 1993.
150. Vincent, P. A., Rebres, R. A., Lewis, E. P., Hurst, V., and Saba, T. M., Release of ED1 fibronectin from matrix of perfused lungs after vascular injury is independent of protein synthesis, *Am. J. Physiol.*, 265, L485, 1993.

CHAPTER 5

Protein Synthesis in Irradiated Endothelial Cells

Eliot M. Rosen and Itzhak D. Goldberg

CONTENTS

I. INTRODUCTION

The endothelium is a multifunctional tissue that forms the lining of the entire circulatory and lymphatic systems, including the heart. Endothelial cells are a mesenchyme-derived cell type that exhibit many epithelioid characteristics and perform a number of highly specialized functions.[1] For example, endothelial cells actively transport macromolecules to and from tissues; metabolize vasoactive substances; synthesize basement membrane; regulate smooth muscle cell proliferation; and interact with the coagulation, fibrinolytic, and complement systems. As discussed earlier (Chapter 1), the vascular endothelium is a crucial target that may limit the dose of radiation that can be utilized in the treatment of cancer. A thorough understanding of endothelial radiobiology may enable us to devise new strategies to selectively radioprotect the endothelium, allowing higher doses of radiation to the tumor with

1-8493-4840-4/98/$0.00+$.50

the same or reduced normal tissue toxicity. This chapter will describe the effects of radiation on total protein synthesis by vascular endothelial cell monolayers, while other chapters describe the effects of radiation on expression of specific endothelial gene products.

II. MAINTENANCE OF THE INTEGRITY OF THE VASCULAR ENDOTHELIUM

The integrity of the normal, uninjured endothelium is maintained, in part, by (1) specialized intercellular junctions, including gap and tight junctions; (2) cell surface proteins that inhibit cell motility and proliferation; (3) expression of anti-thrombogenic molecules; and (4) down-regulation of cell surface molecules that mediate interaction with leukocytes.[2–4] Loss of endothelial continuity by a denuding injury can be repaired by several mechanisms depending upon the extent of the injury.[5–7] Gaps caused by loss of single cells can be repaired rapidly (in less than 1 hour) by spreading of adjacent cells. Gaps caused by loss of several cells are covered by spreading and migration of nearby cells (several hours), while larger gaps are repaired by migration from the edge of the wound, followed by proliferation (several days), to make up for the cell loss. These mechanisms operate to minimize the amount of time that thrombogenic subendothelial tissues are exposed to platelets and clotting factors.[4]

Ionizing radiation (X-rays, gamma-rays) induces loss of endothelial cells from a confluent monolayer in cell culture and in blood vessels *in vivo*.[7–11] Cell loss creates gaps or discontinuities in endothelial monolayer. These discontinuities could be repaired by the following mechanisms: (1) *spreading* and *migration* of neighboring cells from the edge of a gap; (2) *hypertrophy* of the remaining attached cells; (3) *proliferation* of surviving, reproductively viable cells; and (4) *repopulation* by unirradiated cells from outside the radiation field. Some or all of these mechanisms appear to contribute to repair of endothelial damage following exposure to radiation and other toxins. We have investigated the mechanisms by which irradiated endothelial monolayers recover from cell loss during the early postirradiation period.

III. RESPONSES OF BLOOD VESSEL-DERIVED CELLS TO RADIATION

Cultured endothelial cells exhibit D_0 values (100–150 cGy) and extrapolation numbers (2–8) generally similar to those of cultured vascular smooth muscle cells, fibroblasts, and other cell types.[7,11–14] Nevertheless, the behavior of endothelial monolayers following irradiation is distinctly different from that of smooth muscle cells and fibroblasts. We studied the responses of confluent monolayers of bovine large vessel (aorta and pulmonary artery) endothelial cells to ^{137}Cs gamma-radiation.[10] These cells form cobblestone epithelioid monolayers, take up silver nitrate stain at intercellular boundaries, stain positively for Factor VIII antigen, and express angiotensin

I converting enzyme, an endothelial-selective marker.[15] For comparison, similar studies were performed utilizing confluent cultures of bovine and rat aortic smooth muscle cells.

Within several hours after a dose of gamma-irradiation, we observed disruption of the normal, epithelioid configuration of the endothelial monolayer, with retraction of cell membranes, cell separation, appearance of cells with spindle-shaped morphology, and loss of silver staining. Damaged cells rounded up, as "surviving" cells spread out underneath them and undermined their points of attachment. These rounded cells were ultimately shed into the culture medium, leaving a population of intact cells that appeared significantly larger in cell surface attachment area than unirradiated control cells. Somatic hypertrophy of irradiated endothelial cells could not be explained by the accumulation of cells in a G_2-like state or by the accumulation of polyploid cells. Flow cytometric analysis of propidium iodide-stained nuclei of sham-treated vs. irradiated endothelial cells revealed no differences in the distribution of cells with 2C (G_1/G_0 diploid) and 4C (G_2 diploid or G_1 tetraploid) DNA content. The percentages of cells with 2C and 4C DNA content were 86 and 14%, respectively, for unirradiated cells. The corresponding values for cells that had received 3000 cGy 24 hr earlier were 82 and 18%, respectively. Therefore, hypertrophy occurred predominantly in a diploid population of endothelial cells.

We further investigated the mechanism of hypertrophy of irradiated endothelial cells. The increase in cell attachment area was not merely due to increased spreading of irradiated cells; 24 hr after a radiation dose of 3000 cGy, the modal cell volume of the irradiated cells (1944 μm^3) was approximately twice that of unirradiated control cells (980 μm^3). The increase in cell volume was correlated with a similar increase in cell protein content (Table 5.1). Cell loss, cell hypertrophy, and accumulation of cell protein were dose and time dependent. These phenomena were detectable 24 hr after a dose of 400 cGy and were maximal 72 hr after a dose of 3000 cGy. The degree of cellular hypertrophy was directly correlated with the degree of cell loss up to about 80% cell loss. The maximal cell hypertrophy was about four- to fivefold, when the cell loss factor was 80%, indicating a limit to the amount of cell mass that could be accumulated.

Table 5.1 Cell Loss and Hypertrophy of Bovine Aortic Endothelial Cells 24 hr After Irradiation

Dose (cGy)	Cell Density (Cells/cm²)	Protein Content (μg/10⁶ cells)	Clonogenic Survival (% of Control)
0	$(1.95 \pm 0.12) \times 10^5$	149 ± 1	100
400	$(1.39 \pm 0.04) \times 10^5$	202 ± 3	7.4 ± 0.9
1200	$(1.12 \pm 0.11) \times 10^5$	241 ± 2	0.5 ± 0.02
3000	$(1.13 \pm 0.03) \times 10^5$	271 ± 2	—

Note: Confluent cultures were sham-treated or irradiated with ^{137}Cs gamma-rays at a dose rate of 108 cGy/min at 25°C. Values represent means ± standard errors (n = 3–4). All statistical comparisons of irradiated (400, 1200, 3000 cGy) vs. control (0 cGy) cultures were significant ($p < 0.02$, two-tailed *t*-test).

In principle, multiple mechanisms could account for the accumulation of protein in irradiated endothelium, including: (1) increased protein synthesis; (2) decreased protein degradation; (3) pinocytosis of protein from the medium; and (4) phagocytosis of nearby cells. New protein synthesis (mechanism 1) clearly contributes to somatic hypertrophy of endothelial cells, since protein synthesis rates were significantly higher in irradiated cultures than in control cultures, even at doses as low as 400 cGy (Table 5.2). While protein synthesis rates were clearly elevated following irradiation, DNA synthesis rates were decreased. We measured incorporation of ^{14}C-leucine into cell protein and ^{3}H-thymidine into cell DNA during a 1-hour pulse given 1 day after treatment of confluent endothelial monolayers with 0 cGy or 3000 cGy. The ^{14}C-leucine counts were $(7.4 \pm 1.0) \times 10^3$ and $(13.0 \pm 0.1) \times 10^3$ cpm/10^6 cells for control and irradiated cells, respectively. The corresponding values for ^{3}H-thymidine counts were $(16.2 \pm 0.4) \times 10^3$ and $(3.2 \pm 0.5) \times 10^3$ cpm/10^6 cells, respectively. In contrast to the behavior of endothelial cells, somatic hypertrophy was *not* observed in irradiated cultures of vascular smooth muscle cells. Even at a dose of 3000 cGy, no significant cell loss, protein accumulation, or increased protein synthesis was found in irradiated cultures of rat or bovine aortic smooth muscle cells, as compared with unirradiated control cultures (Table 5.3).

It is unlikely that pinocytosis of serum proteins (mechanism 3) contributed significantly to the hypertrophy of irradiated endothelial cells. The same degree of cell hypertrophy was observed whether the experiment was performed in standard culture medium containing 10% fetal calf serum or in serum-free medium (Table 5.4). This does not rule out the possibility that pinocytosis of cell proteins released from radiation-damaged cells contributes to cell hypertrophy. However, this latter possibility is rendered unlikely on the basis of the experiment described below.

To assess the contribution of decreased protein turnover (mechanism 2) to radiation-induced hypertrophy, cells were cultured in medium containing ^{14}C-leucine (0.1 μCi/ml) for 3 days, in order to label cell proteins. This dose of label had little or no effect on the cell number. The cultures were then washed, and the rate of loss of label from acid-precipitatable cell protein was determined. The rate of loss of label from cell protein is a measure of the protein degradation rate. The protein degradation rates of cultures that had received 0, 400, and 3000 cGy were 38, 22, and 0% per day, respectively. Thus, complete cessation of protein degradation could only account for an increase in protein content of 38% per day, far less than the 300 to 400% increases observed in less than 2 days after a dose of 3000 cGy. If phagocytosis of cell debris (mechanism 4) contributed significantly to cell hypertrophy, we would expect a significant increase in the amount of label per cell in the protein turnover experiment. However, after 3000 cGy, increases in label per cell were 25% or less. After lower doses, there was no increase in label per cell. This finding rules against the possibility that phagocytosis of cell debris or pinocytosis of protein released from damaged cells are significant contributors to hypertrophy. Therefore, increased protein synthesis and, to a lesser extent, decreased protein degradation, appear to be the predominant mechanisms underlying cell hypertrophy.

Table 5.2 Increase in Endothelial Cell Protein Synthesis Following Irradiation

Parameter	0 cGy (Sham)	400 cGy	3000 cGy
	Day 1		
Cell density (cells/cm^2)	$(1.59 \pm 0.17) \times 10^5$	$(0.98 \pm 0.04) \times 10^5$	$(0.88 \pm 0.04) \times 10^5$
Protein content (μg/10^6 cells)	194 ± 1	268 ± 3	300 ± 14
Protein synthesis (cpm/10^6 cells)	$(2.54 \pm 0.30) \times 10^4$	$(3.45 \pm 0.59) \times 10^4$	$(3.72 \pm 0.3) \times 10^4$
	Day 3		
Cell density (cells/cm^2)	$(1.69 \pm 0.17) \times 10^5$	$(0.66 \pm 0.06) \times 10^5$	$(0.35 \pm 0.03) \times 10^5$
Protein content (μg/10^6 cells)	182 ± 1	344 ± 4	632 ± 6
Protein synthesis (cpm/10^6 cells)	$(1.70 \pm 0.11) \times 10^4$	$(3.87 \pm 0.30) \times 10^4$	$(10.4 \pm 0.30) \times 10^4$
	Day 5		
Cell density (cells/cm^2)	$(1.66 \pm 0.06) \times 10^5$	$(0.70 \pm 0.06) \times 10^5$	$(0.25 \pm 0.01) \times 10^5$
Protein content (μg/10^6 cells)	150 ± 3	276 ± 9	676 ± 13
Protein synthesis (cpm/10^6 cells)	$(1.51 \pm 0.10) \times 10^4$	$(2.96 \pm 0.24) \times 10^4$	$(7.26 \pm 0.20) \times 10^4$

Note: Confluent cultures of bovine aortic endothelial cells in 35-mm dishes were exposed to ^{137}Cs gamma-rays on Day 0. Cell number, protein content, and protein synthesis rates were determined at different times after irradiation. Values represent mean ± standard error of triplicate determinations. The protein synthesis rate was determined by pulse-labeling cultures with 10 μCi of ^{35}S-methionine per dish for 1 hr and precipitating cell protein using trichloroacetic acid.

Table 5.3 Response of Aortic Smooth Muscle Cells to Ionizing Radiation

Cell Strain	Dose (cGy)	Cell Density (cells/cm^2)	Protein Content (μg/10^6 cells)	Protein Synthesis (cpm/10^6 cells/hr)
SMS1	0	$(1.67 \pm 0.29) \times 10^5$	431 ± 4	$(1.97 \pm 0.18) \times 10^4$
	3000	$(1.51 \pm 0.06) \times 10^5$	405 ± 6	$(2.22 \pm 0.07) \times 10^4$
SMS63c	0	$(1.35 \pm 0.13) \times 10^5$	276 ± 6	$(2.05 \pm 0.15) \times 10^4$
	3000	$(1.25 \pm 0.08) \times 10^5$	303 ± 12	$(2.06 \pm 0.07) \times 10^4$

Note: SMS1 is a diploid strain of rat aortic smooth muscle cells; SMS63c is a strain of fetal bovine aortic smooth muscle cells. Cells were irradiated with ^{137}Cs gamma-rays on Day 0, and measurements were made on Day 7 (SMS1) or on Day 2 (SMS63c). All values are means ± standard errors (n = 3-5). To measure protein synthesis, cells were pulse-labeled for 1 hr with (^{14}C)-leucine (1 μCi/ml), and the incorporation of label into trichloroacetic acid-precipitatable protein was determined, as described before.

Table 5.4 Hypertrophy of Irradiated Bovine Aortic Endothelium in Serum-Free Medium

Culture Medium	Dose (cGy)	Cell Density (cells/cm²)	Protein Content (μg/10⁶ cells)
MEM + 10% serum	0	$(1.37 \pm 0.19) \times 10^5$	206 ± 5
	3000	$(0.46 \pm 0.08) \times 10^5$	498 ± 46
Serum-free MEM	0	$(1.16 \pm 0.29) \times 10^5$	228 ± 8
	3000	$(0.50 \pm 0.02) \times 10^5$	464 ± 29

Note: Cells were washed and the culture medium was replaced with fresh MEM (Minimal Eagle's medium) with or without 10% fetal calf serum immediately prior to irradiation. Cell density and protein content were measured 24 hr after sham treatment (0 cGy) or 3000 cGy of ^{137}Cs gamma-radiation. Values are means ± standard errors (n = 3).

IV. SIGNIFICANCE OF THE ENDOTHELIAL HYPERTROPHIC RESPONSE

In contrast to the behavior of vascular smooth muscle cells and fibroblasts, irradiated cultures of vascular endothelial cells undergo dose-dependent cell loss and compensatory hypertrophy of the remaining attached cells. The ability of the endothelium to rapidly reestablish its continuity following injury is essential *in vivo*, in order to minimize the exposure of the thrombogenic subendothelium to platelets, clotting factors, and smooth muscle growth factors.[3–6] Somatic cellular hypertrophy may be an expression of the tendency of a partially denuded endothelial monolayer to reestablish continuity under circumstances in which cell proliferation is inhibited or abolished. Cell hypertrophy associated with the accumulation of cell protein but no change in DNA content also occurs during *in vitro* cellular senescence of endothelial cells.[15] Increases in endothelial cell surface attachment area, cell volume, and protein content occurred in association with decreases in the ^{3}H-thymidine labeling index and the population doubling time. Increases in endothelial cell surface area also occur in human aortas during *in vivo* aging.[16] Similarities between aging- and radiation-induced endothelial hypertrophy include loss of proliferative capacity, accumulation of protein but not DNA, and a maximum increase in cell protein content of four- to fivefold. Similar hypertrophic responses would be expected following exposure to any toxin that induces endothelial cell loss and inhibits endothelial cell proliferation.

V. IMPLICATIONS FOR *IN VIVO* VASCULAR RADIOBIOLOGY

Clinically, the consequences of irradiation of large arteries and veins appear to be much less severe than that of irradiation of microvessels (e.g., capillaries). Radiobiologic studies suggest that large vessel- and microvessel-derived endothelial cells exhibit similar intrinsic radiosensitivity. However, in large vessels, a scaffold of subendothelial tissue (basement membrane and smooth muscle) may facilitate re-endothelialization of the vessel lumen. As discussed above, re-endothelialization may be due to a combination of cell hypertrophy, proliferation of the remaining clonogenic cells, and repopulation by cells from outside the irradiated field. The latter may include cells derived from unirradiated endothelium adjacent to the radiation field and from circulating endothelial cells.

The major consequence of irradiation of large arteries in humans appears to be accelerated development of atherosclerotic lesions. Survivors of Hodgkin's disease and other cancers who received moderate doses of cervicothoracic radiation (4000 to 5000 cGy) as part of their treatment exhibit an increased incidence of transient ischemic attacks and myocardial infarction.[17–20] Angiography performed many years after radiation often reveals narrowing and plaque formation in vessels within the radiation field. Experimental studies utilizing a variety of different animal species have revealed accelerated atherogenesis in irradiated vessels.[21–24] Damage to the endothelium is thought to play a crucial role in spontaneous atherogenesis, and is likely to contribute significantly to radiation-induced atherogenesis.[25] Irradiation of moderate to large arteries in experimental animals leads to endothelial injury and endothelial cell depletion, with subsequent recovery by 3 to 12 months, depending on the fractionation and total dose.[8,26] Interestingly, vascular smooth muscle cell depletion was smaller in magnitude than endothelial cell depletion, and occurred much later (12 to 18 months versus 3 months).[26] These *in vivo* findings appear to correlate with the lack of cell loss from irradiated smooth muscle cultures *in vitro*. Although the relative contributions of endothelial cell hypertrophy, clonogenic survival, and repopulation by unirradiated cells have not been determined, hypertrophy of irradiated endothelial cells has been documented *in vivo*.[27]

VI. MOLECULAR MECHANISM OF RADIATION-INDUCED ENDOTHELIAL DAMAGE

The hypertrophic response that we described appears to be triggered by radiation-induced cell loss, a phenomenon that is not directly related to clonogenic survival (Table 5.1). This response is *not* observed in smooth muscle cells. Radiation-induced cell retraction, associated with depolymerization of F-actin microfilament stress fibers, was observed in pulmonary microvascular endothelium at doses lower than 100 cGy.[28] The molecular mechanism of this phenomenon has not been established, but various studies suggest possible contributory mechanisms. Alterations in eicosanoid synthesis following irradiation have been well established (see Chapter 10). 12(*S*)-hydroxyeicosatetraenoic acid, a lipoxygenase metabolite of arachidonic acid, has been found to induce retraction and cytoskeletal rearrangement in cultured endothelial cells via a protein kinase C-dependent pathway.[29]

Radiation induces the release of various growth factors and cytokines from endothelial cells (see Reference 12 and Chapters 7 and 9). The increased growth factor expression appears to result from *de novo* synthesis and secretion rather than from cell lysis. Preliminary studies from our laboratory (unpublished data) suggest that radiation induces endothelial cell expression of scatter factor (hepatocyte growth factor), an angiogenic cytokine that induces profound morphologic alterations in endothelial and epithelial cells.[30–33] These alterations, which include cell retraction and conversion from an epithelioid to a spindle morphology, are similar to the morphologic alterations observed within the first 24 hours after irradiation of a confluent endothelial monolayer. In an interesting study, conditioned medium from irradiated monolayers of bovine pulmonary artery endothelium was found to stimulate protein synthesis two- to threefold in irradiated fibroblasts, without increasing DNA synthesis.[34] Protein synthesis was also stimulated in unirradiated fibroblasts, but to a lesser extent. Protein synthesis was increased in less than 10% of the 1100 proteins that could be resolved by two-dimensional gel electrophoresis. These findings suggest that factors released by irradiated endothelial cells are capable of inducing a marked increase in the synthesis of selected proteins.

VII. CONCLUSION

Following irradiation, cells are shed from a confluent endothelial monolayer, creating gaps in the monolayer. To compensate for the cell loss, the remaining attached cells undergo hypertrophy, in part by new protein synthesis and in part by a reduction in the rate of protein turnover. These mechanisms, which operate over a time frame of several days *in vitro*, may reflect an early vascular homeostatic response before clonal proliferation of surviving cells or repopulation by cells from outside of the irradiated field can compensate for the cell loss. Elucidation of the molecular events leading to radiation-induced disruption of endothelial sheets and stimulation of protein synthesis may suggest methods to ameliorate tissue damage.

REFERENCES

1. Thilo-Korner, D. G. S. and Freshney, R. I., Eds., *The Endothelial Cell — A Pluripotent Control Cell of the Vessel Wall*, Karger, Basel, 1983.
2. Heimark, R. I. and Schwartz, S. M., The role of membrane-membrane interactions in the regulation of endothelial cell growth, *J. Cell Biol.*, 100, 1934–1940, 1985.
3. Schwartz, S. M., Dynamic maintenance of the endothelium, in *The Endothelial Cell — A Pluripotent Control Cell of the Vessel Wall*, Thilo-Korner, D. G. S. and Freshney, R. I., Eds., Karger, Basel, 1983, 113.
4. Sakariassen, K. S., Bolhuis, P. A., and Sixma, J. J., Human blood platelet adhesion to artery subendothelium is mediated by factor VIII-von Willebrand Factor bound to the subendothelium, *Nature*, 279, 636, 1979.
5. Sholley, M. M., Gimbrone, M. A., and Cotran, R. S., Cellular migration and replication in endothelial regeneration: a study using irradiated endothelial cultures, *Lab. Invest.*, 36, 18, 1977.
6. Hirsch, E. Z., Chisolm, G. M., and White, H. M., Re-endothelialization and maintenance of endothelial integrity in longitudinal denuded tracks in the tracks in the thoracic aorta of rats, *Atherosclerosis*, 46, 287, 1983.
7. Fischer, J. J., Proliferation of rat aortic endothelial cells following X-irradiation, *Radiat. Res.*, 92, 405, 1982.
8. Fonkalsrud, E. W., Sanchez, M., Zerubavel, R., and Mahoney, A., Serial changes in arterial structure following radiation therapy, *Sur. Gynecol. Obstet.*, 145, 395, 1977.
9. Johnson, L. K., Longenecker, J. P., and Fajardo, L. F., Differential radiation response of cultured endothelial cells and smooth myocytes, *Anal. Quant. Cytol.*, 4, 188, 1982.

10. Rosen, E. M., Vinter, D. W., and Goldberg, I. D., Hypertrophy of cultured bovine aortic endothelium following irradiation, *Radiat. Res.*, 117, 395, 1989.
11. Rubin, D. B., Drab, E. A., Ward, W. F., and Bauer, K. D., Cell cycle changes and cytotoxicity in irradiated cultures of bovine aortic endothelial cells, *Radiat. Res.*, 108, 206, 1986.
12. Witte, L., Fuks, Z., Haimovitz-Friedman, A., Vlodavsky, I., Goodman, D. S., and Eldor, A., Effects of irradiation on the release of growth factors from cultured bovine, porcine, and human endothelial cells, *Cancer Res.*, 49, 5066, 1989.
13. Rosen, E. M., Goldberg, I. D., Myrick, K. V., and Levenson, S. E., Radiation survival of vascular smooth muscle cells as a function of age, *Int. J. Radiat. Biol.*, 48, 71, 1985.
14. Rosen, E. M., Goldberg, I. D., Shapiro, H. M., Zoller, L. C., Myrick, K. V., Levenson, S. E., and Halpin, P. A., Growth kinetics as a function of ploidy in diploid, tetraploid, and octaploid smooth muscle cells derived from the normal rat aorta, *J. Cell Physiol.*, 125, 512, 1985.
15. Rosen, E. M., Mueller, S. N., Noveral, J. P., and Levine, E. M., Proliferative characteristics of clonal endothelial strains, *J. Cell Physiol.*, 107, 123, 1981.
16. Repin, V. S., Dolgov, V. V., Zaikina, O. E., Novikov, I. D., Antonov, A. S., Nikolaeva, M. A., and Smirnov, V. N., Heterogeneity of endothelium in human aorta. A quantitative analysis by scanning electron microscopy, *Atherosclerosis*, 50, 35, 1984.
17. Cohn, K. E., Stewart, J. R., Fajardo, L. F., and Hancock, E. W., Heart disease following radiation, *Medicine*, 46, 281, 1967.
18. Dollinger, M. R., Lavine, D. M., and Foye, L. V., Myocardial infarction due to post-irradiation fibrosis of the coronary arteries, *J. Am. Med. Assoc.*, 195, 316, 1966.
19. McReynolds, R. A., Gold, G. L., and Roberts, W. C., Coronary heart disease after mediastinal irradiation for Hodgkin's disease, *Am. J. Med.*, 60, 39, 1976.
20. Silverberg, G. D., Britt, R. H., and Goffinet, D. R., Radiation-induced carotid artery disease, *Cancer*, 41, 130, 1978.
21. Amromin, G. D., Gildnehorn, H. L., Solomon, R. D., Nadkarni, B. B., and Jacobs, M. L., The synergism of X-irradiation and cholesterol-fat feeding on the development of coronary artery lesions, *J. Atheroscler. Res.*, 4, 325, 1964.
22. Artom, C., Lofland, H. B., and Clarkson, T. B., Ionizing radiation, atherosclerosis, and lipid metabolism in pigeons, *Radiat. Res.*, 26, 165, 1965.
23. Lindsay, S., Entenman, C., Ellis, E. E., and Geraci, C. L., Aortic atherosclerosis in the dog after localized aortic irradiation with electrons, *Circ. Res.*, 61, 1962.
24. Lindsay, S., Kohn, H. I., Dakin, R. L., and Jew, J., Aortic artheriosclerosis in the dog after localized aortic X-irradiation, *Circ. Res.*, 10, 51, 1962.
25. Ross, R., The pathogenesis of atherosclerosis, *N. Engl. J. Med.*, 314, 488, 1986.
26. Hirst, D. G., Denekamp, J., and Hobson, B., Proliferation studies of the endothelial and smooth muscle cells of mouse mesentery after irradiation, *Cell Tissue Kinet.*, 13, 91, 1980.
27. Narayan, K. and Cliff, W. J., Morphology of irradiated microvasculature: a combined *in vivo* and electron-microscopic study, *Am. J. Pathol.*, 106, 47, 1982.
28. Kantak, S. S., Diglio, C. A., and Onoda, J. M., Low dose radiation-induced endothelial cell retraction, *Int. J. Rad. Biol.*, 64, 319, 1993.
29. Tang, D. G., Timar, J., Grossi, I. M., Renaud, C., Kimler, V. A., Diglio, C. A., Taylor, J. D., and Honn, K. V., The lipoxygenase metabolite, 12(*S*)-HETE, induces a protein kinase C-dependent cytoskeletal rearrangement and retraction of microvascular endothelial cells, *Exp. Cell Res.*, 361, 1993.
30. Stoker, M., Gherardi, E., Perryman, M., and Gray, J., Scatter factor is a fibroblast-derived modulator of epithelial cell mobility, *Nature*, 327, 238, 1987.

31. Rosen, E. M., Meromsky, L., Setter, E., Vinter, D. W., and Goldberg, I. D., Purification and migration-stimulating activities of scatter factor, *Proc. Soc. Exp. Biol. Med.*, 195, 34, 1990.
32. Grant, D. S., Kleinman, H. K., Goldberg, I. D., Bhargava, M., Nickoloff, B. J., Polverini, P., and Rosen, E. M., Scatter factor induces blood vessel formation *in vivo*, *PNAS U.S.A.*, 90, 1937, 1993.
33. Naidu, Y. M., Rosen, E. M., Zitnik, R., Park, M., Goldberg, I., Polverini, P. J., and Nickoloff, B. J., Role of scatter factor (hepatocyte growth factor) in the pathogenesis of AIDS-related Kaposi's sarcoma, *PNAS U.S.A.*, in press.
34. Flavin, M. P., Parton, L. A., and Bowman, C. M., Conditioned medium from irradiated bovine pulmonary artery endothelial cells stimulates increased protein synthesis by irradiated bovine fibroblasts *in vitro*, *Exp. Lung Res.*, 16, 405, 1990.

CHAPTER 6

Signaling in the Radiation Response of Endothelial Cells

Adriana Haimovitz-Friedman and Zvi Fuks

CONTENTS

I. INTRODUCTION

Radiation-induced changes in capillaries and small caliber arterioles constitute a basic injury in the pathogenesis of chronic radiation damage to the skin,[1] heart,[2] lung,[3] liver,[4] kidney,[5] and the central nervous system.[6] Histopathological studies have shown that microvascular damage is predominant during the early phases of radiation injury to these tissues,[7] while tissue atrophy and fibrosis, developing secondary to microvascular degeneration and ischemia, characterize the late radiation damage syndrome.[8–11] Ultrastructural studies have demonstrated that endothelial cells are the most sensitive structures to radiation in the vessel wall,[7,9] exhibiting a spectrum of

1-8493-4840-4/98/$0.00+$.50

dose- and time-dependent lesions, ranging from early reversible endothelial cell edema and vesicle formation to advanced degeneration and cell death.[7–11]

Radiation produces in endothelial cells a variety of biochemical and morphological abnormalities,[10] but the damage most frequently monitored as a quantitative marker of radiation sensitivity is the loss of the clonogenic capacity. Endothelial cells that survive the lethal effects of radiation maintain their clonogenic competence, and when cultured after irradiation at low cell densities, each surviving cell produces a discrete colony of progeny cells. The number of colonies can be determined before and after irradiation, thus enabling calculations of the surviving fraction. The fraction of surviving cells is dose dependent and is inversely related to the radiation dose. When plotted on semilogarithmic coordinates, clonogenic dose-survival curves consist of an exponential component (the "linear component") at the high-dose range and a gradually curving component at the low-dose range, referred to as the "shoulder region."[12–15] The slope of the linear component (D_0) is believed to represent the inherent radiation sensitivity of a tested cell population, while the width of the shoulder (D_q) represents the capacity to repair sublethal or potentially lethal radiation lesions.[12–14] Although vascular endothelial cells *in vivo* are mostly quiescent and exhibit a low incidence of cycling cells, with labeling index values of less than 1%,[16] clonogenic dose-survival evaluations can still be carried out *in vivo* if angiogenic stimuli are produced. Physical or chemical stimulation of the rat subcutaneous air pouch or of the dog and rabbit cornea result in an angiogenic response of capillary neovascularization, with each capillary regarded as a clone of a single endothelial cell.[17–19] The density of the capillary network can be measured, and it serves as an indicator of the clonogenic capacity of microvascular endothelial cells.

In vitro studies on the sensitivity of endothelial cells to radiation[20–25] have disclosed a wide range of sensitivities, with D_0 values of the clonogenic dose survival curves ranging from 101 to 165 cGy and D_q values between 64 and 230 cGy (Table 6.1). In comparison, the D_0 of radiation dose-effect curves *in vivo* have ranged between 170 and 265 cGy, and D_2–D_1 values of split dose experiments in the range of 180 to 290 cGy.[17–19] Although reservations can be raised regarding comparisons between *in vivo* and *in vitro* data, these observations suggest that when irradiated *in vivo*, endothelial cells may be more radioresistant and capable of repairing more radiation damage than would be appreciated from the *in vitro* experiments. The broad range of the D_q values *in vitro*, and the differences between the *in vitro* and *in vivo* data, suggest that in addition to cell type specificities, the repair function of radiation lesions in endothelial cells may be sensitive to variations in microenvironmental conditions.

This review summarizes observations on factors that affect the sensitivity of endothelial cells to radiation *in vitro* and *in vivo*. The response of endothelial cells to radiation was indeed found to be sensitive to several microenvironmental factors, including angiogenic factors and components of the extracellular matrix. Signaling pathways associated with modulation of the radiation response by growth factors may serve as a model for regulatory mechanisms of radiation resistance *in vitro* and *in vivo*, with potential clinical implications.

Table 6.1 Parameters of Radiation Dose Survival Curves of Endothelial Cells Published in the Literature

Author (Ref.)	Cells	Culture Surfaces	FGF	D_0 (cGy)	D_q (cGy)
Kwock (20)	BPAEC	Plastic	None	161	71
Rhee (21)	BAEC	Plastic	None	101	65
Fuks (22)	BAEC	Plastic	bFGF	94	135
Hei (23)	HUVEC	Gelatin	aFGF	165	130
Sholley (24)	HUVEC	Gelatin	ECGF	146	115
Martin (25)	RAEC	Irrad. RAEC	CM	120	230
Fuks (22)	BAEC	HR9-bFGF/ECM	bFGF	107	144
	BAEC	BAEC/ECM	bFGF	89	217
	BAEC	HR9/ECM	bFGF (& 72 hr)	107	29

Note: The abbreviations used are: BPAEC, bovine pulmonary artery endothelial cells; HUVEC, human umbilical vein endothelial cells; BAEC, bovine aortic endothelial cells; RAEC, rabbit aortic endothelial cells; FGF, fibroblast growth factor; ECGF, endothelial cell growth factor; CM, conditioned medium; ECM, extracellular matrix.

II. EFFECTS OF AUTOLOGOUS MATRIX SIGNALS ON THE RADIATION RESPONSE OF ENDOTHELIAL CELLS

One of the microenvironmental variables of the cell culture conditions used to generate the dose-survival data listed in Table 6.1 is the type of culture dish surface upon which endothelial cells were seeded to assess the clonogenic capacity after irradiation. In some studies, the cells were cultured postradiation either on top of the plastic surfaces of standard culture dishes, while in others gelatin-coated dishes, or dishes coated with a variety of natural extracellular matrices were used. The latter conditions more closely resemble the normal endothelial cell microenvironment *in vivo*. Endothelial cells rest *in vivo* upon basement membranes that serve as anatomical support structures for cell anchorage and cytoskeletal orientation, and provide regulatory physiological stimuli for a variety of normal cellular functions.[26–28] When cultured *in vitro*, endothelial cells synthesize and secrete a membranous layer of extracellular matrix (ECM), which closely resembles in composition and supramolecular structure the vascular subendothelial basement membrane *in vivo*.[26] Such ECM-coated culture dishes were found to support endothelial cell proliferation and differentiation and a variety of other cellular functions.[29,30]

Fuks et al.[22] reported on the effects of ECM on the clonogenic competence of endothelial cells after X-ray exposure. When irradiated plateau phase bovine aortic endothelial cells (BAEC) were assayed for colony formation on top of uncoated plastic of standard culture dishes, the dose survival curves showed D_0 values of 107 ± 6 cGy and D_q of 63 ± 28 cGy (Figure 6.1). When assayed in dishes precoated with the ECM produced *in vitro* by BAEC (designated as BAEC/ECM) the curves showed a similar D_0 (106 ± 2 cGy), but the D_q was increased to 194 ± 8 cGy ($p < 0.05$). This observation indicated that matrix components confer in endothelial cells an improved capacity to repair radiation lesions and to restore the clonogenic capacity. However, when the HR9-bFGF/ECM produced by a mouse endodermal tumor cell line was used, a decreased repair capacity was noted, with a D_q value of 156 ± 30 cGy ($p < 0.05$ compared to BAEC/ECM). The HR9-bFGF/ECM was reported to differ in composition and structure from the BAEC/ECM[31] and to confer different biological responses in cells seeded on top of it.[32] These observations suggested a high specificity of the repair function of irradiated endothelial cells to interaction with autologous matrix-based molecules. This notion is, however, inconsistent with studies reporting that the radiosensitivity of human ovarian carcinoma cells cultured on the Matrigel ECM was not different from the sensitivity of the same cells assayed on plastic.[33] It should, nonetheless, be noted that Matrigel is a reconstituted basement membrane-like gel,[34] and being of mouse origin it is xenogeneic to human ovarian carcinoma cells. Indeed, the experiments by Fuks et al.[22] demonstrated that the xenogeneic combination of BAEC and the mouse HR9-bFGF/ECM exhibited radiosensitivity characteristics not significantly different from those observed on plastic, but both were associated with a decreased repair capacity as compared to the autologous BAEC/ECM (Figure 6.1).

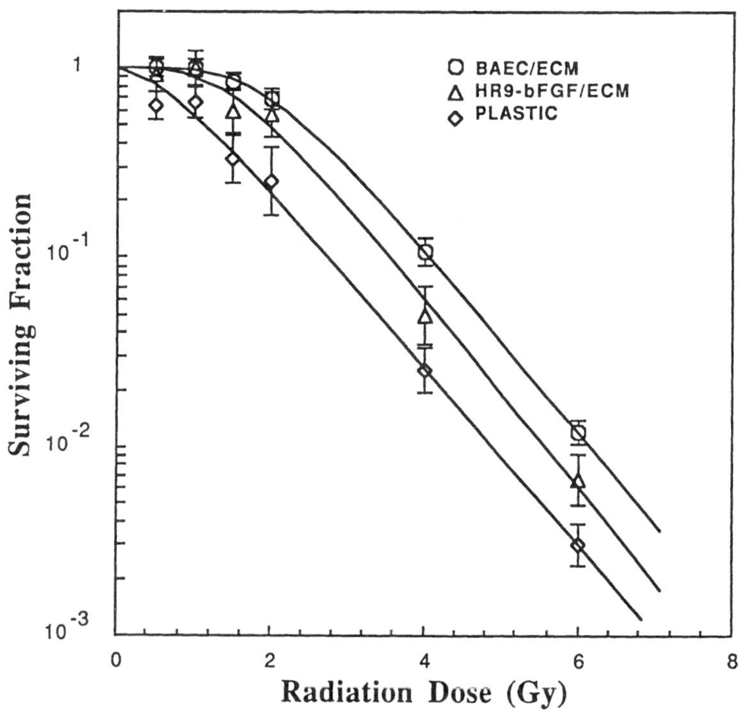

Figure 6.1. Radiation dose survival curves of plateau-phase BEAC plated for colony formation assays on either ECM or on regular tissue culture plastic. The cells were irradiated in suspension and seeded for colony formation assays on top of the uncoated plastic surfaces of standard tissue culture dishes (Δ), or in culture dishes pre-coated with the BAEC/ECM (○), or with the HR9-bFGF/ECM (□). (From Fuks, Z. et al., *Eur. J. Cancer*, 28/A: 725-731, 1992. With permission.)

The sensitivity and specificity of the repair function to components of autologous ECM are consistent with other known effects of matrix on mammalian cells. The tissue specificity of the ECM is dependent on the relative concentrations and supramolecular assembly of collagens, elastin, structural glycoproteins (i.e., laminin, fibronectin, entactin, nidogen, and thromboplastin) and proteoglycans in each type of matrix.[35] Anchorage of cells to basement membranes is mediated via focal adhesion plaques that bind the cell membrane to specific domains of ECM molecules and also serve as structural links between the ECM and the cytoskeleton.[36] These highly specialized structures also contain regulatory proteins that mediate transmission of ECM stimuli to the cytoplasm by chemical signaling similar to that triggered by soluble growth factors.[37–39] ECM-cellular linkages were found to control both the shape, spatial orientation and tissue distribution of normal cells[29,30] and to affect the cytoskeletal organization.[40] In addition, they were shown to regulate cytodifferentiation and cell migration during embryogenesis,[28] to induce the expression of differentiation genes in mesenchymal and epithelial cells,[27,28,41,42] and to modulate the response of cells to hormones and growth factors.[29,41] The specificity of the matrix functions to the autologous cell type was demonstrated by studies that showed that the BAEC/ECM promoted both proliferation and normal differentiation of BAEC *in vitro*, while the ECM produced by human fibroblasts supported BAEC proliferation, but failed to support differentiation functions.[43] The high specificity of the repair function of radiation damage in endothelial cells to autologous matrix indicates that signaling via membrane receptors is most probably associated with this function. However, the specific molecules associated with the initiation of this signaling and whether it leads to induction or to enhanced expression of repair enzymes remain unknown.

III. EFFECT OF BASIC FIBROBLAST GROWTH FACTOR (bFGF) ON ENDOTHELIAL CELL KILL BY RADIATION

Another variable microenvironmental condition that may affect the response of endothelial cells to X-irradiation *in vitro* is the use of growth factors in the culture media to support clonal growth (Table 6.1). Endothelial cells express a variety of plasma membrane receptors for factors associated with angiogenesis, coagulation, inflammation, and regulation of the vascular tone.[44] Several mitogenic growth factors and hematopoietic cytokines (i.e., basic and acidic FGF, PDGF, VEGF, angiogenin, erythropoietin, and GM-CSF) induce in endothelial cells receptor-dependent proliferative responses, while TGFβ, PF-4, IL-1, and TNF confer antiproliferative signals.[45] The most potent and universal angiogenic factor is basic fibroblast growth factor (bFGF).[45,46] Other growth factors confer more limited angiogenic and proliferative effects, correlating with patterns of expression of plasma membrane receptors for these factors. For example, while PDGF-β receptors are expressed in human and rat microvascular endothelial cells[47,48] and in human glioma endothelial cells,[49] PDGF receptors were not found on human umbilical vein endothelial cells (HUVEC)[50] or on bovine aortic and rat heart endothelial cells.[51] Some types of endothelial cells exhibit an absolute dependence of *in vitro* clonal growth on the presence of FGF in the culture media.[52] Rogelj et. al.[53] demonstrated that when BAEC cells were plated on top of uncoated plastic surfaces in the absence of bFGF, no colonies could be detected after 7 to 10 days in culture. Addition of bFGF resulted in colony growth stimulation,[53] with plating efficiencies of 10 to 30%.[18] When BAEC were seeded on top of several types of extracellular matrix that contained matrix-bound bFGF, the addition of exogenous bFGF to the culture medium was not required to elicit clonal growth.[22,53] Because of the high sensitivity of the repair functions to microenvironmental conditions, Haimovitz-Friedman et al. postulated that variations in the postradiation application of bFGF may affect the recovery patterns of endothelial cells from radiation-induced damage.[54]

A recent study by Fuks et al.[55] in fact demonstrated that when bFGF was added to cultures of bovine aortic endothelial cells (BAEC) immediately after irradiation, the survival curves disclosed significant sensitivity to ionizing radiation, with a D_0 value of 83.7 ± 4.3 cGy. At the low-dose range, however, the cells exhibited a remarkable ability to overcome the lethal effects of radiation, producing a wide shoulder region (D_q value of 205 ± 3.1 cGy). The colony formation assay in these experiments was carried out in culture dishes precoated with the bFGF-free HR9/ECM.[55] This matrix does not support the clonal growth of BAEC unless bFGF is added to the culture medium.[53,54] However, the addition of bFGF can be postponed for several days, because while attached to this matrix, BAEC maintain their viability and exhibit a normal clonal efficiency even if the addition of bFGF is delayed for as many as 96 hours.[54] When the addition of bFGF to the clonogenic assay cultures was delayed for 72 hours after irradiation, the cells exhibited a significant decrease capacity for recovery from radiation damage, as evidenced from the minimal shoulder region (D_q of 34.4 ± 2.5 cGy; $p < 0.05$ as compared to the 0 time curve), although there was no significant change in the slope of the linear component of the curve.

The ability to modify the shoulder of the dose-survival curve has traditionally attracted a great deal of attention because the dose range in which the shoulder phenomenon occurs corresponds with the dose range commonly used for the management of human tumors. In his classical publications on the response of mammalian cells to radiation, Elkind conceptualized the shoulder phenomenon as representing an ability of irradiated cells to accumulate sublethal quantities of radiation lesions and to subsequently repair them over the course of time.[22,25,56] The prevailing hypothesis on the mechanisms of radiation-induced cell kill identifies the cell nucleus as the primary target for the lethal effects of ionizing radiation and heterologous DNA double-strand breaks as the most common type of radiation lesions that lead to mammalian cell death.[57,58] The direct interaction of the DNA with X-ray photons, or with reactive oxygen intermediates produced in the cell by the radiolysis of water,[57,58] creates a variety of DNA lesions that are mostly repaired; this is because mammalian cells are proficient in the capacity to recombine DNA breaks and to repair other types of DNA damage.[59–61] However, when the load of DNA double-stranded breaks exceeds the repair capacity of the cell, or when structurally complex lesions are produced, the cell may sustain potentially lethal damage.[56,62] Thus, residual DNA double-strand breaks that the cell fails to repair are regarded as those that eventually lead to radiation-induced cell death.[60,61] According to these concepts on radiation damage and its repair, the data on the radioprotective effect of bFGF quoted above

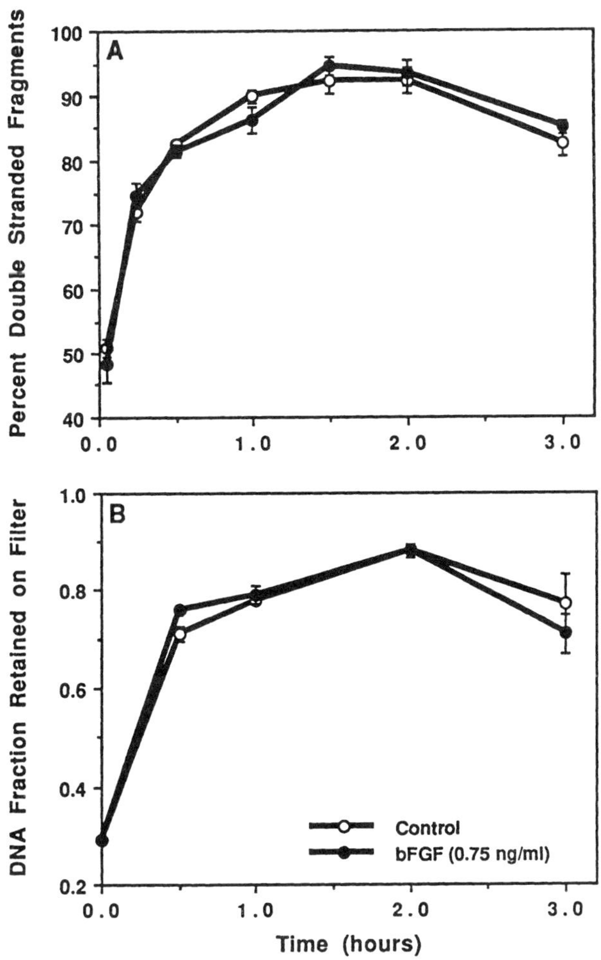

Figure 6.2. Effects of bFGF on repair of radiation-induced DNA breaks. Plateau-phase BAEC were exposed to a single dose of 20 Gy in the presence and absence of 0.75 ng/ml bFGF. The DNA was extracted at the indicated time points and evaluated for DNA breaks. (A) shows the repair of radiation-induced, single-strand breaks following a dose of 20 Gy, using the nondenaturing neutral DNA elution method.[63] The repair of radiation-induced DNA single-strand breaks was expressed as a time-dependent increase in double-stranded fragments produced by sonication of the extracted DNA, calculated at each time point as radioactive counts of the double-stranded fragments fraction divided by the overall radioactive counts. (B) shows the repair of radiation-induced, double-strand breaks by the neutral filter elution method.[64] The fraction of intact or repaired DNA retained on the filter is shown as a function of time after irradiation.

would suggest that bFGF, if provided shortly after irradiation, induces or enhances the repair of radiation-induced DNA double-strand breaks in endothelial cells.

To explore this possibility, Fuks et al.[55] measured the effect of bFGF on DNA double-strand break repair in irradiated BAEC, using the nondenaturing neutral DNA elution method.[63] Figure 6.2A shows that neither the initial yield of breaks nor the kinetics of the final level of repair was affected by bFGF stimulation, with the repair process completed within 90 to 120 minutes after irradiation regardless of the presence or absence of bFGF.[55] Similarly, there was no effect of bFGF on the rejoining of radiation-induced DNA single-strand breaks, measured by the alkali unwinding technique,[64] with the repair reaching near maximum levels at about 30 minutes after irradiation regardless of bFGF stimulation (Figure 6.2B). However, because of the significant decrease in the clonogenic survival associated with the delay in bFGF stimulation shown in Figure 6.2, it was assumed that either the structural or the functional integrity of the DNA was impaired in bFGF-deprived cells.[55] This issue was addressed in experiments that used the terminal transferase nick-end translation assay[65] to identify DNA breaks. The studies revealed that following completion of the repair of primary radiation-induced DNA breaks, a secondary wave of DNA fragmentation became apparent in bFGF-deprived BAEC. The nature of the secondary DNA fragmentation process was evaluated by agarose gel electrophoresis (Figure 6.3) and disclosed a typical "DNA ladder" of oligonucleosomal fragments characteristic of apoptosis.[55] Staining of irradiated cells with the DNA-binding fluochrome *bis*benzimide trihydrochloride (Hoechst-33258) revealed in the nuclei of irradiated cells extensive morphological features of apoptosis, including nucleoplasmic condensation, nuclear budding, and formation of apoptotic bodies (Figure 6.4A). The frequency of apoptotic cells was radiation dose dependent, and these changes inhibited in bFGF-treated cells (Figure 6.4B). Flow cytometry analysis showed that the apoptotic process in these cells involved exclusively a subpopulation of G_0-G_1 cells and progressed as an interphase phenomenon without requiring a transition through the M phase of the cell cycle. Basic FGF stimulation rescued the cells by inhibiting the apoptotic pathway, thus creating a shoulder pattern in the clonogenic dose survival curve. On the other hand, bFGF did not affect the inherent sensitivity to radiation killing due to unrepaired primary DNA double-strand breaks, as demonstrated by the lack of bFGF effect on the slope (D_0) of the linear component of the dose survival curve.

Apoptosis (programmed cell death) is a major physiological mechanism involved in the regulation of embryonal development, growth and differentiation, and in the wear-and-tear maintenance of adult mammalian tissues.[66,67] In addition, programmed cell death serves as a mechanism for induction of cell kill by cytokines,[68,69] chemical agents,[70,71] radiation,[72,73] and heat.[74] Apoptosis is conceptualized as a preprogrammed pathway of sequential biochemical events that are only partially known, which eventually lead to activation of a calcium-magnesium-dependent endonuclease that cleaves the nuclear chromatin at selective internucleosomal linker sites.[65] Signals generated at the membrane of the affected cell activate neighboring intact cells and infiltrating macrophages to phagocytize the dying cell and its disintegrating nucleus.[75] Ionizing radiation induces programmed cell death in thymic, lymphoid, and hematopoietic cells[76,77] and in stem and undifferentiated progenitors of testicular, intestinal, renal, neuronal, and oligodendrocytic cell lineages.[77] Mature and differentiated progenies of non-lymphoid mammalian cells are believed to undergo radiation-induced apoptosis only rarely,[73] although it has been reported in the epithelium of intestinal crypts,[77] the ascini of the parotid and lacrimal glands,[78] and in several experimental murine tumors.[79,80] BAEC belong to the latter category because these

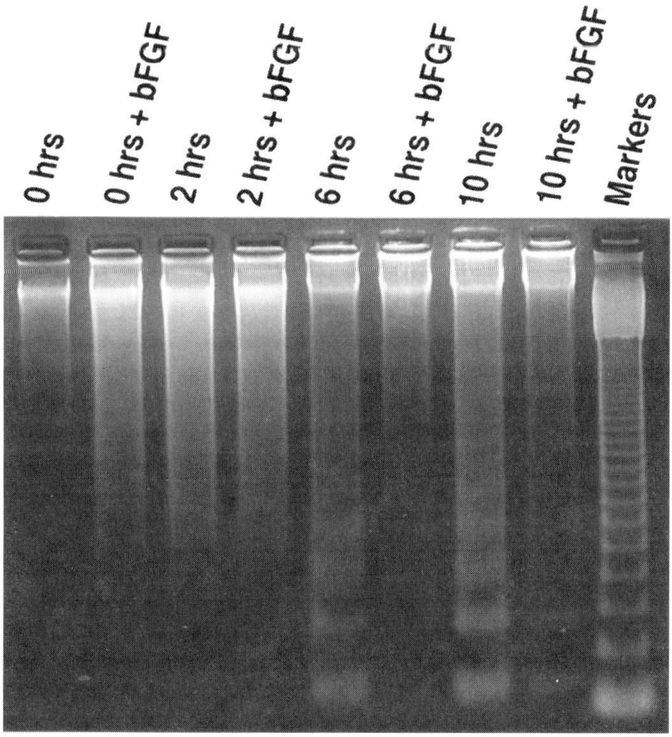

Figure 6.3. Agarose gel electrophoresis of DNA extracted from plateau-phase BAEC irradiated with a single dose 500 cGy and incubated after irradiation with (0.75 ng/ml) or without bFGF.

cells were shown to be morphologically and functionally differentiated and to express phenotypic characteristics similar to those expressed by adult aortic endothelial cells *in vivo*.[52,81]

The data suggesting that the bFGF-mediated inhibition of apoptosis creates a shoulder pattern in the clonogenic dose-survival curve of endothelial cells provide a possible alternative to the current hypothesis on the nature of the shoulder phenomenon in mammalian cells. The proposed model[55] postulates that the preprogrammed pathway of apoptosis may be triggered by radiation exposure more frequently than has been appreciated in the past. Activation of cell-type specific anti-apoptotic protection may rescue apoptosis-sensitive subpopulations of irradiated cells, thus creating a shoulder in the dose survival curve. The shoulder region may, therefore, represent a dose range in which the potential for the interphase mode of apoptosis may prevail over the postmitotic mechanism of cell death, whereas postmitotic cell death from unrepaired DNA double-strand breaks may be the predominant mechanism of radiation-induced cell kill at the higher dose range. The relative contribution from each mode of cell death may differ not only with dose, but may vary from one cell type to another, relative to their inherent and inducible capacities to overcome each of these types of potentially lethal radiation damage.[55]

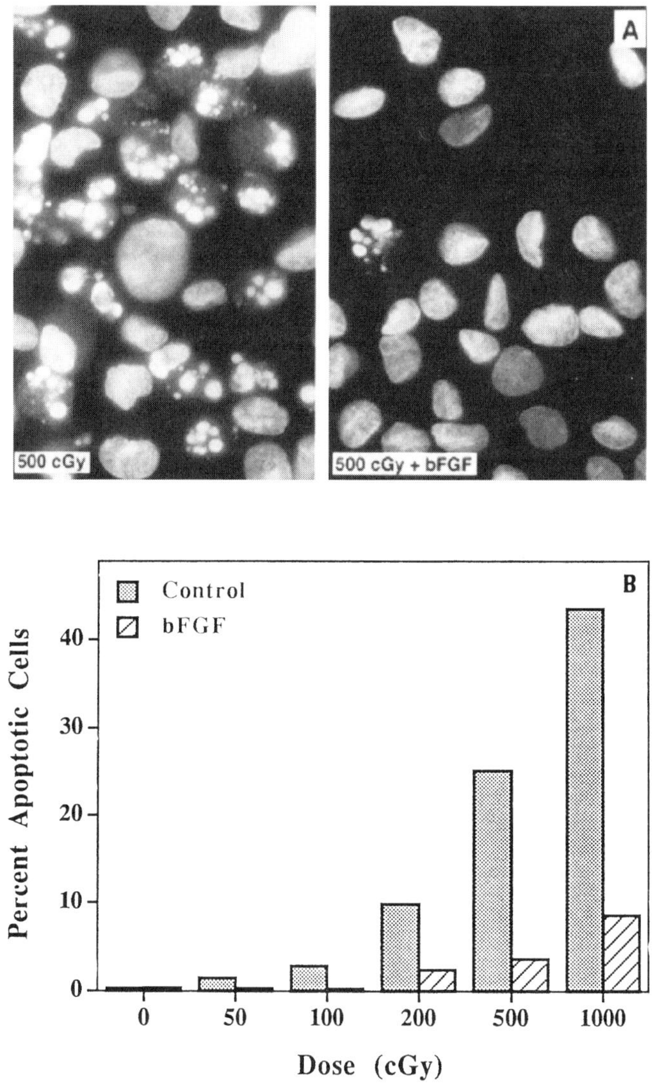

Figure 6.4. Morphological changes associated with radiation-induced apoptosis in plateau-phase BAEC. (A) shows cells irradiated with a single dose of 500 cGy and incubated with or without bFGF (0.75 ng/ml) for 8 hr. The cells were then fixed with paraformaldehyde and stained with Hoechst 33258. Note that the majority of the irradiated control cells show chromatin changes typical of apoptosis, while only a single cell with apoptotic changes can be seen among the bFGF-treated cells. (B) shows the effect of bFGF (0.75 ng/ml) on the percentage of cells exhibiting morphological changes of apoptosis 8 hr after exposure to increasing doses of irradiation. A total of 500 cells were counted for each data point. (From Fuks, Z. et al., *Cancer Res.*, 54, 2591, 1994. With permission.)

IV. SIGNALING IN THE RADIATION PROTECTIVE EFFECT CONFERRED BY bFGF

The biological effects of bFGF are mediated via binding to its high-affinity transmembrane receptors, and activation of its tyrosine kinase domain.[82,83] The downstream signaling from the activated receptor involves several pathways. The best-characterized pathway involves transphosphorylation and activation of the γ-isoform of phospholipase C (PLCγ) by the activated receptor tyrosine kinase.[83] The activated PLCγ hydrolyzes membrane phosphatidylinositol to generate inositol 1,4,5-triphosphate and diacylglycerol,[83,84] which in turn leads to mobilization of cellular Ca^{2+} and to activation of protein kinase C (PKC).[84] In some cell types, bFGF signaling was found to rapidly activate p44-mitogen-activated protein (MAP) kinase and S6 protein kinase (pp90rsk),[85,86] possibly involving upstream two other kinases, Raf-1 and mitogen-activated protein kinase-kinase.[87,88] In endothelial cells, Haimovitz-Friedman et al.[89] reported that bFGF stimulation induces a rapid turnover of phosphoinositides, translocation of the α-isotype of PKC from the cytosol to the membrane, and its activation within seconds after bFGF stimulation. Peak stimulation to a 3.5-fold increase in PKC activity was noted at 30 seconds, and it was down-regulated to nearly baseline levels by 90 seconds after bFGF stimulation. Exposure of the cells to a dose of 200 cGy without concomitant stimulation with bFGF did not result in a PKC activation.[89]

The involvement of the bFGF receptor tyrosine kinase in the bFGF effect to inhibit radiation-induced apoptosis was demonstrated by Haimovitz-Friedman et al.,[89] who showed that the anti-apoptotic protection of bFGF in irradiated BAEC was completely abrogated by the tyrosine kinase inhibitor tyrophostin AG213. The role of bFGF-activated PKC in mediating this effect was suggested by experiments that showed that nonspecific PKC activation by short-term exposure to low doses of the phorbol ester TPA mimicked the radioprotective effect of bFGF, while pre-treatment with the PKC inhibitor H-7 abrogated the bFGF effect, as was observed after depletion of cellular PKC by overnight preincubation with high-dose TPA.[89] Agarose gel electrophoresis of DNA extracted from irradiated BAEC showed that both TPA and bFGF inhibited the apoptotic degradation of DNA induced in these cells by exposure to 500 cGy, and that these radioprotective effects could be reversed by pretreatment with the PKC inhibitor H-7 (Figure 6.5). These data demonstrated that PKC activation is both required and sufficient for the rescue of endothelial cells from programmed cell death induced by radiation in endothelial cells.

The rescue of endothelial cells from radiation-induced apoptosis by bFGF is analogous to the prevention of apoptosis by hematopoietic cytokines. Several hematopoietic cytokines were shown to inhibit apoptosis via PKC in serum- or growth factor-deprived and in glucocorticosteroid-treated lymphoid and hematopoietic cells.[90–93] PKC activation was also reported to mediate the down-regulation of radiation-induced apoptosis in thymic and lymphoid cells.[94–97] In the experiments reported by Haimovitz-Friedman et al.,[89] bFGF or TPA stimulation were required to elicit the activation of PKC, because exposure to radiation alone in the absence of

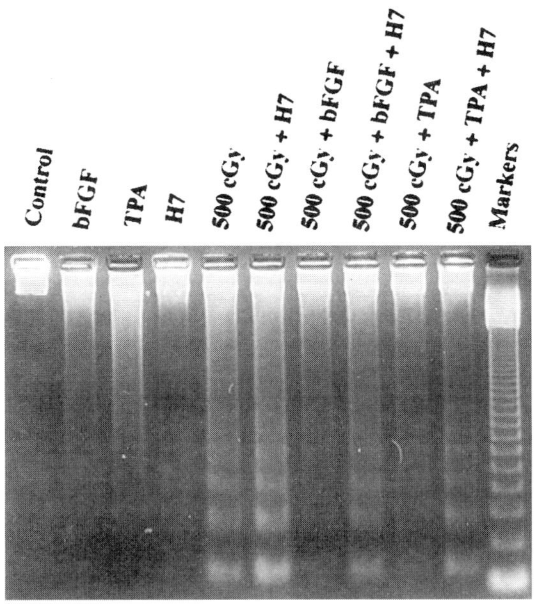

Figure 6.5. Effects of bFGF and TPA on apoptotic DNA degradation in irradiated endothelial cells. Plateau-phase BAEC were irradiated with a single dose of 500 cGy while still adherent to the original culture dishes as contact inhibited monolayers. In some cultures, the phorbol ester TPA (30 ng/ml) was added to the medium immediately after irradiation and removed 30 minutes later by replacement of the medium. Basic FGF (1 ng/ml) and/or H-7 (20 ng/ml) were added to some of the culture immediately after irradiation. The DNA was extracted from the cultures 10 hr after irradiation and submitted to agarose gel electrophoresis to detect apoptotic DNA degradation, as previously described.[38] (From Haimovitz-Friedman, A. et al., *Cancer Res.*, 54, 2591, 1994. With permission.)

bFGF failed to activate PKC. However, other recent studies have demonstrated that in several cell types, radiation activated PKC directly, without an apparent involvement of specific growth factors in eliciting this response.[98–101] Hallahan et al. reported that a spontaneous activation of PKC within 1 minute after radiation exposure in HL-60[98] and RIT-3 cells,[99] while Kim et al.[100] found that radiation spontaneously activated the ε isoform of PKC in human lung adenocarcinoma A549 cells. Uckun et al.[101] reported radiation-induced activation of PKC in B-lymphocyte precursor cells. PKC activation in these cells was abolished by protein tyrosine kinase inhibitors, and it was suggested that as yet unidentified tyrosine kinases activated by direct radiation effects mediate PKC activation in these cells.[107] These data suggest cell-type specificities and multiple metabolic pathways may be involved in the PKC response after irradiation.

Whether growth factor-independent mechanisms of PKC activation play a role in the protection of cell other than thymic, lymphatic, hematopoietic, and endothelial cells against radiation-induced apoptosis is unknown. Experiments reported by Hallahan et al.[102] and by Kim et al.[100] showed that inhibition of radiation-induced activation of PKC by staurosporine enhanced the radiation killing of human tumor cells. In both studies, the increase in radiation killing of staurosporine-treated cells was not associated with a measurable difference in the rejoining of DNA double-strand breaks as compared to the staurosporine untreated control cells. Although the investigators did not demonstrate that apoptosis was involved in the enhanced radiation killing of PKC-inhibited cells, the resemblance of these observations to the bFGF-deprivation effect on irradiated endothelial cells[55,89] is striking. Taken together, these data suggest that direct activation of PKC by irradiation, or its activation by growth factors and cytokines, may play an important role in the homeostatic control of radiation resistance in many cell types, possibly by providing a protective mechanism against radiation-induced apoptosis.

V. SIGNALING IN RADIATION-INDUCED APOPTOSIS IN ENDOTHELIAL CELLS

The biochemical regulation of programmed cell death has recently attracted a great deal of attention because of the potential to affect the outcome of cancer therapy by modulation of the apoptotic pathway. Primary lesions to the DNA are regarded by many investigators as the signals that trigger chemotherapy- and radiation-induced apoptosis.[70, 73,103] Experiments to explore the role of signals originating from radiation-damaged DNA in apoptosis were performed using pretreatment of cells with bromodeoxyuridine (BrdUrd). This halogenated pyrimidine is known to sensitize mammalian cells to radiation killing because it replaces thymidine during DNA synthesis, thus creating defective DNA with increased propensity for DNA breaks upon radiation exposure.[104] Warters[105] reported that pretreatment of murine T-hybridoma cells with BrdUrd enhanced both X-ray-induced DNA strand breaks and apoptosis to a similar extent. In contrast, Akagi et al.[106] reported that pretreatment of Molt-4 cells with BrdUrd had no effect on the incidence of apoptotic cell death, although it did sensitize the cells to the postmitotic modes of radiation-induced cell kill. Several investigators have suggested that one of the biological roles of the p53 protein may be to interact with damaged DNA and thus trigger the biochemical cascade of radiation-induced apoptosis.[107–112] p53-Mediated apoptosis in irradiated cells was found to be associated with suppression of the cyclin-Cdk2 kinase complex by p53-activated p21WAF1/CIP1 suppressor gene,[117,118] but further details on the involvement of p53 in the triggering radiation-induced apoptosis are unknown.

In contrast to these observations, other studies have suggested that the apoptotic cascade may be triggered by extranuclear signals. Oxygen radical damage to the plasma membrane has been reported to induce apoptosis in resting B lymphocytes exposed to gamma-irradiation.[113] Macklis et al.[114] reported that binding of ^{212}Bi-radiolabeled anti-Thy 1.2 monoclonal antibody to cell surface receptors in EL-4 cells resulted in significant radiation-induced cell kill via activation of the apoptotic mechanism. Dosimetry evaluations of the range of migration of the particle irradiation emitted from the ^{212}Bi-radionuclide indicated that the majority of the radiation was deposited in the cell membrane. Uckun et al.[101] reported that extranuclear protein tyrosine kinases regulate radiation-induced apoptosis in B-lymphocyte precursors. However, although highly suggestive, none of these studies provided conclusive evidence for an extranuclear triggering mechanism. Hence, the question of whether radiation-induced apoptosis is initiated by signals generated at the cell membrane, nucleus, or both, remains open.

The approach to study mechanisms by which radiation-induced signals generated at the plasma membrane may initiate the apoptotic pathway has been derived from investigations on the induction of programmed cell death by tumor necrosis factor α (TNFα). Obeid[115] and Jarvis[116] have recently reported that the sphingomyelin pathway constitutes the early events in the apoptotic cascade of TNFα-induced apoptosis. The sphingomyelin pathway is a signal transduction pathway that mediates the signaling of several cytokines, such as TNFα and Interleukin-1β.[117] This pathway has similarities to the phosphoinositide signal transduction pathway.[117,118] Sphingomyelin (*N*-acylsphingosin-1-phosphocholine) is a phospholipid preferentially concentrated in the outer leaflet of the plasma membrane of most mammalian cells and consists of a long sphingoid-base backbone, a fatty acid, and a phosphocholine head group.[117,118] The central lipid in the sphingomyelin pathway is ceramide, produced by hydrolysis of sphingomyelin by a specific sphingomyelin-directed phospholipase-C, designated as sphingomyelinase. Two general forms of sphingomyelinase have been identified. The neutral sphingomyelinase is distinguished from its acidic counterpart by Mg^{2++} dependence, DAG independence, a different pH maximum, and association with the plasma membrane.[117,118]

Hydrolysis of sphingomyelin to yield ceramide has been identified as the initial step in a signaling pathway that mediates the induction of IL-2 transcription by IL-1β in EL-4 murine T-helper cells,[119] HIV-1 replication in TNFα-stimulated J22HL-60 cells,[120] TNFα-induced apoptosis in U937 monoblastic and HL-60 promyelocytic leukemia cell lines, and in L292/LM and WEHI-164/13 human fibrosarcoma cell lines.[115,116] Ceramide serves as a lipid second messenger in the sphingomyelin signaling pathway, activating a 97-kDa proline-directed serine/threonine kinase, termed ceramide-activated protein kinase.[120,121] The downstream signaling events activated by ceramide are only partially known, but phosphorylation of the p42 isoform of MAP kinase,[122] nuclear translocation of NFκB,[123–125] and activation of a serine/threonine phosphatase, which is similar to the type 2A protein phosphatase,[126] are known.

Activation of the sphingomyelin pathway by TNFα is mediated via its 55-kDa low-affinity receptor[127] and is tightly coupled to receptor activity, as activation of the sphingomyelin cascade can be induced by TNF or by a stimulatory effect of an antibody against the 55-kDa receptor in isolated membranes of promyelocytic HL-60 cells.[128] Further, incubation of HL-60 cell extracts with *Staphylococcus aureus* sphingomyelinase generates ceramide and recapitulates the effect of TNF on kinase activation, while other phospholipases, including phospholipases A2, C, and D, are ineffective.[128] The rapid kinetics of activation in isolated plasma membranes and the cell-free reconstitution of the cascade indicate that the sphingomyelin pathway is a primary membrane signaling system for TNFα and is independent of nuclear signals in the initial phases of this metabolic pathway. It is of interest to note that there is a striking similarity between the effects of TNF and ionizing radiation on endothelial cells. Robaye et al.[129] reported that TNF induces apoptotic DNA degradation in BAEC, similar to that induced by X-rays.[55] Activation of the sphingomyelin pathway in mediating TNF-induced apoptosis was found to be inhibited by PKC activation,[115,116] similar to the effect of bFGF in BAEC.[89] Furthermore, bFGF protects BAEC against TNFα-induced cell death in a similar fashion to its effect on irradiated BAEC.

Because of the similarities between the effects of radiation and TNF to induce apoptosis, Haimovitz-Friedman et al.[130] postulated that the triggering of apoptosis by radiation in BAEC may be generated by primary cell membrane signals and may be mediated via the sphingomyelin signaling pathway. The studies by Haimovitz-Friedman et al.[130] showed that treatment with synthetic ceramide analogs, or *Staphylococcus aureus* sphingomyelinase, mimicked the effect of radiation to induce apoptosis in endothelial cells. In fact, ionizing radiation, like TNF, induced rapid sphingomyelin hydrolysis to ceramide by the action of a neutral sphingomyelinase (Figure 6.6). The generation of ceramide was maximal within seconds after irradiation, and was sensitive to low-dose irradiation with a calculated ED_{50} of ~100–200 cGy. Phorbol esters abolished sphingomyelin hydrolysis to ceramide and blocked radiation-induced apoptosis in BAEC, further suggesting that ceramide generated by radiation exposure may act as a second messenger in the pathway of radiation-induced apoptosis.[130] The most compelling evidence that ceramide is a critical and obligatory element of the apoptotic cascade of radiation-induced apoptosis in endothelial cells was provided by experiments showing that when ceramide elevation was restored by addition of exogenous C2-ceramide, the phorbol ester effect to inhibit apoptosis was overcome (Figure 6.7).

The fact that radiation closely mimicked the effect of TNF to generate ceramide and induce apoptosis raised the question whether, similar to TNF, radiation acts directly on the plasma membrane to produce ceramide. The rapid kinetics of ceramide

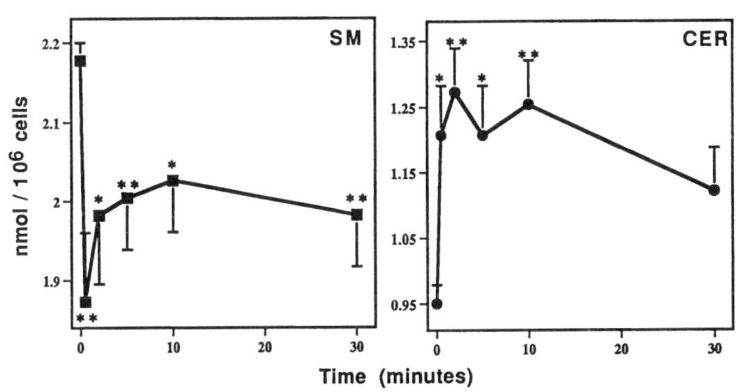

Figure 6.6. Effect of radiation on sphingomyelin (SM) and ceramide (CER) levels in endothelial cells. Plateau-phase BAEC were irradiated with a single dose of 1000 cGy. Sphingomyelin was extracted and resolved by TLC as previously described.[93] Ceramide measurements were performed by the diacylglycerol kinase assay.[95] The level of sphingomyelin decreased from 2.2 to 1.9 pmol/10^6 cells ($p < 0.001$), and concomitantly the ceramide level increased from 0.95 to 1.3 pmol/10^6 cells ($p < 0.025$). These changes occurred within 30 seconds of irradiation and sustained for 30 minutes. The kinetics and magnitude of these effects were similar to those reported for TNFα in human leukemia HL-60 and U937 cells.[78,79] One star (*) indicates $p < 0.025$; two stars (**) indicate $p < 0.005$. (From Haimovitz-Friedman, A. et al., *J. Exp. Med.*, 180, 525, 1994. With permission.)

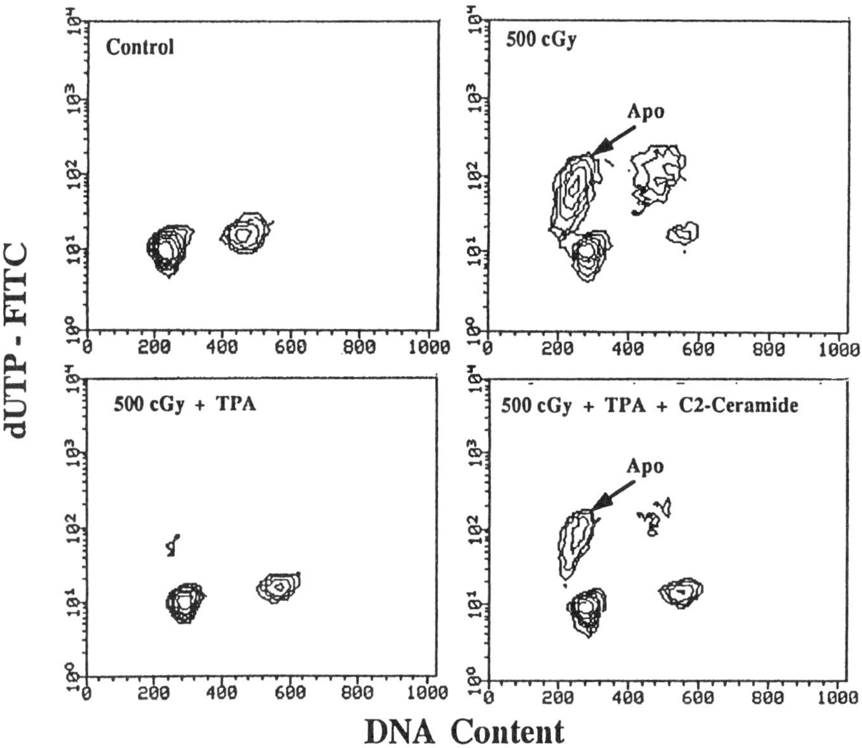

Figure 6.7. Restoration of apoptotic DNA fragmentation by C2-ceramide in endothelial cells treated with ionizing irradiation and TPA detected by flow cytometry. Plateau-phase BAEC were incubated for 5 minutes with or without 50 ng/ml of TPA. The cells were then irradiated with a single dose of 500 cGy and incubated for additional 30 minutes at 37°C. The cells were then washed and incubated for additional 24 hours at 37°C in DMEM with or without 50 μ*M* C2-ceramide. Flow cytometry was then performed, using concomitant terminal transferase staining of apoptotic DNA breaks and propidium iodide staining of total cellular DNA content. (From Haimovitz-Friedman, A. et al., *J. Exp. Med.*, 180, 525, 1994. With permission.)

generation after radiation exposure did not exclude a possibility of a very rapid transfer of nuclear-generated signals to the plasma membrane. To address this issue, Haimovitz-Friedman et al.[130] irradiated fresh supernates of nuclei-free membranes prepared from BAEC, and demonstrated that radiation-induced hydrolysis of sphingomyelin to ceramide occurred in these nuclei-free extracts (Figure 6.8). The magnitude of this effect was similar to that observed after irradiation of intact BAEC, and a concomitant reduction in sphingomyelin levels in the membrane preparations was also observed. These studies indicated that a direct effect of ionizing radiation on the membrane is sufficient to produce the critical lipid ceramide that transduces the apoptotic signals without requiring the presence or involvement of the cell nucleus. The data provide the first conclusive evidence that radiation can generate apoptotic signaling at the cell membrane and offer an alternative to the prevailing

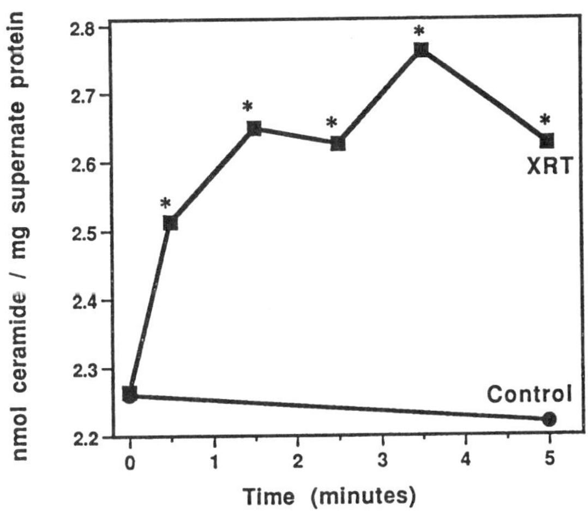

Figure 6.8. Effect of ionizing radiation (XRT) on ceramide level in nuclei-free membrane preparations. Plateau-phase BAEC were disrupted at 4°C by a tight-fitting Dounce homogenizer and the nuclei-free membrane fraction was irradiated with a dose of 1000 cGy at 4°C. To measure the effect of radiation on ceramide, samples of irradiated and control membranes were incubated at 37°C in a reaction mixture containing ATP and sphingomyelinase assay buffer.[94] Ceramide was quantified by the diacylglycerol kinase assay as described.[95] The level of ceramide increased within 30 seconds from a basal level of 2.25 to 2.50 nmol/mg supernate protein ($p < 0.001$). The magnitude of this effect was similar to that observed after irradiation of intact BAEC, shown in Figure 6.7. A concomitant reduction was observed in the sphingomyelin levels (data not shown). One star (*) indicates $p < 0.001$. (From Haimovitz-Friedman, A. et al., *J. Exp. Med.*, 180, 525, 1994. With permission.)

paradigm on the mechanism of the lethal effects of radiation, which identifies the nucleus as the primary and direct target for the lethal effects of ionizing irradiation.

VI. THE ROLE OF APOPTOSIS IN RADIATION-INDUCED DAMAGE TO THE MICROVASCULAR ENDOTHELIUM *IN VIVO*

Studies performed on the sensitivity of the endothelium of blood vessels have shown that capillaries represent the most radiation-sensitive sections of the vascular system.[8–11] Electron microscopy studies of the irradiated rabbit myocardium[2] demonstrated that, whereas the endothelium of capillaries exhibited severe damage, endothelial cells of large vessels and of the endocardium, even when present in the same tissue sections, seemed to be intact. The variations in sensitivity of the various sections of the vascular tree to radiation correlate with the patterns of distribution of angiogenic factors in the basement membranes of blood vessels. Immunohistochemical studies carried out by Cordon-Cardo et al.[131] demonstrated that bFGF is ubiquitously detected in the basement membranes of large- and intermediate-size blood vessels. However, capillaries were found to have significant segments of their basement membranes with diminished or absent bFGF deposits. It was, therefore suggested that bFGF-deficient segments of the microvascular endothelium may represent vulnerable sites for damage by radiation exposure.[54,55]

Based on these findings, Fuks et al.[55] postulated that intravenous injections of bFGF may saturate bFGF-deficient sections of the capillary endothelium and would serve as radioprotective therapy against radiation-induced damage to the microvasculature *in vivo*. Pharmacokinetic studies in C_3H/HeJ mice showed that intravenously injected ^{125}I-bFGF disappears from the circulation within 1 to 2 minutes after injection. Biotin-conjugated bFGF injected intravenously was shown by immunohistochemical staining to be bound to the endothelium of blood vessels in the lung and brain up to 60 minutes after injection with minimal, if any, leakage into the perivascular space. The rapid clearance of bFGF and the patterns of its intravascular binding suggest that injected bFGF binds to the heparan sulfate proteoglycan that coats the intraluminal surface of the vascular endothelium,[132] known for its bFGF binding affinity.[30]

Having established the availability of intravenously injected bFGF to the vascular endothelium, Fuks et al.[55] tested its effect on the development of the radiation pneumonitis in C_3H/HeJ mice receiving bilateral whole-lung irradiation. Intravenous injections of bFGF were given at doses of 400 ng each, immediately before and immediately after irradiation, and at 1 and 2 hours thereafter, but no additional maintenance therapy was given subsequently. The actuarial survival in bFGF-treated animals was significantly improved at each dose within the range of 1500 to 2500 cGy as compared to control bFGF-untreated mice. The actuarial survival at 180 days of mice receiving a dose of 2250 cGy was 35%, compared to 57% in bFGF-treated mice ($p = 0.001$). At 60 days after irradiation, the cumulative mortality from radiation pneumonitis in the control group was 29/45 (64%), compared to 12/42 (28%) in bFGF-treated mice ($p < 0.0001$). These data suggested that bFGF therapy, if given immediately before and after irradiation, affects an early and critical radiation-induced event that triggers the pathogenesis of radiation pneumonitis, which in bFGF-untreated animals leads to death weeks to months later on. To further investigate the nature of this putative event, *in situ* terminal transferase nick-end staining for the presence of apoptotic nuclei was performed, and it revealed extensive apoptotic changes in the nuclei of the capillary endothelium. Other pulmonary tissue elements, including the pneumocytes, the bronchial epithelium, and the endothelium of large- and intermediate-size blood vessels, did not show similar apoptotic changes. Treatment with intravenous bFGF abrogated the radiation-induced apoptotic changes in the capillary endothelium, although examination of thymic tissue present in the same sections showed apoptotic changes in more than 80% of the thymocytes. These data indicated that the therapeutic effect of bFGF in mice receiving lung irradiation may be associated with inhibition of programmed cell death, selectively induced in the pulmonary capillary endothelium by the radiation exposure.

VII. CONCLUSION

The demonstration that radiation induces programmed cell death in endothelial cells *in vitro* and *in vivo*, and that PKC activation via bFGF blocks radiation-induced apoptosis, may have broad implications for the understanding of mechanisms of radiation resistance in mammalian cells. The studies in endothelial cells indicate the existence of a balance between radiation induction of apoptosis via the sphingomyelin pathway and its down-regulation by natural suppressor mechanisms through PKC. High resting levels of PKC activity, its stimulation by hormones or growth factors, or PKC activation as a direct effect of radiation without the involvement of growth factors or cytokines, may provide an anti-apoptotic mechanism *in vitro* and *in vivo* and constitute a generic mechanism of radiation resistance. This hypothesis may provide an explanation for the frequent appearance of a shoulder in the dose survival curves of irradiated mammalian cells, and for the observation that PKC inhibitors sensitize the radiation killing of several cell types. PKC activation as a mechanism of radiation resistance may be of special relevance at the clinically applicable dose range, where primary radiation-induced DNA double-strand breaks are less prevalent. At higher dose levels, the postmitotic mechanisms of cell death from DNA double-strand breaks and chromosomal aberrations may prevail, while interphase programmed cell death may contribute to the overall effect of radiation, albeit with a decreased relative impact on the overall survival.

The existence of *in vivo* correlates to the *in vitro* phenomena in irradiated endothelial cells supports the concept that the interphase mode of apoptotic cell death may represent a fundamental element of the pleiotropic effects of radiation in nonlymphoid mammalian cells. It is becoming increasingly apparent that molecular events mediated by autocrine, paracrine, and endocrine growth and inhibitory factors play a major role in tissue responses to ionizing radiation.[133] Recent studies have provided an insight into mechanisms by which messages exchanged between cells via paracrine loops may mediate specific cellular and tissue responses that lead to the eventual development or prevention of radiation damage at the tissue level.[135] Understanding the molecular basis and the temporal evolution of apoptosis and anti-apoptotic effects during a course of fractionated irradiation may have important clinical implications. The evolution of radiation injury in normal and tumor tissues is thus placed within a cellular and molecular context that provides a background for testable experimental hypotheses at dose ranges relevant to clinical radiotherapy. As mechanisms of radiation-induced apoptosis in normal and tumor tissues are unraveled, it may become possible to actively intervene in specific pathways to favorably alter the therapeutic ratio in the radiation treatments of several types of human cancers.

REFERENCES

1. Verola, O., Lafaix J. L., Daburon, F., and Brocheriou, C., Vascular damage after acute local irradiation: a light and electron microscopy study, *Br. J. Radiol.*, Suppl. 19, 104, 1986.
2. Fajardo, L. F. and Stewart, J. R., Pathogenesis of radiation induced myocardial fibrosis, *Lab. Invest.*, 29, 244, 1973.
3. Philips, T. L., Benak, S., and Ross, C., Ultrastructural and cellular effects of ionizing radiation, *Front. Radiat. Ther. Oncol.,* 6, 21, 1972.

4. Jaenke, R. S., Robbins, M. E., Bywaters, T., Whitehouse, E., Rezvani, M., and Hopewell, J. W., Capillary endothelium. Target site of renal radiation injury, *Lab. Invest.*, 68, 396, 1993.
5. Ingold, J.A., Reed, G. B., Kaplan, H. S., and Bagshaw, M. A., Radiation hepatitis, *Am. J. Roentgenol.*, 93, 200, 1965.
6. Calvo W., Hopewell, J. W., Reinhold, H. S., and Yeung, T. K., Time and dose related changes in the white matter of the rat brain after single doses of X-rays, *Br. J. Radiol.*, 61, 1043, 1988.
7. Hopewell, J. W., Calvo, W., Jaenke, R., Reinhold, H. S., Robbins, M. E., and Whitehouse, E. M., Microvasculature and radiation damage, *Recent Results Cancer Res.*, 130, 1, 1993.
8. Law, M. P., Radiation-induced vascular injury and its relation to late effects in normal tissue, *Adv. Radiat. Biol.*, 9, 37, 1981.
9. Hopewell, J. W., Campling, D., Calvo, W., Reinhold, H. S., Wilkinson, J. H., and Yeung, T. K., Vascular irradiation damage: its cellular basis and likely consequences, *Br. J. Cancer*, 53 (Suppl. VII), 181, 1986.
10. Fajardo, L. F., The unique physiology of endothelial cells and its implications in radiobiology, *Front. Radiat. Ther. Oncol.*, 23, 96, 1989
11. Reinhold, H. S., Fajardo, L. F., and Hopewell, J. W., The vascular system, *Adv. Rad. Biol.*, 14, 177, 1990.
12. Elkind, M. M. and Sutton, H. G., X-ray damage in recovery in mammalian cells in culture, *Nature,* 184, 1293, 1959.
13. Curtis, S. B., Lethal and potentially lethal lesions induced by radiation — A unified repair model, *Radiat. Res.,* 106, 252, 1986.
14. Chadwick, K. H. and Leenhouts, H. P., A molecular theory of cell survival, *Phys. Med. Biol.,* 18, 78, 1973.
15. Elkind, M. M., Repair Processes in the treatment and induction of cancer with radiation, *Cancer,* 65, 2165, 1990.
16. Hobson, B. and Denekamp, J., Endothelial proliferation in tumours and normal tissues: continuous labelling studies, *Br. J. Cancer*, 49, 405, 1984.
17. Reinhold H. S. and Binsman G. H., Radiosensitivity of capillary endothelium, *Br. J. Rad.,* 46, 54, 1973.
18. Van den Brenk, H. A., Sharpington, C., Orton, C., and Stone, M., Effects of X-radiation on growth and function of the repair blastema (granulation tissue). II. Measurements of angiogenesis in the Selye pouch in the rat, *Int. J. Rad. Biol.*, 25, 277, 1974.
19. Fike, J. R., Gillette, E. L., and Chow, D. J., Repair of sublethal radiation damage by capillaries, *Int. J. Rad. Oncol. Biol. Phys.*, 5, 339, 1979.
20. Kwock, L., Douglas, W. H., Lin, P. S., Baur, W. E., and Fanburg, B. L., Endothelial cell damage after gamma irradiation *in vitro* impaired uptake of alpha-aminoisobutyric acid, *Am. Respir. Dis.,* 125, 95, 1982.
21. Rhee, J. G., Lee, I., and Song, C. W., The clonogenic response of bovine aortic endothelial cells in culture to radiation, *Radiat. Res.,* 106, 182, 1986.
22. Fuks, Z., Vlodavsky, I., Andreeff, M., McLoughlin, M., and Haimovitz-Friedman, A., Effects of extracellular matrix on the response of endothelial cells to radiation *in vitro*, *Eur. J. Cancer,* 28/A, 725, 1992.
23. Hei, T. K., Marchese, M. J., and Hall, E. J., Radiosensitivity and sublethal damage repair in human umbilical cord vein endothelial cells, *Int. J. Radiat. Oncol. Biol. Phys.,* 13, 879, 1987.

24. Sholley, M. M. and Wilson, J. D., Proliferation and migration of irradiated endothelial cells, in D.B. Rifkin and M. Klagsbrun (Eds.), *Angiogenesis, Mechanism and Pathology*, Cold Spring Harbor, NY: Cold Spring Harbor Laboratory, 1987, 139.
25. Martin, D. F. and Fisher, J. J., Radiation sensitivity of cultured rabbit aortic endothelial cells, *Int. J. Radiat. Oncol. Biol. Phys.,* 10, 1903, 1984.
26. Vlodavsky, I., Lui, G. M., and Gospodarowicz, D., Morphological appearance, growth behavior and migratory activity of human tumor cells maintained on extracellular matrix vs. plastic, *Cell,* 19, 607, 1980.
27. Bissel, N. J. and Barcellos-Hoffman, H., The influence of extracellular matrix on gene expression: is structure the message?, *J. Cell. Sci.*, Suppl. 8, 327, 1987.
28. Hay, E. D., Extracellular matrix, cell skeleton, and embryonic development, *Am. J. Med. Genet.*, 34, 14, 1989.
29. Vlodavsky, I., Folkman, J., Sullivan, R., Fridman, R., Sullivan, R., Ishai-Michaeli, R., Sasse, J., and Klagsbrun, M., Endothelial cell-derived basic fibroblast growth factor: synthesis and deposition into subendothelial extracellular matrix, *Proc. Natl. Acad. Sci. U.S.A.,* 84, 2292, 1987.
30. Bashkin, P., Klagsbrun, M., Doctrow, S., Svahn, C. M., Folkman, J., and Vlodavsky, I., Basic fibroblast growth factor binds to subendothelial extracellular matrix and is released by heparanase and heparin-like molecules, *Biochem.*, 28, 1737, 1989.
31. Kramer, R. H. and Vogel, K. G., Selective degradation of basement membrane macromolecules by metastatic melanoma cells, *J. Natl. Cancer Inst.*, 72, 889, 1984.
32. Gospodarowicz, D., Lepine J., Massoglia, S., and Wood, I., Comparison of the ability of basement membranes produced by corneal endothelial and mouse-derived endodermal PF-HR-9 cells to support the proliferation and differentiation of bovine kidney tubule epithelial cells *in vitro, J. Cell Biol.,* 99, 947, 1984.
33. Kofstad, E. K. and Sutherland, R. M., Radiation sensitivity of human ovarian carcinoma cell lines *in vitro*: effects of growth factors and hormones, basement membrane, and intercellular contact, *Int. J. Rad. Oncol. Biol. Phys.,* 15, 921, 1988.
34. Kleinman, H. K., McGarvey, M. L., Hassell, J. R., Star, V. L., Cannon, F. B., Laurir, G. W., and Martin, G. R., Basement membrane complexes with biological activity, *Biochemistry,* 25,312, 1986.
35. Timple, R., Structure and biological activity of basement membrane proteins, *Eur. J. Biochem.,* 180, 487, 1989.
36. Burridge, K., Fath, K., Kelly, T., Nuckols, G., and Turner, C., Focal adhesions: transmembrane junctions between the extracellular matrix and the cytoskeleton, *Ann. Rev. Cell Biol.,* 4, 487, 1988.
37. McDonald, L. A., Receptors for extracellular matrix components, *Am. J. Phys.,* 257, L331, 1987.
38. Laken, S., Leach, K., and Klauck, T., Association of type 3 protein kinase C with focal contacts in rat embryo fibroblasts, *J. Cell Biol.,* 109, 697, 1989.
39. Nuckols, G. H., Turner, C. E., and Burridge, K., Functional studies of the domains of talin, *J. Cell Biol.,* 110, 1635, 1990.
40. Ben-Ze'ev, A., The cytoskeleton in cancer cells, *Biochem. Biophys. Acta,* 780, 197, 1985.
41. Bissell, M. J. and Ram, T. G., Regulation of functional cytodifferentiation and histogenesis in mammary epithelial cells: role of extracellular matrix, *Envir. Health Persp.,* 80, 61, 1989.

42. Fujita, M., Spray, D. C., Choi, H., Saez, J., Jefferson, D. M., Hertzberg, E., Rosenberg, L. C., and Ried, L. M., Extracellular matrix regulation of cell-cell communication and tissue specific gene expression in primary liver cultures, *Prog. Clin. Biol. Res.*, 226, 333, 1986.
43. Fridman, R., Fuks, Z., Ovadia, H., and Vlodavsky, I., Differential structural requirements for the induction of cell attachment, proliferation and differentiation by the extracellular matrix, *Exp. Cell Res.*, 157, 181, 1985.
44. Fajardo, L. F., The complexity of endothelial cells, *Am. J. Clin. Pathol.*, 92, 241, 1989.
45. Bikfalvi, A. and Han, Z. C., Angiogenic factors are hematopoietic growth factors and vice versa, *Leukemia*, 8, 523, 1994.
46. Folkman, J. and Shing, Y., Angiogenesis, *J. Biol. Chem.*, 267, 10931, 1992.
47. Beits, J. G., Kim, I. S., Calabresi, P., and Frackelton, A. R., Human microvascular endothelial cells express receptors for platelet-derived growth factors, *Proc. Natl. Acad. Sci. U.S.A.*, 88, 2021, 1991.
48. Smits, A., Hermansson, M., Nister, M., Karnushina, I., Heldin, C. H., Westermark, B., and Funa, K., Rat brain capillary endothelial cells express functional PDGF B type receptors, *Growth Factors*, 2, 1, 1989.
49. Plate, K. H, Breier, G., Farrell, C. L., and Risau, W., Platelet-derived growth factor receptor-beta is induced during tumor development and upregulated during tumor progression in endothelial cells in human gliomas, *Lab. Invest.*, 67, 529, 1992.
50. Heldin, C. H., Wasteson, A., and Westermark, B., Interaction of platelet derived growth factor with its fibroblast receptor. Demonstration of ligand degradation and receptor modulation, *J. Biol. Chem.*, 257, 4216, 1982.
51. Kazlauska, A. and DiCorleto, P. E., Cultured endothelial cells do not respond to platelet-derived growth protein in an autocrine manner, *Biochem. Biophys. Acta*, 846, 405, 1985.
52. Gospodarowicz, D., Moran, J., Brown, D., and Birdwell, C. R., Clonal growth of bovine vascular cells: Fibroblast growth factor as a survival agent, *Proc. Natl. Acad. Sci. U.S.A.*, 73, 4120, 1976.
53. Rogelj, S., Klagsbrun, M., Atzmon, R., Kurokawa, M., Haimovitz-Friedman, A., Fuks, Z., and Vlodavsky I., Basic fibroblast growth factor is an extracellular matrix component required for supporting the proliferation of vascular endothelial cells and the differentiation of PC12 cells, *J. Cell. Biol.*, 109, 823, 1989.
54. Haimovitz-Friedman, A., Vlodavsky, I., Chaudhuri, A., Witte, L., and Fuks, Z., Autocrine effects of fibroblast growth factor in radiation damage repair in endothelial cells, *Cancer Res.*, 51, 2552, 1991.
55. Fuks, Z., Persaud, R. S., Alfieri, A., McLoughlin, M., Schwartz, J. L., Seddon, A., Cordon-Cardo, C., and Haimovitz-Friedman, A., Basic fibroblast growth factor protects endothelial cells from radiation-induced programmed cell death *in vitro* and *in vivo*, *Cancer Res.*, 54, 2591, 1994.
56. Elkind, M. M., Repair processes in radiobiology, *Radiat. Res.*, 100, 425, 1984.
57. Ward, J. F., DNA damage produced by ionizing radiation in mammalian cells: identities, mechanisms of formation and repairability, *Prog. Nucleic Acid. Mol. Biol.*, 35, 95, 1988.
58. Radford, I. R., Evidence for a general relationship between the induced level of DNA double-strand breakage and cell-killing after X-irradiation of mammalian cells, *Int. J. Radiat. Biol.*, 49, 611, 1986.
59. Bradford, J. S., Sublethal damage, potentially lethal damage, and chromosomal aberrations in mammalian cells exposed to ionizing radiation, *Int. J. Radiat. Oncol. Biol. Phys.*, 21, 1457, 1991.

60. Steele, G. G., McMillan, T. J., and Peacock, J. H., The picture has changed in the 1980s, *Int. J. Rad. Biol.,* 56, 525, 1989.
61. McMillan, T. J., Residual DNA damage: what is left over and how does this determine cell fate, *Eur. J. Cancer,* 28, 267, 1992.
62. Brenner, D. J. and Ward, J. F., Constraints on energy depletion and target size of multiply damaged sites with DNA double-strand breaks, *Int. J. Rad. Biol.*, 61, 737, 1992.
63. Bradley, M. D. and Kohn, K. W., X-ray induced DNA double strand break production and repair in mammalian cells as measured by neutral filter elution, *Nucleic Acid Res.,* 7, 793, 1979.
64. Jorgensen, T. J., Prasad. S. C., Brennan, T. P., and Dritschilo, A., Constraints to DNA unwinding near radiation-induced strand breaks in Ewing's sarcoma cells, *Radiat. Res.,* 123, 320, 1990.
65. Gavrieli, Y., Sherman, Y., and Ben-Sasson, S. A., Identification of programmed cell death *in situ* via specific labeling of nucleic DNA fragmentation, *J. Cell Biol.,* 119, 493, 1992.
66. Wyllie, A. H., Glucocorticoid-induced thymocyte apoptosis is associated with endogenous endonuclease activation, *Nature,* 284, 555, 1980.
67. Williams, G. T. and Smith, C. A., Molecular regulation of apoptosis: genetic controls on cell death, *Cell,* 74, 777, 1993.
68. Laster, S. M., Wood, J. G., and Gooding, L. R., Tumor necrosis factor can induce both apoptic and necrotic forms of cell lysis, *J. Immunol.,* 141, 2629, 1988.
69. Oberhammer, F. A., Pavelka, M., Sharma, S., Tiefenbacher, R., Purchio, A. F., Bursch, W., and Schulte-Hermann, R., Induction of apoptosis in cultured hepatocytes and in regressing liver by transforming growth factor beta 1, *Proc. Natl. Acad. Sci. U.S.A.,* 89, 5408, 1992.
70. Eastman, A. and Barry, M. A., The origin of DNA breaks: a consequence of DNA damage, DNA repair, or apoptosis?, *Cancer Invest.*, 10, 229, 1992.
71. Hickman, J. A., Apoptosis induced by anticancer drugs, *Cancer Metast. Rev.,* 11, 121, 1992.
72. Urmansky, S. R., Apoptotic cell death in the radiation-induced cell death of lymphocytes, in *Apoptosis: The Molecular Basis of Cell Death.* Tomei, L. D. and F. O. Cope, Eds., New York, Cold Spring Harbor Laboratory Press, 1991.
73. Allan, D. J., Radiation-induced apoptosis: its role in a MADCaT (mitosis-apoptosis-differentiation-calcium toxicity) scheme of cytotoxicity mechanisms, *Int. J. Radiat. Biol.*, 62, 145, 1992.
74. Harmon, B. V., Corder, A. M., Collins, R. J., Gobe, G. C., Allen, J., Allan, D. J., and Kerr, J. F., Cell death induced in a murine mastocytoma by 42-47 degrees C heating *in vitro*: evidence that the form of death changes from apoptosis to necrosis above a critical heat load, *Int. J. Radiat. Biol.,* 58, 845, 1990.
75. Duvall, E., Wyllie, A. H., and Morris, R. G., Macrophage recognition of cells undergoing programmed cell death (apoptosis), *Immunology,* 56, 351, 1985.
76. Yamada, T. and Ohyama, H., Radiation-induced interphase death of rat thymocytes is internally programmed (apoptosis), *Int. J. Radiat. Biol.*, 53, 65, 1988.
77. Potten, C. S., Extreme sensitivity of some intestinal crypt cells to X and gamma irradiation, *Nature*, 269, 518, 1977.
78. Stephens, L. C., Schultheiss, T. E., Price, R. E., Ang, K. K., and Peters, L. J., Radiation apoptosis of serous ascinar cells of salivary and lacrimal glands, *Cancer,* 67, 1539, 1991.

79. Stephens, L. C., Ang, K. K., Schultheiss, T. E., Milas, L., and Meyn, R. E., Apoptosis in irradiated murine tumor, *Radiat. Res.*, 127, 308, 1991.
80. Stephens, L. C., Hunter, N. R., Ang, K. K., Milas, L., and Meyn, R., E., Development of apoptosis in irradiated murine tumors as a function of time and dose, *Radiat. Res.*, 135, 75, 1993.
81. Vlodavsky, I., Johnson, L. K., Greenburg, G., and Gospodarowicz, D., Vascular endothelial cells maintained in the absence of fibroblast growth factor undergo structural and functional alterations that are incompatible with their *in vivo* differentiated properties, *J. Cell Biol.*, 83, 468, 1979.
82. Jaye, M., Schlessinger, J., and Craig, C. A., Fibroblast growth factor receptor tyrosine kinase: molecular analysis and signal transduction, *Biochim. Biophys. Acta*, 1135, 185, 1992.
83. Ullrich, A. and Schlessinger, J., Signal transduction by receptors with tyrosine kinase activity, *Cell*, 61, 203, 1990.
84. Nishizuka, Y., Intracellular signalling by hydrolysis of phospholipids and activation of protein kinase C, *Science*, 258, 607, 1992.
85. Kahan, C., Seuwen, K., Meloche, S., and Pouyssegur, J., Coordinate, biphasic activation of p44 mitogen-activated protein kinase and S6 kinase by growth factors in hamster fibroblasts. Evidence for thrombin-induced signals different from phosphoinositide turnover and adenylylcyclase inhibition, *J. Biol. Chem.*, 267, 13369, 1992.
86. Vouret-Craviari, V., Van-Obberghen-Schilling, E., Scimeca, J. C., Van-Obberghen, E., and Pouyssegur, J., Differential activation of p44mapk (ERK1) by alpha-thrombin and thrombin-receptor peptide agonist, *Biochem. J.*, 289, 209, 1993.
87. Graves, L. M., Northrop, J. L., Potts, B. C., Krebs, E. G., and Kimelman, D., Fibroblast growth factor, but not activin, is a potent activator of mitogen-activated protein kinase in Xenopus explants, *Proc. Natl. Acad. Sci. U.S.A.*, 91, 1662, 1994.
88. Lange-Carter, C. A., Pleiman, C. M., Gardner, A. M., Blumer, K. J., and Johnson, G. L., A divergence in the MAP kinase regulatory network defined by MEK kinase and Raf, *Science*, 260, 315, 1993.
89. Haimovitz-Friedman, A., Balaban, N. A., McLoughlin, M., Ehleiter, D., Michaeli, J., Vlodavsky, I., and Fuks, Z., Protein kinase C mediates basic fibroblast growth factor protection of endothelial cells against radiation-induced apoptosis, *Cancer Res.*, 54, 2591, 1994.
90. Lotem, J., Cragoe, E. J., and Sachs, L., Rescue from programmed cell death in leukemic and normal myeloid cells, *Blood*, 78, 953, 1991.
91. Vazquez, A., Auffredon, M. A., Chaouchi, N., Taieb, J., Sharma, S., Galanaud, P., and Leca, G., Differential inhibition of interleukin 2- and interleukin 4-mediated human B cell proliferation by ionomycin: a possible role for apoptosis, *Eur. J. Immunol.*, 21, 2311, 1991.
92. Zubiaga, A. M., Munoz, E., and Huber, B. T., IL-4 and IL-2 selectively rescue Th cell subsets from glucocorticoid-induced apoptosis, *J. Immunol.*, 149,107, 1992.
93. Rajotet, D., Haddad, P., Haman, A., Cragoe, E. J., and Hoang, T., Role of protein kinase C and the Na^+/H^+ antiport in suppression of apoptosis by granulocyte macrophage colony-stimulating factor and interleukin-3, *J. Biol. Chem.*, 267, 9980, 1992.
94. Tomei, L. D., Kanter P., and Wenner C. E., Inhibition of radiation-induced apoptosis *in vitro* by tumor promoters, *Biochem. Biophys. Res. Commun.*, 155, 324, 1988.
95. Kizaki, H., Tadakuma, T., Okada, C., Maramatsu, J., and Ishimura, Y., Activation of a suicide process of thymocytes through DNA fragmentation by calcium and Phorbol esters, *J. Immunol.*, 143, 1790, 1989.

96. McConkey, D. J., Hartzell, P., Jondal, M., and Orrenius, S., Inhibition of DNA fragmentation in thymocytes and isolated thymocyte nuclei by agents that stimulate protein kinase C, *J. Biol. Chem.*, 264, 13399, 1989.
97. Story, M. D., Stephens, L. C., Tomasovic, S. P., and Meyn, R. E., A role for calcium in regulating apoptosis in rat thymocytes irradiated in vitro, *Int. J. Radiat. Biol.*, 61, 243, 1992.
98. Hallahan D. E., Virudachalam, S., Sherman, M. A., Huberman, E., Kuff, D. W., and Weichselbaum, R. R., Tumor necrosis factor gene expression is mediated by protein kinase C following activation by ionizing radiation, *Cancer Res.*, 51, 4565,1991.
99. Hallahan, D. E., Virudachalam, S., Grdina, D., and Weichselbaum, R. R., The isoquinoline sulfonamide H7 attenuates radiation-mediated protein kinase C activation and delays the onset of X-ray-induced G2 arrest, *Int. J. Radiat. Oncol. Biol. Phys.*, 24, 687, 1992.
100. Kim, C. Y., Giaccia, A. J., Strulovici, B., and Brown, J. M., Differential expression of protein kinase C epsilon protein in lung cancer cell lines by ionising radiation, *Br. J. Cancer,* 66, 844, 1992.
101. Uckun, F. M., Schieven, G. L., Tuel-Ahlgren, L. M., Dibirdik, I., Myers, D. E., Ledbetter, J. A., and Song, C. W., Tyrosine phosphorylation is a mandatory proximal step in radiation-induced activation of the protein kinase C signaling pathway in human B-lymphocyte precursors, *Proc. Natl. Acad. Sci. U.S.A.,* 90, 252, 1993.
102. Hallahan, D. E., Virudachalam. S., Schwartz, J. L., Panje, N., Mustafi, R., and Weichselbaum, R. R., Inhibition of protein kinases sensitizes human tumor cells to ionizing radiation, *Radiat. Res.,* 129, 345, 1992.
103. Radford, I. R. and Murphy, T. K., Radiation response of mouse and myeloid cell lines. III. Different signals can lead to apoptosis and may influence sensitivity to killing by DNA double-strand breaks, *Int. J. Radiat. Biol.,* 65, 229, 1994.
104. Kinsella, T. J., Dobson, P. P., Mitchell, J. B., and Fornace, A. J., Enhancement of X–ray induced DNA damage by pre-treatment with halogenated pyrimidine analogs, *Int. J. Radiat. Oncol. Biol. Phys.*, 13, 733, 1987.
105. Warters, R. L., Radiation-induced apoptosis in a murine T-cell hybridoma, *Cancer Res.,* 52, 883, 1992.
106. Akagi, Y., Ito, K., and Sawada., S., Radiation-induced apoptosis and necrosis in Molt–4 cells: a study of dose-effect relationships and their modification, *Int. J. Radiat. Biol.*, 64, 47, 1993.
107. Clarke, A. R., Purdie, C. A., Harrison, D. J., Morris R. G., Bird, C. C., Hooper, M. L., and Wyllie, A. H., Thymocyte apoptosis induced by p53-dependent and independent pathways, *Nature,* 362, 849, 1993.
108. Lowe, S. W., Schmitt, E. M., Smith, S. W., Osborne, B. A., and Jacks, T., p53 is required for radiation-induced apoptosis in mouse thymocytes, *Nature,* 62, 847, 1993.
109. Lowe, S. W., Ruley, H. E., Jacks, T,. and Housman, D. E., p53-dependent apoptosis modulates the cytotoxicity of anticancer agents, *Cell,* 74, 957, 1993.
110. Yonish-Rouach, E., Grunwald, D., Wilder, S., Kimchi, A., May, E., Lawrence, J. J., May, P., and Oren, M., p53-mediated cell death: relationship to cell cycle control, *Mol. Cell Biol.*, 13, 1415, 1993.
111. El-Deiry, W. S., Harper, J. W., O'Connor, P. M., Velculescu, V. E., Canman, C. E., Jackman, J., Pietenpol, J. A., Burrell, M., Hill, D. E., Wang, Y. et al., WAF1/CIP1 is induced in p53-mediated G1 arrest and apoptosis, *Cancer Res.* 54, 1169, 1994.
112. Dulic, V., Kaufmann, W. K., Wilson, S. J., Tlsty, T. D., Lees, E., Harper, J. W., Elledge, S. J., and Reed, S. I., p53-dependent inhibition of cyclin-dependent kinase activities in human fibroblasts during radiation-induced G1 arrest, *Cell,* 76,1013, 1994.

113. Ashwell, J. D., Schwartz, R. H., Mitchell, J. B., and Russo, A., Effect of gamma radiation on resting B lymphocytes. I. Oxygen-dependent damage to the plasma membrane results in increased permeability and cell enlargement, *J. Immunol.,* 136, 3649, 1986.
114. Macklis, R. M., Lin, J. Y., Beresford, B., Atcher, R. W., Hines, J. J., and Humm, J. L., Cellular kinetics, dosimetry and radiobiology of α-particle radioimmunotherapy: induction of apoptosis, *Radiat. Res.,* 130, 220, 1992.
115. Obeid, L. M., Linardic, C. M., Karolak, L. A., and Hannun, Y. A., Programmed cell death induced by ceramide, *Science,* 259, 1769, 1993.
116. Jarvis, D. W., Kolesnick, R. N., Fornari, F. A., Traylor, R. S., Gewirtz, D. A., and Grant, S., Induction of apoptotic DNA degradation and cell death by activation of the sphingomyelin pathway, *Proc. Natl. Acad. Sci. U.S.A.,* 91, 73, 1994.
117. Kolesnick, R. and Golde, D. W., The sphingomyelin pathway in tumor necrosis factor and interleukin-1 signaling, *Cell,* 77, 325, 1994.
118. Hannun, Y. A., The sphingomyelin cycle and the second messenger function of ceramide, *J. Biol. Chem.,* 269, 3125, 1994.
119. Mathias, S., Younes, A., Kan, C.-C., Orlow, I., Joseph, C., and Kolesnick, R. N., Activation of the sphingomyelin pathway in intact EL4 cells and in a cell-free system by IL-1β, *Science,* 259, 519–522, 1993.
120. Rivas, C. I., Golde, D. W., Vera, J. C., and Kolesnick, R. N., Involvement of the sphingomyelin pathway in autocrine tumor necrosis factor signaling for human immunodeficiency virus production in chronically infected HL-60 cells, *Blood,* 15, 83, 2191, 1994.
121. Mathias, S., Dressler, K. A., and Kolesnick, R. N., Characterization of a ceramide-activated protein kinase: Stimulation by tumor necrosis factor α, *Proc. Natl. Acad. Sci. U.S.A.,* 88, 10009, 1991.
122. Joseph, C. K., Byun, H. S., Bittman, R., and Kolesnick, R. N., Substrate recognition by ceramide-activated protein kinase. Evidence that kinase activity is proline-directed, *J. Biol. Chem.,* 268, 20002, 1993.
123. Raines, M. A., Kolesnick, R. N., and Golde, D. W., Sphingomyelinase and ceramide activate mitogen-activated protein kinase in myeloid HL-60 cells, *J. Biol. Chem.,* 268, 14572, 1993.
124. Schutze, S., Potthoff, K., Machleidt, T., Berkovic, D., Wiegmann, K., and Kronke, M., TNF activates NF-kappa B by phosphatidylcholine-specific phospholipase C-induced "acidic" sphingomyelin breakdown, *Cell,* 71, 765, 1992.
125. Yang, Z., Costanzo, M., Golde, D. W., and Kolesnick, R. N., Tumor necrosis factor activation of the sphingomyelin pathway signals nuclear factor kappa B translocation in intact HL-60 cells, *J. Biol. Chem.,* 268, 20520, 1993.
126. Dobrowsky, R. T., Kamibayashi, C., Mumby, M. C., and Hannun, Y. A., Ceramide activates heterotrimeric protein phosphatase 2A, *J. Biol. Chem.,* 268, 15523, 1993.
127. Yanaga, F. and Watson, S. P., Tumor necrosis factor alpha stimulates sphingomyelinase through the 55 kDa receptor in HL-60 cells, *FEBS Lett.*, 314, 297, 1992.
128. Dressler, K. A., Mathias, S., and Kolesnick, R. N., Tumor Necrosis Factor-α activates the sphingomyelin signal transduction pathway in a cell-free system, *Science,* 255, 1715, 1992.
129. Robaye, B., Mosselmans, R., Fiers, W., Dumont, J. E., and Galand, P., Tumor necrosis factor induces apoptosis (programmed cell death) in normal endothelial cells *in vitro, Am. J. Pathol.,* 138, 447, 1991.

130. Haimovitz-Friedman, A., Kan, C.-C., Ehleiter, D., Persaud, R. S., Alfieri, A., McLoughlin, M., Fuks, Z., and Kolesnick, R. N., Ionizing radiation acts on cellular membranes to generate ceramide and induce apoptosis, *J. Exp. Med.*, 180, 525, 1994.
131. Cordon-Cardo, C., Vlodavsky, I., Haimovitz-Friedman, A., Hicklin, D., and Fuks, Z., Expression of basic fibroblast growth factor in normal human tissues, *Lab. Inv.*, 63, 832, 1990.
132. Simionescu, M., Simionescu, N., Silbert, J. E., and Palade, G. E., Differential microdomains on the luminal surface of the capillary endothelium. II. Partial characterization of their anionic sites, *J. Cell. Biol.*, 90, 614, 1981.
133. Fuks, Z. and Weichselbaum, R. R., Radiation tolerance and the new biology: growth factor involvement in radiation injury to the lung [editorial comment], *Int. J. Radiat. Oncol. Biol. Phys.*, 24, 183, 1992.
134. Rubin, P., Finkelstein, J., and Shapiro, D., Molecular biology mechanisms in radiation induction of pulmonary injury syndrome: interrelationship between the alveolar macrophage and septal fibrosis, *Int. J. Radiat. Oncol. Biol. Phys.*, 24, 93, 1992.

CHAPTER 7

Effect of Ionizing Radiation on Endothelial Cell Plasma Membrane Processes

Lester Kwock, A. William Blackstock, and Mitchell Friedman

CONTENTS

I. INTRODUCTION

Several studies have proposed that the vascular system plays a major role in the observed radiation effects of ionizing radiation on normal tissue and is the initiating

1-8493-4840-4/98/$0.00+$.50

event in the development of the late effects of radiation; namely, fibrosis and tissue atrophy.[1–5] This proposal is based on the histological appearance of several tissues after receiving therapeutic doses of radiation. In these cases, the normal parenchymal cells in the tissues show no early changes due to radiation, whereas the vascular compartment of the tissue shows progressive changes over a period of several months. Histological studies in irradiated liver,[6] kidney,[7] heart,[8] and brain[9] have shown that marked changes in the vasculature of these tissues appear to precede atrophy of the parenchymal cells in these irradiated tissues. These observations have led to the hypothesis that "impaired vascular function underlies tissue atrophy... after irradiation."[2] According to this hypothesis, the primary effect of irradiation is to injure the microvascular endothelium, thereby inducing an increase in transvascular permeability to plasma proteins, with subsequent edema and fibrin deposition in the interstitial spaces and blood vessel walls. The newly deposited fibrin is eventually replaced by collagen fibers, which results in tissue atrophy and fibrosis.

The major histological effects observed in the irradiated blood vessel studies cited above were (1) retraction of adjoining endothelial cells, (2) areas of hypertrophic attached endothelial cells, and (3) areas of denuded endothelial cells indicating sloughing of highly damaged endothelial cells. These changes indicate major morphological changes in the normal surface membrane architecture of the endothelium that can induce changes in the normal plasma membrane processes in endothelial cells. The purpose of this brief review is to suggest that one of the critical targets in the early effects of ionizing radiation on the vasculature is the plasma membrane of endothelial cells.

II. MORPHOLOGICAL AND FUNCTIONAL CHARACTERISTICS OF THE PLASMA MEMBRANE ENDOTHELIUM: EFFECT OF IONIZING RADIATION

The vascular endothelium serves as a vital multifunctional tissue. The endothelial cells that cover the entire luminal side of blood vessels are tightly bound to each other to form a smooth continuous cellular plasma membrane surface. This membrane surface (1) is nonthrombogenic, (2) plays a role in the active transport of macromolecules and nutrients to and from the parenchymal tissue compartment, (3) interacts with the coagulation, fibrinolytic, and complement systems, and (4) is involved in the transport with subsequent intracellular metabolism of vasoactive substances.[10]

Essential to the maintenance of this continuous smooth surface is the interaction of cell surface membrane components between endothelial cells. These endothelial cell surface membrane components are composed of acidic mucopolysaccharides and form the intercellular junctions between endothelial cells.[11] These junctions allow selected substances to pass through and come into contact with the basement membrane of blood vessels. The basement membrane is acellular in nature but is a dynamic structure that acts as an additional semipermeable barrier to the parenchymal compartment of tissues. The integrity of this barrier is maintained by the endothelial and supporting connective tissue stromal cells.

By recognizing the complexity of the role that the plasma membrane plays in the functioning of the endothelial cell, it can be appreciated that even small, permanent alterations induced by radiation on the vascular endothelium in tissues can have widespread deleterious effects on those tissues. For instance, in the case of the pulmonary vascular endothelium, we and others have shown that either acute high doses or fractionated doses of ionizing radiation can lead to early changes in surface morphology and function of the vascular endothelium.[12–15]

A. *In Vivo* Morphological Studies of Irradiated Microvasculature

In a fractionated 30 Gy dose study (5 × 6 Gy) in rat lungs performed in our laboratory,[12] we found that within 4 to 6 weeks, a 70% decrease in 99mTc-MAA vascular lung perfusion is observed in the irradiated rat lung lobes as compared to the unirradiated contra lateral lungs. Microradiographs of irradiated lung sections at 4 to 6 weeks after exposure to the fractionated doses of radiation demonstrated a widespread loss of capillary filling with small arteries and arterioles demonstrating a "sausage-like" segmented appearance (Figure 7.1). Histological examination of tissue lung sections showed areas in irradiated small arteries or arterioles that appeared to be denuded of endothelium and areas where endothelial cells appeared to be bulging into the lumen to the vessels (Figure 7.2). Similar microradiographic and morphologic changes have been observed in small arteries and arterioles in irradiated kidneys, brain, and hamster cheek pouches.[13–15] The "sausage-like" segmentation pattern found in these studies was attributed to damage to the endothelial cells lining the lumen of the vessels with subsequent hyperplasia of either endothelial or smooth muscle cells and the vasoconstrictive effects of 5-hydroxytryptamine (5-HT).

B. *In Vitro* Morphological Studies of Irradiated Endothelial Cells (EC)

With the ability to isolate endothelial cells, studies examining the effects of ionizing radiation have been performed that have provided information about early changes occurring in irradiated EC. Endothelial cells isolated from different tissues and grown in plastic culture dishes, form confluent monolayers and exhibit a normal cobblestone appearance by inverted phase contrast microscopy (Figure 7.3A). However, when EC are exposed to either a fractionated (5 × 6 Gy) or an acute large dose of radiation, alterations in attachment, morphology, and discontinuity of the monolayer is observed.[12,16,17] In the case of the fractionated radiation studies (Figure 7.3B), 24 hr after the initial 6-Gy dose, a small number of cells are observed floating in the medium. Between the second and fourth 6-Gy fractionated dose, the density of attached cells in the irradiated monolayers compared to the sham-exposed control monolayers appears to decrease (Figure 7.3C). In many areas, EC hypertrophy is observed and the overall cobblestone appearance of the monolayers is clearly altered (Figure 7.4C). Discontinuities within the monolayer are observed after the fourth 6 Gy dose (total dose of 24 Gy) (Figures 7.3D and 7.4B). By 2 to 3 days after the final 6-Gy dose (total dose of 30 Gy), the areas of discontinuity of the monolayer have increased and the irradiated cells appear to be highly vacuolated.

A)
B)

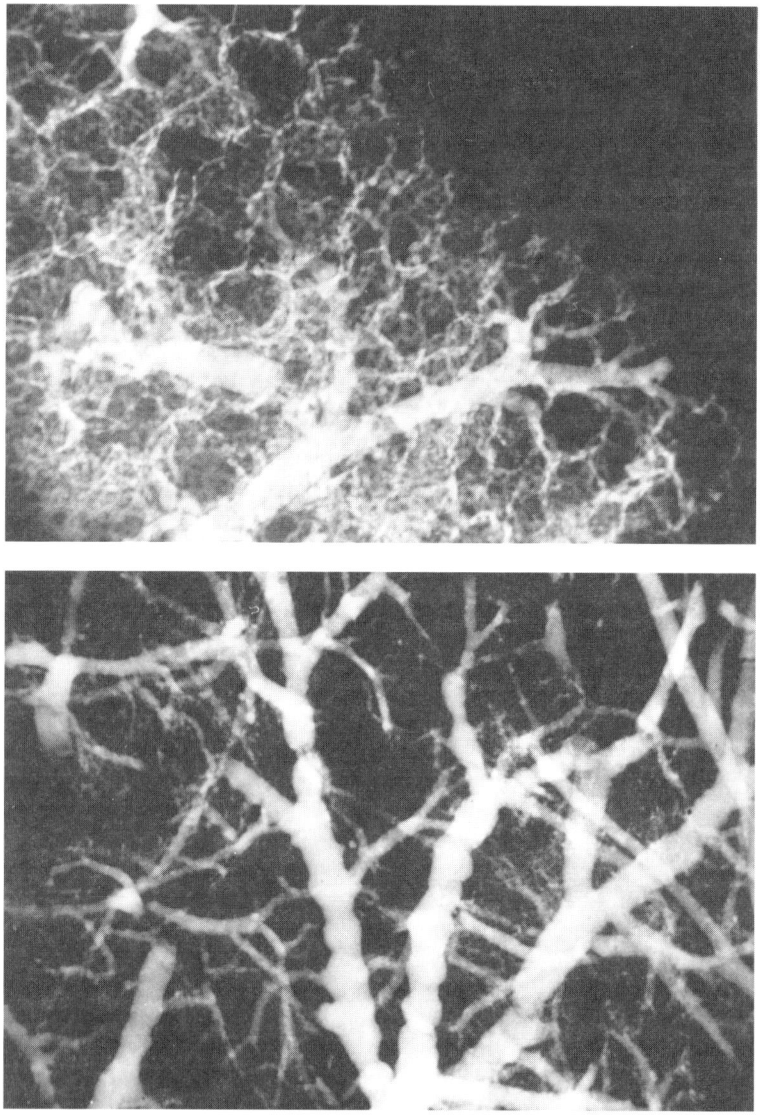

Figure 7.1. A: Microradiograph of unirradiated contralateral lung lobe (6 weeks post-irradiation, total dose = 30 Gy). B: Microradiograph of irradiated lung lobe. Note loss of filling in the capillary beds and beaded appearance of the small arteries (6 weeks post-irradiation, total dose = 30 Gy).

In a quantitative acute dose study (4 to 30 Gy) on confluent vascular EC by Rosen et al.[16] the cultures showed a dose-dependent cell loss, with a compensatory hypertrophy of the remaining attached cells. Rosen suggested that the somatic cellular hypertrophy is a response of the denuded endothelial monolayer to reestablish the surface continuity of the endothelium without replication following

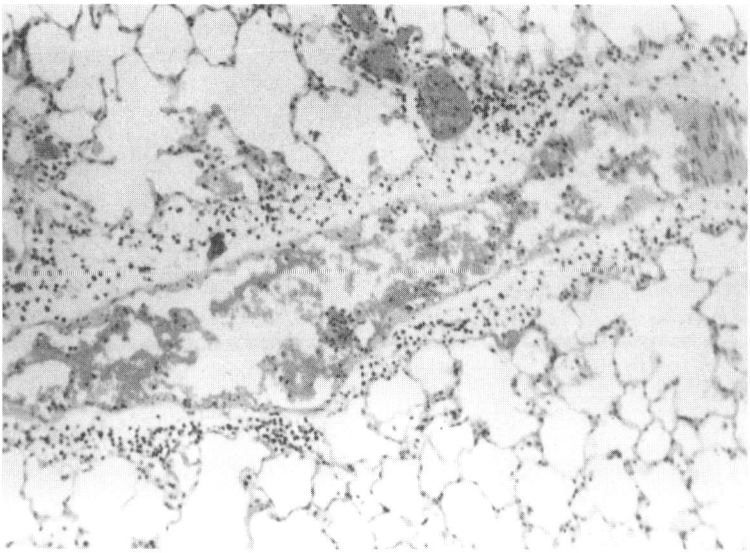

Figure 7.2. Histological lung section (40×) taken from a 30-Gy lung 29 days postirradiation. Note the mixture of blood and barium in the vessel and the striking undulating appearance of the vessel intima.

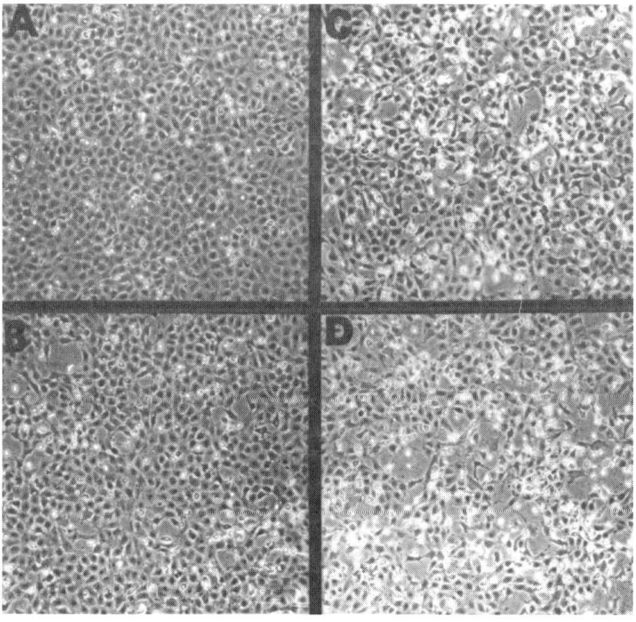

Figure 7.3. Effect of ionizing radiation on isolated endothelial cells. A: normal EC prior to irradiation; B: 6-Gy exposed EC; C: 12-Gy exposed EC; D: 24-Gy exposed EC.

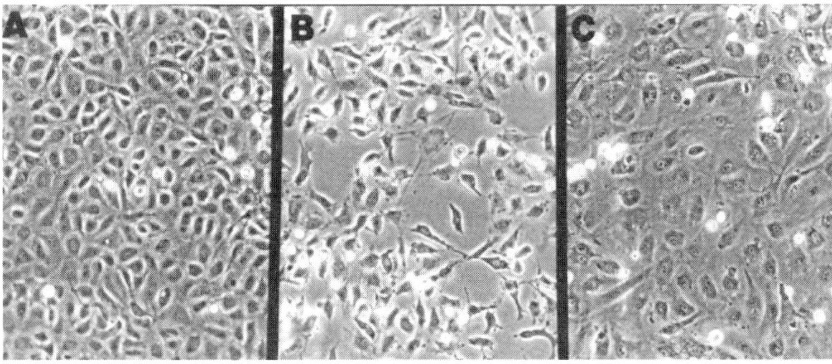

Figure 7.4. Effect of ionizing radiation on 30-Gy exposed EC. A: Sham-irradiated control; B: area of 30-Gy exposed EC showing discontinuities in monolayer; C: hypertrophic appearing areas in EC exposed to 30 Gy.

injury to minimize exposure of the thrombogenic subendothelium to platelets, clotting factors, and smooth muscle cell mitogens.

A major consequence of cell detachment is a loss in the continuity of the endothelium due to loss of intercellular junctions formed by plasma membrane surface components. This loss in intercellular junctions, if not repaired, leads to increased edema and transvascular permeability to plasma proteins. In an albumin permeability radiation study of EC monolayers, Friedman et al.[18] showed that radiation increases endothelial albumin transfer rates. The increased albumin transfer rates were associated with morphologic evidence of EC monolayer disruption and EC injury. This effect, as presented in the introduction, can lead to tissue atrophy and fibrosis.[2]

Additionally in this study, a dose-dependent alteration in actin filament cytoskeletal organization in cells remaining attached was observed.[18] The cytoskeleton of cells has been implicated in modulating many plasma membrane-associated processes such as cell attachment, cell motility, expression of surface membrane receptors, intracellular transport properties, and endothelial wound-repair processes.[19,20] Thus, changes in the cytoskeleton organization in irradiated EC observed by Friedman et al. suggests that the changes observed in cellular morphology may lead to changes in EC plasma membrane function.

III. EFFECTS OF IONIZING RADIATION ON EC PLASMA MEMBRANE FUNCTION

A. Membrane Transport Processes

5-HT is released into the plasma when platelets are activated and has been shown to be a powerful vasoconstrictor in most vascular beds.[21] Under normal conditions increases in 5-HT are removed from the circulation by pulmonary endothelial cells via a Na^+ and energy-dependent transport process.[21,22] This transport system in the lung has been shown to be highly sensitive to oxygen free radicals[23,24] such as superoxide anion and hydroxyl free radicals, which are the free radical species involved in the indirect effects of ionizing radiation on plasma membrane components.[25] Exposure of lungs to hyperbaric oxygen conditions leads to decreases in the plasma clearance of 5-HT and can be prevented by pretreatment with superoxide dismutase.[24] This would suggest that this transport system in EC is also sensitive to the effects of ionizing radiation.

In *in vitro* endothelial cell culture studies, Vegt et al.,[26] using EC isolated from human umbilical cord, and Kwock et al.,[27] using EC isolated from calf mainstem pulmonary artery, found a radiation dose dependence and early (3–4 hr postirradiation) inhibition of the Na^+-dependent and energy-dependent amino acid transport system. This transport system has been shown to be inhibited by superoxide anion[28] similar to the Na^+-dependent 5-HT transport system. The Na^+-dependent amino acid transport system is involved in several vital cell functions including cell growth and cell repair processes.[29]

Vegt et al. showed that the decrease in Na^+-dependent amino acid transport in EC was due to a decreased affinity of the transport carrier for the amino acid and not to either the number of transport carriers or energy-dependent processes. In both studies, the Na^+-independent amino acid transport system was not impaired by ionizing radiation. Additionally, these investigators showed that glucose transport was transiently increased (3 hr postirradiation), but within 24 hr postirradiation returned to normal transport levels. The effect of ionizing radiation on the Na^+-dependent transport system changes preceded any morphological and viability changes in the EC cultures.

B. Effect of Ionizing Radiation on the Activity of Plasma Membrane-Bound Enzymes

1. Phospholipase A2

Membrane phospholipase A2 activity has been shown to be increased by hydroxyl radicals generated by ionizing radiation.[30] Increased phospholipase A2 activity results in the release of arachidonic acid from membrane phospholipids and the generation of various endoperoxide metabolites such as lipid endoperoxides, thromboxane A2, and other prostaglandin-like metabolites such as prostacyclin. Prostacyclin is a potent inhibitor of platelet aggregation and a vasodilator. It is believed that prostacyclin is involved in the regulation of platelet-vessel wall interactions and vasotone.[31,32]

There have been several studies dealing with the effects of ionizing radiation on EC prostacyclin synthesis. In several studies, decreases in prostacyclin synthesis has been found,[33–35] whereas in other studies increases in prostacyclin have been found.[36–38] Hosoi et al.[39] have suggested that the timing between irradiation and assay of prostacyclin is important in order to understand the differences observed in these prostacyclin studies. They suggest that inhibition of prostacyclin synthesis is an early and transient event that occurs within 90 minutes after irradiation, and that this early decrease in prostacyclin production is due to transient changes in cellular membranes, which can be abrogated by the presence of membrane lipid peroxidation scavengers.

Increased EC synthesis of prostacyclin occurs at times between 4 and 24 hr postirradiation.[36–38] It has been hypothesized[38] that an increase in cyclooxygenase and/or prostacyclin synthetase activity and increased release of arachidonic acid from membrane lipids are the major mechanisms responsible for increased EC prostacylin production induced by ionizing radiation.

These data suggest that the levels of arachidonic acid released from membrane lipid components and the level of membrane lipid peroxidative components regulate the production of prostacylin in EC.

2. *Angiotensin Converting Enzyme (ACE)*

Angiotensin converting enzyme is located on the luminal surface of EC and is an enzyme involved in hydrolyzing angiotensin I into angiotensin II which is a potent vasoconstrictor.[40] In addition, ACE hydrolyzes the potent vasodilator bradykinin into peptides that are not vasoactive but may be chemotactic to neutrophils.[40] Rubin et al.[41] have shown that over a range of radiation doses relevant to radiation therapy, vascular EC exhibit decreased cellular ACE activity. This plasma membrane enzyme dysfunction could be demonstrated before cell EC detachment occurred.

The biological relevance of this finding is unknown at this time. However, decreases in ACE activity certainly will lead to a decreased ability of EC to hydrolyze bradykinin, which may be involved in maintaining the thromboresistant character of the endothelium. Bradykinin has been shown to stimulate prostacyclin production in EC cultures[42] by activation of a phospholipase, which then releases arachidonate from membrane phospholipids for prostacylin synthesis. Additionally, bradykinin has been shown to be involved in stimulating EC growth processes.[43] Thus, the inhibition of ACE by radiation may lead to increased interaction of bradykinin with the EC to maintain its thromboresistant character by stimulating prostacylin synthesis and activation of repair processes.

3. *Transglutaminase*

Endothelial cells have been shown to contain a Ca^{2+}-dependent transglutaminase.[44] This enzyme catalyzes the Ca^{2+}-dependent acyl transfer reaction between γ-carboximide groups of peptide-bound glutamine residues and various primary amines. With protein substrates, transglutaminases can catalyze two nucleophilic displacement reactions leading to either protein cross-linking or to the incorporation of small primary amines into the protein substrate. Studies in a number of laboratories have indicated that transglutaminase is involved in cell growth and differentiation.[45–47] The activity of this enzyme varies with the proliferative state of cells; namely, the activity usually increases with decreasing cell growth.

Recently, Korner et al.[48] have shown that irradiation of EC leads to an increase in the activity of transglutaminase. In confluent irradiated EC cultures, the increased activity was dose dependent up to 20 Gy (twofold increase) and reached a plateau at 24 to 36 hr following irradiation. This increase in transglutaminase activity was attributed to the activation of an inactive form of the enzyme stored in the EC plasma membranes. It was suggested that activation of the membrane-bound transglutaminase leads to cross-linking of adjacent cytoplasmic and membrane proteins, which in turn activates membrane processes involved in programmed cell death; namely, apoptosis.

IV. ROLE OF THE PLASMA MEMBRANE IN RADIATION-INDUCED ENDOTHELIAL CELL PROGRAMMED CELL DEATH: APOPTOSIS

In radiation-induced cell death, the established postulate designates the cell nucleus as the susceptible site for radiation-induced lethality.[49] Heterologous double-strand breaks in the DNA are the most common types of lesions that lead to mammalian cell death.[50–51] The subsequent insult produced in the DNA is either a result of the direct interaction with X-rays, or with the reactive oxygen intermediates generated within the cell by radiation.[51] Yet, the compiled proof for chromosomal DNA being the central target for radiation-induced cell killing remains presumptive.[49] Until recently, little could be assumed about specific membrane-generated controls and their relationship to cell function and death. Clearly, during specific events in the cell cycle, the cell membrane and segments of the nuclear DNA come into close proximity, which at least implies a potential interaction.[49] In this section we will review apoptosis and the role the cell membrane may have in the apoptotic cell death of endothelial cells.

Apoptosis (programmed cell death) takes place under normal physiologic conditions.[52,53] The analysis of whether a cell is apoptotic is based on a number of morphological and biochemical criteria. Apoptosis occurs regularly during tissue turnover, embryogenesis, metamorphosis, and endocrine-dependent tissue atrophy. Cells undergoing apoptosis demonstrate a gradual contraction of cell volume with preservation of the integrity of cytoplasmic organelles.[54–55] The shrinkage in cell volume results in the cell becoming increasingly dense. In alliance with the cell shrinkage are general changes in nuclear morphology. The nuclear chromatin becomes compacted into sharply circumscribed, uniformly dense masses aligned with the nuclear envelope. During the latter phases of apoptosis, the cells break apart into a number of membrane-bound apoptotic bodies containing one or more fragments of the nucleus.[54,56,57] Wyllie in 1980 was the original researcher who described the biochemical events associated with apoptosis.[58] While isolating DNA from apoptotic cells and analyzing the sample using agarose gel electrophoresis, he was able to confirm the distinctive ladder pattern consisting of 180 bp subunits. He also proved the temporal emergence of DNA cleavage with chromatin condensation, suggesting that the resulting DNA fragmentation was responsible for the observed alteration in nuclear morphology. Further data maintains that the fragmentation seen during apoptosis is the result of activation of an endonucleus that cleaves the DNA in specific linker regions between nucleosome cores.[59] The endonuclease responsible for the DNA fragmentation appears to be Ca^{2+}/Mg^{2+} associated, and the induction of apoptosis requires synthesis of specific proteins or "death" genes. It is important to realize that depending upon the cell system, apoptosis can either be dependent upon, independent of, or induced by termination of protein synthesis.[60–61]

The biochemical regulation of programmed cell death has recently attracted a great deal of attention. Many researchers have attempted to understand the physiological mechanisms associated with this type of cell death in hopes that it may provide insights to grasping other parameters of cell growth and differentiation. As stated, primary lesions to the DNA resulting from ionizing irradiation, were felt to trigger the apoptotic response, yet little definitive data are known about the signaling mechanisms involved.[49] Recent reports would suggest that perhaps the signal is initiated at the cell membrane and not the nucleus; and that perhaps the initial signals for the induction of apoptosis are contingent upon membrane-generated events.[62]

Recent experiments performed by Haimovitz-Friedman et al. have demonstrated that ionizing radiation induces changes at the cell membrane associated with the apoptotic response in cultured bovine aortic endothelial cells.[63] The proximal events in the pathway proposed by Haimovitz-Friedman and Fuks involve sphingomyelin. Several researchers have shown that this pathway constitutes the early events in the apoptotic cascade of TNF-α-induced programmed cell death.[64,65] The sphingomyelin pathway is a signal transduction pathway that negotiates the signaling of a number of cytokines. In these systems, activation of cell surface receptors stimulates a plasma membrane neutral sphingomyelinase that then hydrolyzes sphingomyelin to generate ceramide and phosphocholine.[63]

Ceramide then functions as a second messenger, initiating a number of downstream events. In an effort to ascertain whether ceramide initiated the apoptotic response in endothelial cells, Haimovitz-Friedman and Fuks utilized the synthetic ceramide analog N-acetylsphingosine (C2-ceramide). Treatment with synthetic ceramide analogs has been shown to initiate the apoptotic pathway in several types of cells.[64,65] The researchers at Memorial Sloan Kettering were able to demonstrate induction of programmed cell death in bovine aortic endothelial cells (BAEC) using exogenous ceramide.[63] Haimovitz-Friedman and Fuks were further able to show that exposing BAEC to ionizing radiation leads to the production of elevated levels of ceramide; following doses of radiation as low as 0.5 Gy increased the level of ceramide to 115% above control. The maximal effect observed following irradiation was 155% above control at a dose of 10 Gy.[64] The authors then tested their observation in similar experiments with nuclei-free membranes and again demonstrated an increase in ceramide formation following irradiation; this in part would indicate that the cellular production of ceramide was a membrane-associated event and not merely a rapid transfer of nuclear-generated signal to the plasma membrane. To further determine if a true relationship between irradiation, induction of ceramide formation, and apoptosis existed, further experiments using 12-O-tetradecanoylphorbol-13-acetate (TPA) to abolish the radiation induced sphingomyelin hydrolysis were performed; this resulted in the ensuing inhibition of the apoptotic degradation of the DNA. Later incubations of these same BAEC in exogenous ceramide demonstrated a reappearance of the classical ladder formation consistent with apoptosis via gel electrophoresis. These data provide commanding evidence that radiation-induced generation of plasma membrane ceramide is a necessary event in radiation-induced apoptosis in endothelial cells.[63]

The capacity of endothelial cells to undergo apoptosis following irradiation appears to be complicated by the multiple controlling points in the apoptotic cascade. Further work reported by Haimovitz-Friedman et al. implicates the role of basic fibroblast growth factor (bFGF) as a protector to the lethal effects of ionizing radiation in endothelial cells *in vitro*.[64] Later experiments by Fuks support this observation but indicate the effect is not exclusively mediated via an effect on the repair of DNA breaks, but rather through inhibition of interphase apoptosis.[65] These researchers observed a significant increase in clonogenic survival in endothelial cells incubated in culture media containing bFGF immediately after irradiation. In studies directed at assessing apoptosis, endothelial cells in G_0-G_1 exposed to ionizing radiation displayed inhibition of apoptosis by bFGF stimulation. The mechanism, as proposed by these authors, is associated with the activation of membrane-bound protein kinase C.

Protein kinase C (PKC) was recognized in 1977 as a proteolytically activated protein kinase that is found ubiquitously in tissues and organs.[68] In most tissues, this enzyme is retrieved mainly from the soluble fraction as an inactive form and is seemingly translocated to membranes in a Ca^{2+}-dependent fashion when cells are stimulated.[67] Immunohistocytochemical analysis using monoclonal antibodies against the enzyme reveal the enzyme to be absent or poorly represented in the nucleus, suggesting PKC is closely associated with the cell membrane.[69]

There has recently been an accumulation of data implicating PKC and the cell membrane as the target for directing cellular responses to ionizing radiation. It is known that the radiation-mediated expression of immediate early/transcription factors jun/AP-1 and Egr-1/zif-268 in mammalian cells is abolished with the down-regulation of protein kinase C.[70] Further, an attenuated gene induction following irradiation[71] has been observed in cell variants deficient in PKC-mediated signal transduction. The phosphorylation potential of PKC increases fivefold within 15 seconds of irradiation, and PKC-specific substrates are subsequently phosphorylated.[71] And finally, Woloschak et al. have reported that PKC gene expression is enhanced soon after irradiation, further implicating PKC, at least indirectly, as a target for directing cellular responses to ionizing radiation.[72] Recent data reported by Uckun et al. provide more direct evidence. Using a human B lymphocyte system, the authors were able to show, following irradiation, the stimulation of phosphatidylinositol turnover, and the subsequent downstream activation of covalent modification of multiple serine-specific protein kinases, including protein kinase C — thus providing additional support that protein kinase C and the cell membrane are important steps in the generation of radiation-induced cellular events.[73]

These data are important if we are to assume that PKC and the cell membrane are involved in the apoptotic process in endothelial cells. Activation of PKC has been implicated in the mechanism for cytokine-mediated protection against glucocorticoid-induced apoptosis.[74–76] Thymic and lymphoid cells have also been shown to be protected by PKC activation against radiation-induced apoptosis.[77,78] In the endothelial cell system, Haimovitz-Friedman has shown that PKC activation mediates the protection conferred in endothelial cells by basic fibroblast growth factor (bFGF) against radiation-induced apoptotic DNA degradation.[79] It has been reported that the cellular effects of FGF are mediated via a ligand binding to high-affinity transmembrane receptors,[80] and that this binding activates the receptor by triggering autophosphorylation of its tyrosine kinase domain.[81] The downstream signaling resulting from the activated receptor involves binding of phosphorylated receptor tyrosine residues to catalytic SH_2 domains of the γ isoform of phospholipase C (PLC), thus activating the latter enzyme. Activated PLCγ hydrolyzes membrane phosphatidylinositol to generate inositol 1,4,5-triphosphate and diacylglycerol, which leads to the mobilization of cellular Ca^{2+} stores and the ultimate activation of PKC.[82] Haimovitz-Friedman has shown the involvement of PKC in the radioprotective effect conferred by bFGF by demonstrating that nonspecific PKC activation by short-term exposure to phorbol ester, 12-O-tetradecanoylphorbol-13-acetate (TPA; 30 ng/ml), mimicked the radioprotective effect of bFGF. Further, inhibition of PKC abrogated the radioprotective effects of bFGF.[79] The above data strongly identify the role of PKC in mediating the anti-apoptotic effects in endothelial cells by bFGF following irradiation.

V. CONCLUSION

The microvasculature is a ubiquitous organ system that plays a major role in the pathogenesis of radiation damage to normal tissues and is regarded as the dose-limiting factor in tissues in the therapeutic management of tumors. The initial expression of radiation injury to the microvasculature, as we have attempted to describe in this brief review, are changes in the structure and function of endothelial cell plasma membrane processes that if not repaired, can lead to impairment of the normal tissue microvascular environment. For complete regeneration and repair of the endothelium, both migration and replication must occur.[83] If migration occurs without subsequent replication or the replication process is impaired, then the regeneration process leads to a weakness within the endothelium. These cells will eventually detach due to natural cellular processes and either are not replaced, or become susceptible to inflammatory responses due to changes in membrane functions or structures involved in maintaining the nonthrombogenic surface of the endothelium. This effect, depending on the extent of radiation damage to the endothelial cells, can either lead to or amplify the later sequelae of radiation damage observed in tissues as hypothesized by Law.[2]

REFERENCES

1. Rubin, P. and Casarett, G. W., Clinical Radiation Pathology as applied to curative radiotherapy, *Cancer*, 22, 767, 1968.
2. Law, M. P., Radiation-induced vascular injury and its relation to late effects in normal tissues, *Adv. Radiat. Biol.*, 9, 37, 1981.
3. Phillips, T. L., An ultrastructural study of the development of radiation injury in the lung, *Radiology*, 87, 49, 1966.

4. Adamson, I. Y. R. and Bowden, D. H., Endothelial injury and repair in radiation-induced fibrosis, *Am. J. Pathol.*, 112, 224, 1983.
5. Ts'ao, C., Ward, W. F., and Port, C. D., Radiation injury in rat lung, *Radiat. Res.*, 96, 284, 1983.
6. Reed, G. B. and Cox, A. J., The human liver after radiation: a form of veno-occlusive disease, *Am. J. Pathol.*, 48, 597, 1966.
7. Casarett, G. W., Pathology of single intravenous doses of polonium, *Radiat. Res.*, 5(Suppl.), 246, 1964.
8. Fajardo, L. F. and Stewart, J. R., Capillary injury preceding radiation-induced myocardial fibrosis, *Radiology*, 101, 429, 1971.
9. Hopewell, J. W. and Wright, E. A., Modification of subsequent freezing damage by previous local brain irradiation, *Exp. Neurol.*, 26, 160, 1970.
10. Thilo-Korner, D. G. S. and Freshney, R. I., Eds., *The Endothelial Cell: A Pluripotent Control Cell of the Vessel Wall*, Karger, Basel, 1983.
11. Heimark, R. I. and Schwartz, S. M., The role of membrane-membrane interactions in the regulation of endothelial cell growth, *J. Cell Biol.*, 100, 1934, 1985.
12. Kwock, L., Davenport, W. C., Clark, R. L., Zaremba, J., Lingle, B., Chaney, E. L., and Friedman, M., The effects of ionizing radiation on the pulmonary vasculature of intact rats and isolated pulmonary endothelium, *Radiat. Res.*, 111, 276, 1987.
13. Scanlon, G. T., Vascular alteration in the irradiated rabbit kidney, *Radiology*, 94, 401, 1970.
14. Hopewell, J. W., The late vascular effects of radiation, *Br. J. Radiol.*, 47, 157, 1974.
15. Hopewell, J. W., The importance of vascular damage in the development of late effects in normal tissues, in *Radiation Biology in Cancer Research*, R. E. Meyn and H. R. Withers, Eds., Raven Press, New York, 1979, 449.
16. Rosen, E. M., Vinter, D. W., and Goldberg, I. D., Hypertrophy of cultured bovine aortic endothelium following irradiation, *Radiat. Res.*, 117, 395, 1989.
17. Lee, S.-L., Douglas, W. H. J., Lin, P.-S., and Fanburg, B. L. Ultrastructural changes of bovine pulmonary artery endothelial cells irradiated *in vitro* with a Cs-137 source, *Tissue Cell*, 15, 193, 1983.
18. Friedman, M., Ryan, U. S., Davenport, W. C., Chaney, E. L., and Kwock, L., Reversible alterations in cultured pulmonary artery endothelial cell monolayer morphology and albumin permeability induced by ionizing radiation, *J. Cell. Physiol.*, 129, 237, 1986.
19. Blose, S. H., The endothelial cytoskeleton, in *The Biology of Endothelial Cells*, E. A. Jaffe, Ed., Martinus Nijhoff, The Netherlands, 1984, 141.
20. Gotlieb, A. I. and Wong, M. K. K., Current concepts on the role of the endothelial cytoskeleton in endothelial integrity, repair, and dysfunction, in *Endothelial Cells*, Vol. II, U. S. Ryan, Ed., CRC Press, Boca Raton, FL, 1987, 82.
21. Pearson, J. D. and Gordon, J. L., Metabolism of serotonin and adenosine, in *Biology of Endothelial Cells*, E. A. Jaffe, Ed., Martinus Nijhoff, The Netherlands, 1984, 330.
22. Strum, J. M. and Junod, A. F., Radioautographic demonstration of 5–hydroxytryptamine uptake by pulmonary endothelial cells, *J. Cell Biol.*, 54, 456, 1972.
23. Block, E. R. and Fisher, A. B., Depression of serotonin clearance by rat lungs during oxygen exposure, *J. Appl. Physiol.*, 42, 33, 1977.
24. Block, E. R. and Fisher, A. B., Prevention of hyperoxic induced depression of pulmonary serotonin clearance by pretreatment with superoxide dismutase, *Am. Rev. Respir. Dis.*, 116, 441, 1977.

25. Petkau, A. and Chelack, W. S., Radioprotective effect of superoxide dismutase on model phospholipid membranes, *Biochim. Biophys. Acta*, 433, 445, 1976.
26. Vegt, G. B., Wassenaar, A. M., Kawilarang-de Haas, E. W. M., Schutte, P. P., van der Linden, M., Di Bon-de Ruijter, M., and Boon, A., Radiation induced changes in the cell membrane of cultured human endothelial cells, *Radiat. Res.*, 104, 317, 1985.
27. Kwock, L., Douglas, W. H. J., Lin, P.-S., Baur, E., and Fanburg, B. L., Endothelial cell damage after gamma irradiation *in vitro*: impaired uptake of alpha-aminoisobutyric acid, *Am. Rev. Respir. Dis.*, 125, 95, 1982.
28. Kwock, L, Lin, P.-S., and Ciborowski, L., Effect of the superoxide anion on Na^+-dependent amino acid transport, *Cell Biol. Int. Rpts.*, 5, 83, 1981.
29. Guidotti, G. G., Borghetti, A. F., and Gazzola, G. C., The regulation of amino acid transport in animal cells, *Biochim. Biophys. Acta*, 515, 329, 1978.
30. Del Maestro, R. F., Thaw, H. H., Bjork, J., Planker, M., and Arfors, K. E., Free radicals as mediators of tissue injury, *Acta Physiol. Scand.*, 492(Suppl.), 43, 1980.
31. Moncada, S. and Vane, J. R., Unstable metabolites of arachidonic acid and their role in haemostasis and thrombosis, *Br. Med. Bull.*, 34, 129, 1978.
32. Moncada, S. and Vane, J. R., Arachidonic acid metabolites and the interactions between platelets and blood-vessel walls, *N. Engl. J. Med.*, 300, 1142, 1979.
33. Allen, J. B., Sagerman, R. H., and Stuart, M. J., Irradiation decreases vascular prostacyclin formation with no concomitant effect on platelet thromboxane production, *Lancet*, 2, 1193, 1981.
34. Ts'ao C. H., Ward, W. F., and Port, C. D., Radiation injury in rat lung. I. Prostacyclin production, arterial perfusion, and ultrastructure, *Radiat. Res.*, 96, 284, 1983.
35. Eldor, A., Vlodavsky, I., Hyam, E., Atzman, R., and Fuks, Z., The effect of radiation on prostacyclin production by cultured endothelial cells, *Prostaglandins*, 25, 263, 1983.
36. Rubin, D. B., Drab, E. A., Ts'ao, C. H., Gardner, D., and Ward, W. F., Prostacyclin synthesis in irradiated endothelial cells cultured from bovine aorta, *J. Appl. Physiol.*, 58, 592, 1985.
37. Hahn, G. L., Menconi, M. J., Cahill, M., and Polgar, P., The influence of gamma irradiation on arachidonic acid release and prostacyclin synthesis, *Prostaglandins*, 25, 263, 1983
38. Friedman, M., Saunders, S., Madden, M. C., Chaney, E. L., and Kwock, L., The effects of ionizing radiation on the pulmonary endothelial cell uptake of alpha-aminoisobutyric acid and synthesis of prostacyclin, *Radiat. Res.*, 106, 171, 1986.
39. Hosi, Y., Yamamoto, M., Ono, T., and Sakamoto, K., Prostacyclin production in cultured endothelial cells is highly sensitive to low doses of ionizing radiation, *Int. J. Radiat. Biol.*, 63, 631, 1993.
40. Ryan, J. W., The metabolism of angiotensin I and bradykinin by endothelial cells, in *Biology of Endothelial Cells*, E. A. Jaffe, Ed., Martinus Nijhoff, The Netherlands, 1984, 317.
41. Rubin, D. B., Drab, E. A., Ward, W. F., Smith, L. J., and Fowell, S. M., Enzymatic responses to radiation in cultured vascular endothelial and smooth muscle cells, *Radiat. Res.*, 99, 420, 1984.
42. Levin, R. I., Weksler, B. B., Marcus, A. J., and Jaffe, E. A., Prostacyclin production by endothelial cells, in *Biology of Endothelial Cells*, E. A. Jaffe, Ed., Martinus Nijhoff, The Netherlands, 1984, 228.

43. Martin, T. W. and Michaelis, K. C., Bradykinin stimulates phosphodiesteratic cleavage of phosphatidylcholine in cultured endothelial cells, *Biochem. Biophys. Res. Commun.*, 157, 1271, 1988.
44. Korner, G., Schneider, D. E., Purdon, M. A., and Bjornsson, D., Bovine aortic endothelial cell transglutaminase: enzyme characterization and regulation of activity, *Biochem. J.*, 262, 633, 1989.
45. Birckbiehler, P. J., Orr, G. R., Patterson, Jr., M. J., Conway, E., and Carter, H. A., Increase in proliferative markers after inhibition of transglutaminase, *Proc. Natl. Acad. Sci., U.S.A.*, 78, 5005, 1981.
46. Jetten, A. M., George, M. A., Pettit, G. R., Herald, C. L., and Rearick, J. J., Action of phorbol esters, bryostatins, and retinoic acid on cholesterol sulfate synthesis: relation to the multistep process of differentiation in human epidermal keratinocytes, *J. Invest. Dermatol.*, 93, 108, 1989.
47. Korner, G. and Bachrach, U., Intracellular distribution of active and inactive transglutaminase in stimulated cultured C6 glioma cells, *J. Cell. Physiol.*, 130, 44, 1987.
48. Korner, G., Deutsch, V. R., Vlodavsky, I., and Eldor, A., Effects of ionizing irradiation on endothelial cell transglutaminase, *FEBS Lett.*, 330, 41,1993.
49. Hall, E. J., Cell survival curves, in *Radiobiology for the Radiologist*, 3rd ed., J.B. Lippincott Company, Philadelphia. 1988, 17.
50. Ward, J. F., DNA damage produced by ionizing radiation in mammalian cells: identities, mechanisms or formation and repairability, *Prog. Nucleic Acid. Mol. Biol.*, 35, 95, 1988.
51. Radford, I. R., Evidence for a general relationship between the induced level of DNA double strand breakage and cell-killing after X-irradiation of mammalian cells, *Int. J. Radiat. Biol.*, 49: 611, 1986.
52. Raff, M. C., Social control on cell survival and cell death, *Nature (London)*, 356, 397, 1992.
53. Vaux, D. L., Toward an understanding of the molecular mechanisms of physiological cell death, *Proc. Natl. Acad. Sci., U.S.A.*, 90, 786, 1993.
54. Wyllie, A. H., Cell death: a new classification separating apoptosis from necrosis, in *Cell Death and Biology and Pathology,* I. D. Bowen and R. A. Lockshin, Eds., Chapman and Hall, London, 1981, 9.
55. Thomas, N. and Bell, P. A., Glucocorticoid-induced cell-size changes and nuclear fragility in rat thymocytes, *Mol. Cell Endocrinol.*, 22, 71, 1981.
56. Wyllie, A. H. and Morris, R. G., Hormone induced cell death. Purification and properties of thymocytes undergoing apoptosis after glucocorticoid treatment. *Am. J. Pathol.*, 109, 78, 1982.
57. Kerr, J. F. R., Wyllie, A. H. and Currie, A. R., Apoptosis: a basic biological phenomenon with wide-ranging implications in tissue kinetics, *Br. J. Cancer*, 26, 239, 1972
58. Wyllie, A. H., Glucocorticoid-induced thymocyte apoptosis associated with endogenous endonuclease activation, *Nature (London)*, 284, 555, 1980.
59. Cohen, J. J. and Duke, R. C., Glucocorticoid activation of a calcium-dependent endonuclease in thymocyte nuclei leads to cell death, *J. Immunol.*, 132, 38, 1984.
60. Martin, S. J., Lennon, S. V., Bonham, A. M., et al., Induction of apoptosis (programmed cell death) in human leukemic HL-60 cells by inhibition of RNA and protein synthesis, *J. Immunol.*, 145, 1859, 1990.
61. Cohen, J. J. Programmed cell death in the immune system, *Adv. Immunol.*, 50, 55, 1990.

62. Ojeda, F., Diehl, H. A., and Folch, H., Radiation induced membrane changes and programmed cell death: possible interrelationships, *Scanning-Microsc.*, 8(3), 645, 1994.
63. Haimovitz-Friedman, A., Kan, C., Ehleiter, D., et al., Ionizing radiation acts on cellular membranes to generate ceramide and initiate apoptosis, *J. Exp. Med.*, 180, 525, 1994.
64. Obeid, L. M., Linardic, L. A., Fornari, R. S., et al., Programmed cell death induced by ceramide, *Science*, 259, 1769 1993.
65. Jarvis, D. W., Kolesnick, F. A., Fornari, R. S., et al., Induction of apoptotic DNA degradation and cell death by activation of the sphingomyelin pathway, *Proc. Natl. Acad. Sci., U.S.A.*, 91:73, 1994.
66. Haimovitz-Friedman, A., Vlodavsky, I., Chaudauri, A., et al., Autocrine effects of fibroblasts growth factor in radiation damage repair in endothelial cells, *Cancer Res.*, 51, 2552, 1991.
67. Fuks, Z., Peraud, R., Alfieri, A., et al., Basic fibroblast growth factor protects endothelial cells against radiation-induced programmed cell death *in vitro* and *in vivo*, *Cancer Res.*, 54, 2582, 1994.
68. Nishizuka, Y., Studies and perspectives of protein kinase C, *Science*, 233, 305, 1986.
69. Kikkawa, U., *Calcium and the Cell*, P. Baker, Ed. (Ciba Foundation) Symposium 122, Pittman, London.
70. Hallahan, D. E., Sukhatme, M. L., Sherman, M. L. et al., Protein kinase-C mediates X-ray inducibility of nuclear signal transducers, egr-1 and c-jun, *Proc. Natl. Acad. Sci., U.S.A.*, 88, 2152, 1991.
71. Hallahan, D. E., Virudachalam, S., Sherman, M. L. et al., Tumor necrosis factor gene expression is mediated by protein kinase C following activation by ionizing radiation, *Cancer Res.*, 51, 4565, 1991.
72. Woloschak, G., Chang-Liu, C., and Sherin-Jones, P., Regulation of protein kinase C by ionizing radiation, *Cancer Res.*, 50, 3963, 1990.
73. Uckun, F. M., Schieven, G. L., Tuel-Ahlgren, L. M., et al., Tyrosine phosphorylation is a mandatory proximal step in radiation-induced activation of the protein kinase C signaling pathway in human B-lymphocyte precursors, *Proc. Natl. Acad. Sci., U.S.A.*, 90, 252, 1993.
74. Lotem, J., Cragoe, E. J., and Sachs, L., Rescue programmed cell death in leukemic and normal myeloid cells, Blood, 78, 953, 1991.
75. Vazquez, A., Auffredon, M. A., Chaouchi, N., et al., Differential inhibition of interleukin 2- and interleukin 4-mediated human B cell proliferation by ionomycin: a possible role for apoptosis, *Eur. J. Immunol.*, 21, 2311, 1991.
76. Zubiaga, A. M., Munoz, E., and Huber, B. T., IL-4 and IL-2 selectively rescue the cell subsets from glucocorticoid-induced apoptosis, *J. Immunol.*, 149, 107, 1992.
77. Tomei, L. D., Kanter, P., and Wenner, C. E., Inhibition of radiation induced apoptosis *in vitro* by tumor promotors, *Biochem. Biophys. Commun.*, 155, 324, 1988.
78. Kizaki, H., Tadakuma, T., Okada, C. et al., Activation of a suicide process of thymocytes through DNA fragmentation by calcium and phorbol esters, *J. Immunol.*, 143, 1790, 1989.
79. Haimovitz-Friedman, A., Balaban, N., McLoughlin, M., et al., Protein kinase C mediates fibroblast growth factor protection of endothelial cells against radiation-induced apoptosis, *Cancer Res.*, 54, 2591, 1994.
80. Jaye, M., Schlessinger, J., and Craig, C. A., Fibroblast growth factor receptor tyrosine kinase: molecular analysis and signal transduction, *Biochim. Biophys. Acta*, 1135, 185, 1992.

81. Ullrich, A. and Schlessinger, J., Signaling transduction by receptors with tyrosine kinase activity, *Cell*, 61, 203, 1990.
82. Nishizuka, Y., Intracellular signaling by hydrolysis of phospholipids and activation of protein kinase C, *Science*, 258, 607, 1992.
83. Sholley, M. M., Gimbrone, M. A., and Cotran, R. S., Cellular migration in endothelial regeneration: a study using irradiated endothelial cultures, *Lab. Invest.*, 36, 18, 1977.

CHAPTER 8

Inflammatory Cytokines, Radiation, and Endothelial Gene Products: A Common Role for Reactive Oxygen Intermediates

Cliff Murray

CONTENTS

I. INTRODUCTION

Recent studies have revealed that the spectrum of cellular effects of ionizing radiation is wider and more subtle than previously realized. From the perspective

1-8493-4840-4/98/$0.00+$.50

of clinical radiotherapy, the arrest of cell division and cell death suffice to explain and justify its continued inclusion as a major part of cancer treatment. However, in the experimental context, where a much wider range of doses is investigated, it is clear that cellular effects can range from subtle up-regulation of gene expression to direct cell death caused by membrane damage. This panoply of effects may account for the complexity of radiation damage, in particular the so-called "late effects." In this chapter, I would like to suggest that the up-regulation of specific genes in endothelial cells by radiation, and similarities with the effects of inflammatory cytokines, owe more to design than chance, as the pathways mediating the effects of inflammatory cytokines and radiation on gene expression converge, with the generation of reactive oxygen intermediates and subsequent activation of transcription.

II. ENDOTHELIAL CELL FUNCTION

A. Barrier Function

Within the fully developed organism, the endothelial cell serves two primary functions. The first is to act as a barrier and selective filter between the vasculature and surrounding tissue. Within different organs, and vascular beds within those organs, there exist differences in barrier function, a prime example of which is the microvasculature of the brain, the so-called "blood-brain barrier." The endothelial cells lining these vessels have particularly tight intercellular junctions, making the lining of the vessel impermeable to all but the smallest ions and hydrophobic molecules.[1] The maintenance of this particular characteristic is thought to depend upon direct physical communication between endothelial cells and astrocytes.[2]

B. Vascular Homeostasis

The other major role of the endothelial cell is to present a nonthrombogenic surface to the blood, preventing unwanted platelet aggregation and/or coagulation within the bloodstream, and maintaining vascular homeostasis generally. The endothelial cell does this in a number of ways, including the expression on the cell surface of thrombomodulin[3] and heparan sulfate proteoglycans,[4] both of which inhibit coagulation. In addition, endothelial cells secrete protein S[5] and the enzyme tissue plasminogen activator.[6] Platelet aggregation is further inhibited by two labile low-molecular-weight endothelial products, prostacyclin (PGI_2)[7] and endothelial-derived relaxing factor (EDRF).[8,9]

C. Immune Accessory Function

In addition to these roles, endothelial cells can function in an immune accessory role, regulating the traffic of leukocytes between blood and surrounding tissue. This they do through the controlled expression of different cell adhesion molecules, which not only serve as signals for the adhesion of circulating leukocytes, but also activate specific functions within the arrested cell.[10] In addition, through the inducible expression of Class II histocompatibility antigens by agents such as interferon-γ (IFN-γ), they may act as antigen-presenting cells.[11]

D. Angiogenesis

Finally, the endothelial cell plays an all-important role in angiogenesis, the construction of new blood vessels, both in a pathological context, such as in diabetic retinopathy, rheumatoid arthritis, and solid tumors, and in normal circumstances, as in the cycling uterus.

Activation of endothelial cells into the "angiogenic" state is brought about through the complex interplay of physical and chemical factors that stimulate a carefully choreographed program of proliferation, proteolytic enzyme release, and migration.[12]

III. CONTROL OF ENDOTHELIAL FUNCTION

A. Paracrine Effects of Cytokines

The endothelial cell is subject to physical, chemical, and biochemical influences from the bloodstream on the luminal surface, and on the abluminal surface from the various elements of the surrounding tissue. In keeping with its critical role in the maintenance of vascular homeostasis, the endothelial cell is able to respond rapidly (within seconds to minutes) to a number of important agonists, such as thrombin and histamine.[13] In general, such rapid responses require neither active transcription nor protein synthesis, but rather rely upon the mobilization of intracellular pools (e.g., calcium ions; von Willebrand Factor). A second class of external influences is mediated by cytokines, a group of distinct polypeptides that bind to specific receptors on the cell surface, and that activate gene expression and protein synthesis *de novo*.[14] To distinguish between these two fundamentally different processes, Pober and Cotran[15] have referred to the effects of agonists such as thrombin and histamine as endothelial *stimulation*, and to the effects induced by many cytokines as endothelial *activation*. The activation of cultured endothelial cells by cytokines probably represents the *in vitro* correlate of the type of changes seen in postcapillary venular endothelial cells during delayed hypersensitivity reactions.[16]

Perhaps the best studied effects of paracrine-acting cytokines on endothelial cells are those of the inflammatory cytokines tumor necrosis factor-α (TNF-α) and interleukin-1β (IL-1). These two cytokines, together with bacterial lipopolysaccharide, constitute an "inflammatory triad," inducing a pro-inflammatory, procoagulant phenotype on endothelial cells *in vitro*. This state is conferred by the modulation of surface expression of a number of important effector molecules: tissue factor, a potent procoagulant, and plasminogen activator inhibitor expression are increased within 2 to 6 hr of exposure to cytokines,[17-20] as is the expression of a number of adhesion molecules,[21] including intercellular adhesion molecule-1 (ICAM-1) and vascular cell adhesion molecule-1 (VCAM-1), both members of the immunoglobulin super-family of genes, and the selectin endothelial leukocyte adhesion molecule-1 (ELAM-1) or E-selectin. ICAM-1 and VCAM-1 are constitutively expressed on the surface of endothelial cells; however, their expression is enhanced by cytokines through increased transcription and protein synthesis *de novo* (for review, see Reference 21). Two other important molecules induced by the inflammatory cytokines, which appear to play a role in leukocyte adhesion, are platelet activating factor (PAF)[22] and interleukin-8 (IL-8);[23] the expression of IL-8 is controlled at the transcriptional level.[23]

While TNFα and IL-1 modulate the expression of a range of gene products, interferon-γ (IFN-γ) plays a unique role in immune inflammation, up-regulating the expression of Class I and II MHC antigens on endothelial cells.[24] Furthermore, it can act additively or synergistically with TNF-α in bringing about the changes associated with immune inflammation.[24]

B. Signal Transduction Pathways

The signal transduction pathways associated with cytokine activation of endothelial cells are complex and poorly understood; however, at least in the case of IL-1 and TNF-α, it is known that activation is linked with the induction of the immediate early response nuclear protooncogenes (ERG) c-*fos* and c-*jun*.[25–27] Low basal expression of c-*fos* and c-*jun* is observed in many cell types, but their expression is rapidly and transiently enhanced by a wide variety of stimuli, in a cycloheximide-independent manner. The protein products of these genes bind to specific sites on DNA as jun-jun homodimers or jun-fos heterodimers, the heterodimeric forms exhibiting more *trans*-activating potential than the dimers. Active dimers interact with a DNA regulatory site referred to as the AP-1 site. It has been suggested that tissue specificity in the regulation of AP-1-dependent genes may be conferred by the existence of different combinations of gene products related to c-*fos* and c-*jun*.[25,26]

Regulated synthesis of transcription factors such as c-*fos* and c-*jun* is often used in the control of *cell-type specific* or *developmentally regulated* genes.[28] Thus, AP-1 binding sites are found in the promoter regions of the genes coding for several proteins particularly associated with endothelial cells, including the cell adhesion molecules ICAM-1,[29] E-selectin,[30] and VCAM-1,[31] and two molecules involved in coagulation/fibrinolysis, namely tissue factor[32] and urokinase/plasminogen activator.[33] Often, AP-1 alone is insufficient for transcriptional activation, and is found in 5′-flanking regions adjacent to other *cis*-acting elements such as the NF-κB binding site.

NF-κB belongs to a class of transcription factors that are constantly present within the cell, but remain inactive until activated by a specific signal. Such transcription factors are often involved in the regulation of *inducible* genes.[28] NF-κB normally resides in the cytoplasm in the form of a complex with an inhibiting subunit, IκB, preventing entry of the transcription factor into the nucleus and thus binding to DNA. IκB is thought to be released from the complex by phosphorylation. Many genes associated with endothelial cell function contain NF-κB and NF-κB-like binding sites, including those for ICAM-1,[29] VCAM-1,[31] E-selectin,[35] urokinase/plasminogen activator,[36] GM-CSF,[37] and MHC I and II.[38,39]

NF-κB itself is known to be activated by a diverse range of stimuli,[37] including TNF-α, IL-1, lipopolysaccharide, phorbol esters, and UV light. This is somewhat surprising given that even the inflammatory cytokines TNF-α and IL-1 do not share a common second messenger pathway; while TNF-α signaling involves G-proteins, IL-1 may act via cAMP and phosphatidylcholine turnover. It has been suggested that these various pathways converge with the generation of reactive oxygen intermediates, which are known to activate NF-κB.[40]

C. Reactive Oxygen Intermediates

Reactive oxygen intermediates (ROI) are produced by all cells as by-products of electron transfer reactions. Such species include singlet oxygen, hydrogen peroxide, superoxide anion (O_2^-), and hydroxyl radical (OH·), which can in turn give rise to secondary reactive products such as lipid peroxides and hypochlorite.[41] Cellular anti-oxidant mechanisms ensure that ROI cause minimal injury; these include the enzymes superoxide dismutase and catalase, and the glutathione/peroxidase system. If the rate of formation of ROI exceeds the capacity of the cell to eliminate them, then oxidative injury will result, in its most extreme form comprising DNA fragmentation and cell lysis.

In some cells, notably phagocytes, the cytotoxic potential of ROI is harnessed for specific purposes. During the "respiratory burst," the flavoprotein NADPH oxidase, found on the plasma membrane of phagocytic cells, including granulocytes and macrophages, catalyzes the one-electron reduction of oxygen to superoxide anion using NADPH as electron donor.[42] Superoxide is converted to hydrogen peroxide, hydroxyl radical, and oxidized halogens such as hypochlorous acid. As well as playing a crucial role in inflammatory reactions and tissue injury, these phagocyte-derived ROI are important mediators of bacterial and tumor cell killing. While the endothelial cell can produce ROI, the source of these molecules is unclear and several enzymes have been suggested, including xanthine oxidase, cytochrome P450, cyclooxygenase, as well as NADPH oxidase.[43]

The first suggestion that NF-κB could be activated by ROI was the demonstration by Staal et al.[44] that the activation of an NF-κB-controlled HIV-1 reporter gene by TNF-α or the phorbol ester PMA was inhibited by the ROI scavenger N-acetyl-L-cysteine (NAC). Studies by others[45,46] indicated that NF-κB activation by a wide range of known activators, including IL-1, LPS, and dsRNA, is inhibited by thiol reagents such as dithiocarbamate and disulfiram, again implicating ROI. In addition, hydrogen peroxide alone was demonstrated to be a very potent activator of DNA binding by NF-κB *in vivo*.[45] Of particular relevance to the endothelial cell are recent studies by Suzuki et al.[47] and Marui et al.[48] In the first study, the authors demonstrated that the LPS- or PMA-induced expression of E-selectin, as assessed by enhanced levels of mRNA, is abrogated by apocynin, a specific inhibitor of NADPH oxidase. The second study demonstrated that transcription and expression of the gene for VCAM-1 are regulated by an anti-oxidant-sensitive mechanism in endothelial cells, leading the authors to suggest that oxidative stress is an important regulatory signal in the endothelial cell. As indicated earlier, the genes for both E-selectin and VCAM-1 are known to contain NF-κB binding sites.

Thus, the evidence is very convincing that ROI play a central role in the activation of NF-κB. Indeed, it appears that this transcription factor is promiscuous in its response to ROI, paying little heed to their source. Other transcription factors, such as AP-1, may also respond to ROI; DNA binding and *trans*-activation by AP-1 occur after treatment of HeLa cells with hydrogen peroxide.[49] This mechanism, however, appears to differ from that of NF-κB, as activation of AP-1 via protein kinase C (PKC) is not blocked by ROI scavengers.

IV. EFFECTS OF RADIATION ON ENDOTHELIAL CELLS

A. Radiation and Transcription Factors

Until recently, the effects of ionizing radiation on cells were regarded as generally deleterious, with excessive damage to genomic DNA leading to induction of mutations, neoplastic transformation, and cell death. It was clear, however, that total dose, quality of radiation, and dose rate also play a part in determining the outcome of the radiation insult and, perhaps not surprisingly, ionizing radiation was shown to induce the expression of DNA repair genes in prokaryotes and lower eukaryotes.[50]

An early demonstration of induction of gene expression in mammalian cells by ionizing radiation was that of Hallahan et al.,[51] who showed that treatment of certain human sarcoma cell lines with 250kV X-rays induces the expression of mRNA for TNF-α, and a subsequent increase in TNF-α protein production. The authors did not speculate on the mechanisms involved but did suggest that, in view of the toxicity of TNF-α for certain tumor cells, this might be a mechanism whereby the cytotoxic effects of radiation were potentiated. In a subsequent study examining the expression of TNF-α in leukemia cells and monocytes after X-irradiation,[52] the same group concluded that enhanced transcription of the TNF-α gene might be mediated by a PKC-dependent mechanism, although specifically which *cis*-acting elements in the TNF-α gene promoter region are involved was not clear; AP-1 and NF-κB motifs have both been described in the 5′ sequence of the human TNF-α gene.[53,54] What is less clear is the chain of events that links initial insult by radiation to the activation of transcription factors.

It has been proposed that radiation triggers a cascade of cytoplasmic signaling events; one of the earliest events, thought to occur within 30 minutes of irradiation, is the activation of PKC, leading to the transcription of immediate early response genes such as c-*jun* and c-*fos*.[55] As mentioned earlier the protein products of these genes give rise to AP-1-binding transcription factors.[25] Recent studies of the effects of radiation on gene expression in human lymphocyte precursors by Uckun et al.[56–58] suggest that activation of PKC, and other serine-specific protein kinases, occurs *downstream* of tyrosine-specific protein kinases that are themselves the targets of radiation, and sensitive to ROI. PKC activation may be further enhanced through the accumulation of diacylglycerol (DAG) generated by ROI-mediated membrane lipid peroxidation.[59] Finally, activation of the transcription factor NF-κB is thought to occur in response to phosphorylation of IκB which detaches from NF-κB and allows NF-κB to enter the nucleus.[34]

NF-κB activation has now been shown to occur in several cell lines in response to ionizing radiation.[57,60,61] In a recent comprehensive study, Wilson et al.[61] showed that while the phorbol ester PMA induced the activation of three important transcription factors (AP-1, NF-κB, and EgrI/Krox24), only NF-κB and, to a lesser extent, AP-1 were activated by ionizing radiation in primary human B-lymphocytes. They concluded that the pleiotropic effects of ionizing radiation are mediated through distinct signaling pathways that may be specific for individual cell types; NF-κB in particular plays a major role in these effects.

B. Radiation and Endothelial Function

The relationship between radiation effects on the vasculature and subsequent damage to normal tissues has been recognized for a long time.[62] Nevertheless, this continues to be the subject of considerable controversy. Early studies, constrained by the complexities of *in vivo* approaches, pointed to the consistent presence of gross blood vessel abnormalities in irradiated normal tissues[63] and attributed late effects of radiation to vascular insufficiency. These changes are now thought to occur as a result of damage to capillary endothelial cells; morphologically they are characterized acutely by swelling, vacuolization and swelling, cell sloughing and increased permeability.[64] This is followed by the formation of platelet thrombi that may occlude the vessel. Ultimately, vascular insufficiency and associated anoxia may lead to fibrosis.[64]

The advent of methods for the routine, large-scale culture of endothelial cells[65] opened a new seam of research possibilities for those interested in radiation pathology. Perhaps the most important implication of the studies that followed was that the long-held view of tissue dysfunction resulting from radiation-induced cell death could now be seriously challenged. Evidence for sublethal effects on endothelial function continues to accumulate. Combining this information with an increased understanding of the role of endothelial cells in inflammatory pathways suggests that we can anticipate a range of responses of endothelial cells, and consequently the normal tissues that they support, to radiation of different doses and qualities.

Several morphological and functional changes observed *in vivo* are mirrored by *in vitro* effects of radiation on cultured endothelial cells. For example, radiation doses as low as 2 Gy induce detachment of bovine aortic endothelial cells from culture dishes;[66] at 5 Gy, human umbilical vein endothelial cells show no change in morphology but do demonstrate sites of converged microfilaments, implying disruption of cytoskeletal organization.[67] This phenomenon is probably related to changes in endothelial monolayer permeability observed after radiation.[68] Woloschak et al.[69] have shown that a variety of different qualities of radiation repress expression of the β-actin gene while up-regulating the expression of γ-actin and α-tubulin. Such changes in the profile of cytoskeletal components may explain several functional observations.

In terms of maintaining homeostasis within the bloodstream, the expression of several key effector molecules by cultured endothelial cells is affected by radiation. The radiation-induced release of von Willebrand Factor (vWF) from Weibel-Palade bodies within the endothelial cell[70] probably occurs in response to changes in intracellular calcium, without involving transcriptional mechanisms. However, important changes have been observed in the synthesis and release of prostanoids such as PGI_2 (prostacyclin) and PGE_2. Prostacyclin in particular plays an important role in maintaining homeostasis, both as a potent vasodilator and as an inhibitor of platelet aggregation.[7] Radiation induces an initial increase followed by a decrease in prostacyclin levels in cultured human endothelial cells,[71] while inducing a general decrease in production by isolated rabbit aortic segments.[72] Clearly, the metabolic interactions are complex, although cyclooxygenase, the enzyme responsible for prostacyclin synthesis, has been identified as an early response gene in endothelial cells.[73] Recent studies have identified two genes coding for cyclooxygenase, COX I and COX II,[74] the second gene representing an inducible form of the enzyme; the transcriptional factors controlling its expression are not known, although it seems likely that its expression is linked to local inflammatory responses.[73]

The best established link between changes in endothelial function as a result of radiation and changes in gene expression comes from the work of Witte et al.,[75] who have demonstrated that single doses of γ-radiation between 20 and 60 Gy induce the synthesis and release of PDGF- and FGF-like growth factors by endothelial cells from a variety of species. This phenomenon occurs within 24 hr of irradiation, and enhanced release persists for up to 72 hr postirradiation. While the transcriptional events involved in this up-regulation have not been deciphered, the enhanced release of both PDGF and bFGF is known to occur as a result of injury or inflammatory stress to endothelial cells.[76,77]

Radiation has been shown to induce the transcription of the IL-1 gene in Syrian hamster embryo cells[78] and the TNF-α gene in a variety of human tumor cell lines and peripheral blood monocytes[51,52] and murine brain cells.[79] Whether similar activation occurs in endothelial cells is unknown. The expression of these cytokines by irradiated endothelial cells might be expected to induce a pro-inflammatory, procoagulant state by working in an autocrine manner.

Another functional change observed in irradiated endothelial cells, and closely linked with the known effects of TNF-α and IL-1, is the enhanced interaction of leukocytes with endothelial monolayers,[80] and with the blood vessel wall of irradiated normal tissues.[81] Interestingly, the opposite effect is observed in irradiated tumor-bearing tissues, where fewer adhesive interactions take place between circulating leukocytes and endothelium.[82] The mechanisms behind these effects are very complex; Dunn et al.[80] first described the production of a trypsin-sensitive neutrophil chemotactic factor by irradiated endothelial cells, followed by increased adhesion to the monolayer. Subsequently, Matzner et al.[83] described another neutrophil chemoattractant that was not trypsin sensitive, but which was associated with the lipoxygenase pathway.

Two other molecules that are potentially implicated in the enhanced adhesion of leukocytes are platelet activating factor (PAF) and P-selectin. These molecules appear to act in concert to mediate the earliest adhesion events,[21] and their expression is stimulated by ROI;[84,85] radiation might therefore be expected to enhance their production. Finally, Lin et al.[67] have demonstrated that radiation up-regulates the expression of the E-selectin gene, which is known to contain NF-κB and AP-1 sites in its promoter region.[30,35] As discussed earlier, the expression of E-selectin is characteristic of endothelial cells, and this molecule is believed to mediate the adhesion of certain types of leukocytes to *activated* endothelial cells. It is clear that enhanced interactions between leukocytes and irradiated endothelial cells may occur by a variety of mechanisms, some of which may be the direct or indirect result of transcriptional activation, and some direct membrane effects.

V. CONCLUSION: PATHWAYS TO LATE RADIATION DAMAGE

It is generally agreed that vascular damage, and more specifically damage to the endothelium, plays an important role in the etiology of late radiation damage. Many of the characteristics of late radiation damage suggest similarities to the effects of other chronic inflammatory stimuli, frequently characterized by obliterative vasculitis and ultimately interstitial fibrosis. From the preceding discussion it should be clear that radiation alone is capable of activating a number of key components in such pathways, including permeability changes, prothrombotic activity, and enhanced leukocyte/endothelial interactions. In some instances, this may be through direct effects on the endothelial cell, such as by the release of lipid-based chemoattractants from the cell membrane, and in other instances through activation of transcription. More complex combinations might also apply; E-selectin, expression of which is directly activated in response to radiation, is also sensitive to activation by TNF-α, itself a candidate for up-regulation by radiation. Thus, in this instance, TNF-α could function in an autocrine manner. Witte et al.[75] have postulated just such an autocrine mechanism in the stimulation of endothelial proliferation by radiation; they also point out that this mechanism would account for the cell proliferation associated with radiation-induced fibrosis. More recently, Wilson et al.[61] have suggested an alternative scheme, in which an acute effect of radiation may be to cause the release of small quantities of pre-formed growth factors from intracellular stores, which then engage specific receptors in an autocrine/paracrine manner, stimulating cellular proliferation.

In conclusion, the relationship between endothelial cells, inflammatory cytokines, and radiation is complex, with a multitude of opportunities for cross-talk and interactions. However, whether induced by "conventional" ligand/receptor interactions at the cell surface, or by insults such as ionizing radiation, ROI would appear to play a common role in the expression of many key molecules involved in pro-inflammatory, procoagulant effects. Furthermore, a number of the genes coding for these molecules, which appear to be linked to cellular stress responses and are induced by inflammatory cytokines, share common transcriptional control elements including NF-κB-binding sites in their promoter regions. So, biochemical pathways are in place that dictate that the major response of the endothelial cell to irradiation is to adopt a pro-inflammatory, procoagulant phenotype, presumably in an attempt to limit and repair damage to the tissue that it supports. It would appear that the endothelial cell may have difficulty in discriminating between certain receptor-based signals and other ROI-generating stimuli; presumably, this situation confers biological advantage upon the tissue in terms of response to injury, and repair.

REFERENCES

1. Reese, T. S. and Karnovsky, M. J., Fine structural localisation of a blood-brain barrier to exogenous peroxidase, *J. Cell Biol.*, 34, 207,1967.
2. De Bault, L. E., Gamma-glutamyl transpeptidase induction mediated by glial foot processes to endothelium contact in coculture, *Brain Res.*, 220, 432, 1981.
3. Esmon, N. L., Thrombomodulin, *Semin. Thromb. Hemost.*, 13, 454, 1987.
4. Marcum, J. A., McKenny, J. B., and Rosenberg, R. D., Acceleration of thrombin-antithrombin complex formation in rat hindquarters via heparin-like molecules bound to the endothelium, *J. Clin. Invest.*, 74, 341, 1984.
5. Stern, D. M., Brett, J., Harris, I. C., and Nawroth, P. P., Participation of endothelial cells in the protein C-protein S anticoagulant pathway, *J. Cell Biol.*, 102, 1971, 1986.
6. Van Hinsbergh, V. W. M., Regulation of the synthesis and secretion of plasminogen activators by endothelial cells, *Haemostasis*, 18, 307, 1988.
7. Moncada, S., Biological importance of prostacyclin, *Br. J. Pharmacol.*, 76, 3, 1982.
8. Furchgott, R. F., Role of endothelium in the responses of vascular smooth muscle to drugs, *Annu. Rev. Pharmacol. Toxicol.*, 24, 175, 1984.
9. Palmer, R. M. J., Ferrige, A. G., and Moncada, S., Nitric oxide accounts for the biological activity of endothelium-derived relaxing factor, *Nature*, 327, 524, 1987.
10. Carlos, T. M. and Harlan, J. M., Membrane proteins involved in phagocyte adherence to endothelium, *Immunol. Rev.*, 114, 5, 1990.
11. Collins, T., Korman, A. J., Wake, C. T., Boss, J. M., Kappes, D. J., Fiers, W., Ault, K. A., Gimbrone, M. A., Strominger, J. L., and Pober, J. S., Immune interferon activates multiple class II major histocompatibility complex genes and the associated invariant chain gene in human endothelial cells and dermal fibroblasts, *Proc. Natl. Acad. Sci., U.S.A.*, 81,4917,1988.
12. Folkman, J. and Klagsbrun, M., Angiogenic factors, *Science*, 235, 442, 1987.
13. Jaffe, E. A., Grulich, J., Weksler, B. B., Hampel, G., and Watanabe, K., Correlation between thrombin-induced prostacyclin production and inositol trisphosphate and cytosolic free calcium levels in cultured endothelial cells, *J. Biol. Chem.*, 262, 8557, 1987.
14. Cohen, S., Physiologic and pathologic manifestations of lymphokine action, *Hum. Pathol.*, 17, 112, 1986.
15. Pober, J. S. and Cotran, R. S., The role of endothelial cells in inflammation, *Transplantation*, 50, 537, 1990.
16. Wilms-Kretschmer, K., Flax, M. H., and Cotran, R. S., The fine structure of the vascular response in hapten-specific delayed hypersensitivity and contact dermatitis, *Lab. Invest.*, 17, 334, 1967.
17. Bevilacqua, M. P., Pober, J. S., Wheeler, M. E., Cotran, R. S., and Gimbrone, M. A., Interleukin 1 (IL-1) activation of vascular endothelium: effects on procoagulant activity and leukocyte adhesion, *Am. J. Pathol.*, 121, 393, 1985.
18. Bevilacqua, M. P., Pober, J. S., Majeau, G. R., Fiers, W., Cotran, R. S., and Gimbrone, M. A., Recombinant tumor necrosis factor induces procoagulant activity in cultured human vascular endothelium: characterization and comparison with the actions of interleukin 1, *Proc. Natl. Acad. Sci., U.S.A.*, 83, 4533, 1986.
19. Nawroth, P. P., Handley, D. A., Esmon, C. T., and Stern, D. M., Interleukin 1 induces endothelial cell procoagulant while suppressing cell-surface anticoagulant activity, *Proc. Natl. Acad. Sci., U.S.A.*, 83, 3460, 1986.

20. Bevilacqua, M. P., Schleef, R. R., Gimbrone, M. A., and Loskutoff, D. J., Regulation of the fibrinolytic system of cultured human vascular endothelium by interleukin 1, *J. Clin. Invest.*, 78, 587, 1986.
21. Bevilacqua, M. P., Endothelial-leukocyte adhesion molecules, *Annu. Rev. Immunol.*, 11, 767, 1993.
22. Bussolino, F., Breviario, F., Tetta, C., Aglietta, M., Mantovani, A., and Dejana, E., Interleukin-1 stimulates platelet activating factor production in cultured human endothelial cells, *J. Clin. Invest.*, 77, 2027, 1986.
23. Strieter, R. M., Kunkel, S. L., Showell, H. J., Remick, D. G., Phan, S. H., Ward, P. A., and Marks, R. M., Endothelial cell gene expression of a neutrophil chemotactic factor by TNF-α, LPS, and IL-1β, *Science*, 243, 1467, 1989.
24. Pober, J. S. and Cotran, R. S., Cytokines and endothelial cell biology, *Physiol. Rev.*, 70, 427, 1990.
25. Angel, P. and Karin, M., The role of jun, fos and the AP-1 complex in cell proliferation and transformation, *Biochim. Biophys. Acta*, 1072, 129, 1991.
26. Abate, C. and Curran, T., Encounters with fos and jun on the road to AP-1, *Semin. Cancer Biol.*, 1, 19, 1990.
27. Colotta, F., Lampugnani, M. G., Polentaretti, N., Dejana, E., and Mantovani, A., Interleukin 1 induces c-fos protooncogene expression in cultured human endothelial cells, *Biochem. Biophys. Res. Commun.*, 152, 1104, 1988.
28. Gerritsen, M. E., and Bloor, C. M., Endothelial cell gene expression in response to injury, *FASEB J.*, 7, 523, 1993.
29. Voraberger, G., Schafer, R., and Stratowa, C., Cloning of the human gene for intracellular adhesion molecule 1 and analysis of its 5′ regulatory region. Induction by cytokines and phorbol ester, *J. Immunol.*, 147, 2777, 1991.
30. Collins, T., Williams, A., Johnston, G. I., Kim, J., Eddy, R., Shows, T., Gimbrone, M. A., and Bevilacqua, M. P., Structure and chromosome location of the gene for endothelial leukocyte adhesion molecule 1, *J. Biol. Chem.*, 266, 2466, 1991.
31. Cybulsky, M. I., Fries, J. W. U., Williams, A. J., Sultan, P., Eddy, R., Byers, M., Shows, T., Gimbrone, M. A., and Collins, T., Gene structure, chromosomal location and basis for alternative mRNA splicing of the human VCAM-1 gene, *Proc. Natl. Acad. Sci., U.S.A.*, 88, 7859, 1991.
32. Mackman, N., Morrisey, J. H., Fowler, B., and Edgington, T. S., Complete sequence of the human tissue factor gene, a highly regulated cellular receptor that initiates the coagulation protease cascade, *Biochemistry*, 28, 1755, 1989.
33. Blasi, F., Riccio, A., and Sabastio, G., Human plasminogen activators, genes, and protein structure, *Horiz. Biochem. Biophys.*, 8, 377, 1986.
34. Baeuerle, P., The inducible transcription activator NF-κB: regulation by distinct protein subunits, *Biochem. Biophys. Acta*, 1072, 63, 1991.
35. Whelan, J., Ghersa, P., van Huijsduijnen, P. H., Gray, J., Chandra, G., Talabot, F., and Delamater, J. F., A NF-κB like factor is essential but not sufficient for cytokine induction of endothelial leukocyte adhesion molecule 1 (ELAM-1) gene transcription, *Nucleic Acids Res.*, 19, 2645, 1991.
36. Novack, U., Cocks, B. G., and Hamilton, J. A., A labile repressor acts through the NF-κB-like binding sites of the human urokinase gene, *Nucleic Acids Res.*, 19, 3389, 1991.
37. Schreck, R. and Baeuerle, P.A., NF-κB as inducible transcriptional activator of the granulocyte-macrophage colony stimulating factor gene, *Mol. Cell. Biol.*, 10, 1281, 1990.

38. Baldwin, A.S. and Sharp, P. A., Two transcription factors, NF-κB and H2TF1, interact with a single regulatory sequence in the class I major histocompatibility complex promoter, *Proc. Natl. Acad. Sci., U.S.A.*, 85, 723, 1988.
39. Israel, A., Kimura, A., Kieran, M., Yano, O., Kanellopoulos, J., Le Bail, O., and Kourilsky, P., A common positive *trans*-acting factor binds to enhancer sequences in the promoter of mouse H-2 and β_2-microglobulin genes, *Proc. Natl. Acad. Sci., U.S.A.*, 84, 2653, 1987.
40. Schreck, R. and Baeuerle, P. A., A role for oxygen radicals as second messengers, *Trends Cell Biol.*, 1, 39, 1991.
41. Halliwell, B. and Gutteridge, J. M. C., *Free Radicals in Biology and Medicine*, 2nd ed., Clarendon Press, London, 1990.
42. Rossi, F., The O_2^- forming NADPH oxidase of the phagocytes: nature, mechanisms of activation and function, *Biochim. Biophys. Acta*, 853, 65, 1986.
43. Freeman, B.A. and Crapo, J. D., Biology of disease: free radicals and tissue injury, *Lab. Invest.*, 47, 412, 1982.
44. Staal, F. J. T., Roederer, M., Herzenberg, L. A., and Herzenberg, L. A., Intracellular thiols regulate activation of nuclear factor κB and transcription of human immunodeficiency virus, *Proc. Natl. Acad. Sci., U.S.A.*, 87, 9943, 1990.
45. Schreck, R., Rieber, P., and Baeuerle, P. A., Reactive oxygen intermediates as apparently widely used messengers in the activation of the NF-κB transcription factor and HIV-1, *EMBO J.*, 10, 2247, 1991.
46. Toledano, M. B. and Leonard, W. J., Modulation of transcription factor NF-κB binding activity by oxidation-reduction *in vitro*, *Proc. Natl. Acad. Sci., U.S.A.*, 88, 4328, 1991.
47. Suzuki, Y., Wang, W., Vu, T. H., and Raffin, T. A., Effect of NADPH oxidase inhibition on endothelial cell ELAM-1 mRNA expression, *Biochem. Biophys. Res. Commun.*, 184, 1339, 1992.
48. Marui, N., Offerman, M. K., Swerlick, R., Kunsch, C., Rosen, C. A., Ahmad, M., Alexander, R. W., and Medford, R. M., Vascular cell adhesion molecule-1 (VCAM-1) gene transcription and expression are regulated through an anti-oxidant-sensitive mechanism in human vascular endothelial cells, *J. Clin. Invest.*, 92, 1866, 1993.
49. Devary, Y., Gottlieb, R., Lau, L. F., and Karin, M., Rapid and preferential activation of the c-*jun* gene during the mammalian UV response, *Mol. Cell. Biol.*, 11, 2804, 1991.
50. Little, J. W. and Mount, D. W., The SOS regulatory system of *Escherichia coli*, *Cell*, 29, 11, 1982.
51. Hallahan, D. E., Spriggs, D. R., Beckett, M. A., Kufe, D. W., and Weichselbaum, R. R., Increased tumor necrosis factor α mRNA after cellular exposure to ionizing radiation, *Proc. Natl. Acad. Sci., U.S.A.*, 86, 10104, 1989.
52. Sherman, M. L., Datta, R., Hallahan, D. E., Weichselbaum, R. R., and Kufe, D. W., Regulation of tumor necrosis factor gene expression by ionizing radiation in human myeloid leukemia cells and peripheral blood monocytes, *J. Clin. Invest.*, 87, 1794, 1991.
53. Economou, J. S., Rhoades, K., Essner, R., McBride, W. H., Gasson, J. C., and Morton, D. L., Genetic analysis of the human tumor necrosis factor α/cachectin promoter region in a macrophage cell line, *J. Exp. Med.*, 170, 321, 1989.
54. Collart, M.A., Baeuerle, P., and Vassali, P., Regulation of tumor necrosis factor alpha transcription in macrophages: involvement of four kB-like motifs and of constitutive and inducible forms of NF-κB. *Mol. Cell. Biol.*, 10, 1498, 1990.
55. Sherman, M. L., Datta, R., Hallahan, D. E., Weichselbaum, R. R., and Kufe, D. W., Ionizing radiation regulates expression of the c-*jun* protooncogene, *Proc. Natl. Acad. Sci., U.S.A.*, 87, 5663, 1990.

56. Uckun, F. M., Tuel-Ahlgren, L., Song, C. W., Waddick, K., Myers, D. E., Kirihara, J., Ledbetter, J. A., and Schieven, G. L., Ionizing radiation stimulates unidentified tyrosine-specific protein kinases in human B-lymphocyte precursors, triggering apoptosis and clonogenic cell death, *Proc. Natl. Acad. Sci., U.S.A.*, 89, 9005, 1992.
57. Uckun, F. M., Schieven, G. L., Tuel-Ahlgren, L., Dibirdik, I., Myers, D. E., Ledbetter, J. A., and Song, C. W., Tyrosine phosphorylation is a mandatory proximal step in radiation-induced activation of the protein kinase C signaling pathway in human B-lymphocyte precursors, *Proc. Natl. Acad. Sci., U.S.A.*, 90, 252, 1993.
58. Park Chae, H., Jarvis, L. J., and Uckun, F. M., Role of tyrosine phosphorylation in radiation-induced activation of *c-jun* protooncogene in human lymphohematopoietic precursor cells, *Cancer Res.*, 53, 447, 1993.
59. O'Brian, C. A., Ward, N. E., Weinstein, B., Ball, A. W., and Marnett, L. J., *Biochem. Biophys. Res. Commun.*, 155, 1374, 1988.
60. Brach, M., Hass, R., Sherman, M. L., Gunji, H., Weichselbaum, R., and Kufe, D., Ionizing radiation induces expression and binding activity of the nuclear factor κB, *J. Clin. Invest.*, 88, 691, 1991.
61. Wilson, R. E., Taylor, S. L., Atherton, G. T., Johnson, D., Waters, C. M., and Norton, J. D., Early response gene signalling cascades activated by ionising radiation in primary human B cells, *Oncogene*, 8, 3229, 1993.
62. Gassmann, A., Zur histologie der roentgenulcera, *Fortschr. Roentgenstr.*, 2, 199, 1899.
63. Rubin, P. and Casarett, G. W., *Clinical Radiation Pathology*, Vols. I and II, Saunders, Philadelphia, 1968.
64. Fajardo, L.F. and Berthrong, M., Vascular lesions following radiation, *Pathol. Annu.*, 23, 297,1988.
65. Jaffe, E. A., Nachman, R. L., Becker, C. G., and Minick, C. R., Culture of human endothelial cells derived from umbilical veins. Identification by morphologic and immunologic criteria, *J. Clin. Invest.*, 52, 2745, 1973.
66. Rubin, D. B., Drab, E. A., Ward, W. F., Smith, L. J., and Fowell, S. M., Enzymatic responses to radiation in cultured vascular endothelial and smooth muscle cells, *Radiat. Res.*, 99, 420, 1984.
67. Lin, P. S., Ho, K. C., Sung, S. J., and Gladding, J., Effect of tumor necrosis factor, heat, and radiation on the viability and microfilament organization in cultured endothelial cells, *Int. J. Hyperthermia*, 8, 667, 1992.
68. Stone, A. M., Rubin, D. B., and Drab, E. A., Increased transendothelial movement of albumen and sucrose is a direct effect of gamma irradiation *in vitro*, *Fed. Proc.*, 46, 418, 1987.
69. Woloschak, G. E., Shearin-Jones, R., and Chang-Liu, C. M., Effects of ionizing radiation on expression of genes encoding cytoskeletal elements: kinetics and dose effects, *Mol. Carcinogenesis*, 3, 374, 1990.
70. Sporn, I. A., Rubin, P., Marder, V. J., and Wagner, D. D., Irradiation induces release of von Willebrand protein from endothelial cells in culture, *Blood*, 64, 567, 1984.
71. Eldor, A., Fuks, Z., Matzner, Y., Witte, L. D., and Vlodavsky, I., Perturbation of endothelial functions by ionizing irradiation: effects on prostaglandins, chemoattractants, and mitogens, *Sem. Thromb. Hemost.*, 15, 215, 1989.
72. Sinzinger, H., Cromwell, M., and Firbas, W., Long-lasting depression of rabbit aortic prostacyclin formation by single-dose irradiation, *Radiat. Res.*, 97, 533, 1984.
73. Maier, J. A., Hla, T., and Maciag, T., Cyclooxygenase is an immediate-early response gene induced by interleukin-1 in endothelial cells, *J. Biol. Chem.*, 265, 10805, 1990.

74. Kujubu, D. A. and Herschman, H. R., Dexamethasone inhibits mitogen induction of the TIS10 prostaglandin synthase/cyclooxygenase gene, *J. Biol. Chem.*, 267, 7991, 1992.
75. Witte, L., Fuks, Z., Haimovitz-Friedman, A., Vlodavsky, I., Goodman, D. S., and Eldor, A., Effects of irradiation on the release of growth factors from cultured bovine, porcine, and human endothelial cells, *Cancer Res.*, 49, 5066, 1989.
76. Harlan, J. M., Thompson, P. J., Ross, R. R., and Bowen-Pope, D. F., Thrombin induces the release of platelet-derived growth factor-like molecules from cultured endothelial cells, *J. Cell Biol.*, 103, 1129, 1986.
77. Hajjar, K. A., Hajjar, D. P., Silverstein, R. L., and Nachman, R. L., Tumor necrosis factor mediated release of platelet derived growth factor from cultured endothelial cells, *J. Exp. Med.*, 166, 235, 1987.
78. Woloschak, G. E., Chang-Liu, C.-M., Shearin-Jones, P., and Jones, C. A., Modulation of gene expression in Syrian hamster embryo cells following ionizing radiation, *Cancer Res.*, 50, 339, 1990.
79. Chiang, C.-S. and McBride, W. H., Radiation enhances tumor necrosis factor α production by murine brain cells, *Brain Res.*, 566, 265, 1991.
80. Dunn, M. M., Drab, E. A., and Rubin, D. B., Effects of irradiation on endothelial cell-polymorphonuclear leukocyte interactions, *J. Appl. Physiol.*, 60, 1932, 1986.
81. Jaenke, R. S., Robbins, M. E., Bywaters, T., Whitehouse, E., Rezvani, M., and Hopewell, J. W., Capillary endothelium: target site of renal radiation injury, *Lab. Invest.*, 68, 396, 1993.
82. Wu, N. Z., Ross, B. A., Gulledge, C., Klitzman, B., Dodge, R., and Dewhirst, M. W., Differences in leukocyte-endothelium interactions between tumour and non-tumour bearing tissues in response to radiation, *Br. J. Cancer*, 69, 883, 1994.
83. Matzner, Y., Cohn, M., Hyam, E., Razin, E., Fuks, Z., Buchanan, M. R., Haas, T. A., Vlodavsky, I., and Eldor, A., Generation of lipid neutrophil chemoattractant by irradiated bovine aortic endothelial cells, *J. Immunol.*, 140, 2681, 1988.
84. Lewis, M. S., Whatley, R. E., Cain, P., McIntyre, T. M., Prescott, S. M., and Zimmerman, G. A., Hydrogen peroxide stimulates the synthesis of platelet activating factor by endothelium and induces endothelial cell dependent adhesion, *J. Clin. Invest.*, 82, 2045, 1988.
85. Patel, K. D., Zimmerman, G. A., Prescott, S. M., McEver, R. P., and McIntyre, T. M., Oxygen radicals induce human endothelial cells to express GMP-140 and bind neutrophils, *J. Cell Biol.*, 112, 749, 1991.

CHAPTER 9

Extracellular Matrix Regulates Endothelial Cell Morphology, Receptor Expression, and Response to Low-Dose Radiation

James M. Onoda, Seema S. Kantak, and Clement A. Diglio

CONTENTS

1-8493-4840-4/98/$0.00+$.50

ABSTRACT

The goals of the studies presented here were to characterize the modulatory effects of contrasting extracellular matrix peptides on low-dose radiation (≤1.0 Gy)-induced pulmonary microvascular endothelial cell (PMEC) retraction, F-actin reorganization, and apical surface expression of the αvβ3 integrin adhesion receptors (i.e., vitronectin receptors). We observed time- and dose-dependent radiation-induced injuries to PMEC monolayers characterized by loss of cell-cell contact (i.e., retraction) mediated by F-actin reorganization, as well as stimulated surface expression of vitronectin receptors. Radiation-induced loss of F-actin microfilament stress fibers reorganization was observed ≥2 hours postirradiation and positively correlated with PMEC retraction. Similarly, radiation-stimulated vitronectin receptor expression was also significantly increased 15 minutes postirradiation, but appeared to be independent of the extent of cellular retraction. F-actin reorganization, cellular retraction, and enhanced vitronectin receptor expression were reversible. Cells recovered and reformed cell-cell contacts as monolayers ≥24 hours postirradiation; concomitantly, the depolymerized microfilaments organized to their pre-irradiated state as microfilament stress fibers and the vitronectin receptors down-regulated to their pre-irradiated levels. The effects of extracellular matrix peptides (i.e., type IV collagen and type I collagen) were manifest as changes in the rate and extent of PMEC F-actin reorganization and cellular retraction, but not the degree of vitronectin receptor expression. PMEC monolayers on type IV collagen were more resistant to radiation-induced F-actin reorganization and demonstrated less cellular retraction in comparison to PMEC monolayers formed on Vitrogen (>98% type I collagen), which demonstrated increased sensitivity to radiation-induced F-actin reorganization and retraction. In contrast, the time course and extent of radiation-stimulated vitronectin receptor expression were similar for PMEC on both substrates, and thus its surface expression may be independent of substrate regulation. We suggest that the quality of cell-cell adhesion and cell-substrate adhesion modulate the morphological radiation sensitivity (i.e., cell retraction and F-actin organization) of endothelial monolayers to radiation, but not the apical expression of adhesion receptors.

I. INTRODUCTION

One of the principal dose-limiting factors in radiation therapy of thoracic and pulmonary cancers is the sensitivity of the microvascular endothelium. Radiation therapy of intrathoracic neoplasms often results in acute lung injury.[1,2] An essential element in the development of radiation-induced injury is the loss of endothelial cell integrity leading to increased vascular permeability.[3,4]

Friedman et al.[3] studied the acute effects of high-dose (6 to 30 Gy) ionizing radiation on endothelial monolayers *in vitro*, and have suggested that radiation-induced alterations in pulmonary endothelial structure and function may play an important role in the induction and progression of lung injury. The radiation dose levels and time points to 48 hours employed in these studies were sufficient for endothelial cell death, denudation, and exposure of the extracellular matrix, which are the seminal events leading to acute pathology at the tissue level. This loss of microvascular integrity is part of the acute inflammatory process and can initiate the postradiation lung injuries of edema and fibrosis.[5,6] The loss of endothelial cell-cell contact and the exposure of the underlying extracellular matrix results in a milieu for the leakage of fluids and proteins from the lumen of blood vessels to the interstitium (i.e., the pathological manifestation of edema). A number of *in vivo* studies have demonstrated gaps in the continuity of the endothelium as causal for high permeability leading to lung edema.[7,9]

Most investigators characterizing radiation-induced endothelial cell injuries *in vitro*, and the destruction of vascular and stromal integrity *in vivo*, have evaluated these structures following exposure to lethal levels of radiation (in order to mimic the lethal effects of radiation on the endothelium in the target volume during therapy).[7–11] Few studies have examined the acute effects of nonlethal effects of radiation on endothelial morphology and apical adhesion receptor expression, and no studies (at the time of this report) have evaluated the modulatory role of extracellular matrix peptides on the radiation sensitivity of endothelial cell monolayers.

We previously described low-dose, radiation-induced, cytoskeletal-mediated morphological changes in microvascular endothelium, and the effects of these changes in endothelial monolayer integrity.[12] We demonstrated that radiation-induced depolymerization of F-actin microfilament stress fibers appeared to be causal for the changes in PMEC morphology. Other investigators have studied the disruption and distribution of microfilaments following exposure to radiation, hormones, drugs, and toxic metabolites and have correlated the breakdown of microfilament bundles with the onset of cytoplasmic retraction[13-15] and increased permeability of the pulmonary microvasculature.[16] The radiation-induced morphological changes described in our previous studies were similar to those reported by these investigators. However, these studies did not describe the role for the interaction between endothelial cells and subendothelial matrix, nor did they seek to determine the possible modulatory effects of extracellular matrix on endothelial cell response to radiation.

Extracellular matrix components affect cellular behaviors that are regulated by changes in cytoskeletal organization (morphology, differentiation, and migration).[17–21] Microfilament organization and the association of focal contact proteins can be regulated by the selection of specific extracellular matrix components for use as the cellular adhesion substrate.[22,23] It has been suggested that the extracellular matrix regulates or modulates the overall organization of the cytoskeleton (via adhesion receptors on the cellular abluminal or basal surface).[24,25] At present, no role for extracellular matrix has been described for the apical or luminal surface expression of adhesion receptors.

Results of experiments conducted by Young et al.[20] and Yost et al.[26] indicated that extracellular matrix molecules profoundly affect the rates of post-injury endothelial cell motility, and coincident with these events are significant alterations in the endothelial cytoskeleton. Thus, the quantitative and qualitative aspects of cell-cell adhesion and cell-substrate adhesion act in concert to mediate the response of endothelial cell monolayers to external agonists (e.g., radiation), but whether apical surface expression of adhesion receptors plays a role in these responses is not known. The acknowledged evidence suggests that the $\alpha v\beta 3$ integrins regulate endothelial cell motility and migration.[24,27,28] They are intimately involved in mediating the adhesion between the endothelial cell abluminal/basal plasma membrane surface and the subcellular matrix. These integrins form part of complexes of great adhesive strength, the focal adhesion plaques.[29,30] The $\alpha v\beta 3$ integrins must be mobilized from the adhesion plaques to permit movement of the endothelial cell plasma membrane — as might occur during migration, cell division, or cellular retraction. Accordingly, we examined the mobilization of the $\alpha v\beta 3$ integrins as a presumptive marker of the rearrangement of the endothelial cell cytoskeleton necessary for retraction to occur.

In the present study, our goal was to determine a role for extracellular matrix collagens as modulatory factors regulating the sensitivity of endothelial cells to radiation. We characterized the effects of nonlethal levels of radiation ($\leq$1 Gy) on contact-inhibited murine pulmonary microvascular endothelial cells (PMEC) seeded on extracellular matrix peptides, (purified) type IV collagen or Vitrogen, which is a mixture of types I and III collagens, although the preparation is >98% type I collagen. Our data suggest that radiation-induced changes in endothelial cell-cell adhesion, morphology, and F-actin organization, but not apical surface expression of the vitronectin adhesion receptor, are regulated by matrix peptides.

II. MATERIALS AND METHODS

A. Cell Lines: Mouse Pulmonary Endothelial Cell Cultures

A pulmonary microvascular endothelial cell line (PMEC) was generously provided by Dr. Clement A. Diglio (Department of Pathology, School of Medicine, Wayne State University). A brief description of his isolation procedure is provided below.

Lungs aseptically removed from C57BL/6J mice were immersed in Ca^{2+} and Mg^{2+} free Hank's Balanced salt solution (HBSS). After this initial wash, fixation of the pleural lining of the lung was done by gently applying 70% ethanol over the lung surface which eliminated mesothelial cell contamination in the developing culture preparation. After rinsing in HBSS, using fine forceps small tissue pinches were obtained close (1 to 2 mm) to the lung periphery, avoiding large vessels. Tissue explants were then treated with 0.1% collagenase (Type II, Worthington, Malvern, PA) in Ca^{2+}, Mg^{2+}-free HBSS for 20 min at 37°C. After collagenase treatment, tissue was carefully removed and plated as explants into 100-mm tissue culture dishes containing Dulbecco's modified Eagle's medium (DMEM) with 20% fetal bovine serum (FBS). After 12 to 14 days in culture, the explants were removed and areas containing the endothelial cells with their typical cobblestone growth pattern were marked and allowed to grow an additional 7 days. Isolated colonies of endothelial cells devoid of spindle cells were selectively trypsinized using cloning penicylinders. Three separate endothelial cell colonies were isolated, grown to confluency, and recloned. One surviving clone was established and routinely subcultured at split ratios 1:2 every 7 to 10 days and was designated PMEC (pulmonary mouse endothelial cell). Endothelial cell characteristics were verified by growth behavior pattern, presence of factor VIII, and prostanoid production according to previously published procedures.[31]

For use in the studies presented here, PMEC were cultured in sterile DMEM + 10% FBS (passages between 14 and 20). They were maintained in a water-saturated incubator at 37°C, 95% room air, and 5% CO_2. The medium was changed every 3 days and cultures were passaged [2.5 m*M* EDTA (Sigma Chemical Co.,

St. Louis, MO) and 0.25% trypsin (Worthington Biochemical Corporation, Freehold, NJ)] when a contact-inhibited monolayer was achieved (approximately once a week) Cultures used for experimental purposes were seeded in T-75 culture flasks and used when confluent to seed cells into wells.

B. Endothelial Cell Retraction Studies

Materials. Substrates: Murine Type IV collagen (Collaborative Biomedical Products, Bedford, MA) and Vitrogen (95-99% type I collagen, with type III collagen, Collagen Corporation, Palo Alto, CA).

Methods. PMEC were grown in 16-mm (Falcon 3047, flat bottom), 24-well plates in alternate rows and incubated at 37°C in a CO_2 incubator. First, substrates (types IV and I collagen) were added to the wells (10 μg/100 μl/well) and left overnight in a sterile laminar flow hood for drying. PMEC were then plated at the appropriate cell density (150,000 cells/well) so that a confluent monolayer is formed in 24 hours. Contact-inhibited monolayers were irradiated (0.125, 0.25, 0.5, and 1.0 Gy) using a Picker X-ray unit (dose rate of 205 cGy/ min, 280 kV, 20 mA, 1.3 mm Cu HVL). Sham and experimental monolayers were subjected to the same conditions during the irradiations. In order to deliver different doses of radiation to the different rows of cells, shielding was used to regulate the dose administered to each sample.

After irradiation, samples were incubated (37°C, 5% CO_2) for the required time (0, 2, 4, 8, 24, and 48 hr). After the respective incubation period, culture medium was aspirated from the wells and 1 ml fixative (1% paraformaldehyde, 2% glutaraldehyde in HBSS, pH 7.4) was added. Cells were fixed for 10 minutes at room temperature. They were then washed once and stored under 0.75 ml HBSS. Samples were observed under the phase-contrast microscope and photographed at a magnification of 400× using Kodak Panchromatic T-Max 400 ASA film.

C. Immunofluorescent Staining of Cytoskeletal Elements

Coverslips (18 mm^2) were coated with 10 μg of a specific substrate and left overnight to dry in a sterile laminar flow hood. After drying, coverslips were transferred to 35-mm Cell Wells™ (6-well, flat-bottom plates, Corning, NY). Pulmonary microvascular endothelial cells were added to the substrate-coated coverslips containing DMEM + 10% FBS at a specific cell density (for each substrate) such that the PMEC formed contact-inhibited monolayers in 48 hours. PMEC were incubated at 37°C for 48 hours post initial seeding until contact-inhibited monolayers were achieved. Monolayers were irradiated with the appropriate doses of radiation (0.125, 0.25, 0.5, and 1.0 Gy). Samples were then terminated at various time intervals following radiation. Lysis squirting was used to expose the F-actin microfilaments.[24] Cells were rinsed with HEPES buffer (10 m*M* HEPES, 100 m*M* KCl, 5 m*M* $MgCl_2$, 3 m*M* EGTA, pH 7.0) and then incubated (10 min) in a 20% dilution of HEPES buffer to induce osmotic swelling and cell lysis visualized by visual inspection using phase-contrast microscopy. After cell lysis, samples were rinsed rigorously (to remove membrane fragments) and fixed with 4% paraformaldehyde in HEPES buffer (room temperature, 30 min). Samples were then washed 3× with HEPES buffer.

Labeling of F-actin (rhodamine phalloidin) was accomplished by a single-step staining using 75 μl of 1:50 rhodamine phalloidin. After labeling, samples were rinsed (5×, HEPES) and mounted in a medium containing 1:2 glycerol:HBSS supplemented with 0.1 g cytifluor (Amersham, Arlington Heights, IL) and 0.2 g Mowiol 4-88 (Calbiochem Corporation, La Jolla, CA 92037) to prevent rapid quenching under fluorescent excitement. Samples were analyzed at 600× magnification under oil using a Nikon Optiphot-2 Microscope. Micrographs were recorded on Kodak T-MAX 400 Panchromatic film.

D. Immunofluorescent Staining of Vitronectin Receptors

Source of Antibodies. Polyclonal antibody directed against αvβ3 (Gibco, Grand Island, NY) was used as the primary antibody for detection of the vitronectin receptor. Fluorescein isothiocyanate conjugated IgG fraction goat anti-rabbit (Cappel/Organon Teknika) was used for indirect immunofluorescence photomicroscopy. Goat whole serum (ICN Biomedical, Costa Mesa, CA) was used as a blocking serum for Fc receptors in the immunofluorescence studies.

Detection of Integrin Receptors Using Fluorescence Microscopy. Endothelial cells (PMEC) were grown in 35-mm petri dishes or on glass coverslips until semiconfluent (~50%). Cells were exposed, *in situ*, to various levels of radiation (0 to 1.0 Gy) and incubated for various lengths of time (0 to 48 hours) postirradiation (37°C). Following incubation, media was removed and samples fixed (3.7% paraformaldehyde, 5% sucrose in PBS, pH 7.4) for 10 minutes at room temperature. Samples were washed (5×) with phosphate buffered saline supplemented with 2 m*M* $MgCl_2$ and 2 m*M* $CaCl_2$ (PBS-Mg/Ca). After removing the last wash, 1:100 dilution of polyclonal antibody to αvβ3 was added to each sample. Antibodies (in a PBS + 4% solution of bovine serum albumin, 0.1 *M* glycine, 0.1% sodium azide, to block nonspecific binding) were incubated for 1 hour at 37°C. Primary antibody was removed and samples rinsed (3×, PBS) before addition of the secondary antibody (goat anti-mouse IgG-FITC) 100 μl/coverslip (10 μg protein/ml), incubated for 30 min, room temperature, under foil). After labeling, samples were rinsed (5×, PBS-Mg/Ca) and mounted in glycerol containing citifluor (Ted Pella, Redding, CA) to prevent photobleaching. Fluorescence micrographs were taken with a Nikon Orthoplan Microscope using Kodak T-MAX 400 Panchromatic film using the 100× oil immersion objective lens.

III. RESULTS

We previously described endothelial cell retraction induced by low, nonlethal levels of radiation.[12] These studies described a role for the reorganization of F-actin microfilaments in the regulation of endothelial cell retraction. These studies employed confluent monolayers of pulmonary microvascular endothelial cells (PMEC) grown in plastic wells or on glass coverslips. We demonstrated that low-dose radiation induced F-actin reorganization, leading to changes in PMEC morphology characterized by loss of cell-cell contract and cellular retraction.

It has become increasingly important to understand the role of cell-matrix interactions, typically mediated by cellular adhesion receptor interaction with substrate ligands (e.g., collagen). While it has been demonstrated that the extracellular matrix plays an active role in the regulation of cellular responses via the activation (or suppression) of intracellular signal transduction pathways leading to rapid as well as long-term cellular responses,[17,18,20,21] many possible regulatory roles for the extracellular matrix have not been examined. One issue not previously explored is whether the various extracellular matrices regulate the agonist-induced expression of integrin adhesion receptors on the apical (luminal) surface of endothelium inasmuch as this is the surface of the endothelial cell not in contact with the extracellular matrix.

Pulmonary microvascular endothelium cells (PMEC) were grown on purified layers of type IV collagen or Vitrogen (>98% type I collagen) for confluent monolayers (Figures 9.1 and 9.2). The plating density was adjusted such that the cells formed confluent monolayers ≤24 hours post plating. Monolayers on type IV collagen (Figure 9.1A) had the typical semicobblestone morphology usually associated with confluent microvascular endothelial cells.[12] The cells were in tight apposition with each other. The morphology of the PMEC monolayer plated on Vitrogen was atypical in appearance (Figure 9.2A). The cells appear more spread and less densely packed than cells on type IV collagen. Moreover, most of the cells on the monolayer were spindle-shaped with an elongated appearance.

A. Analysis of Endothelial Cell Retraction by Phase Contrast Microscopy

Irradiation of monolayers induced retraction, which resulted in a dramatic loss of close cell-cell contact and the formation of gaps between apparently healthy endothelial cells within PMEC monolayers on both type IV collagen and Vitrogen. In order to summarize our observations and yet effectively convey the observed substrate-, time-, and dose-dependent effects of radiation, we present our observations in summary as Tables 9.1 and 9.2. Each table is presented in a checkerboard fashion representing the effects of radiation on PMEC retraction at four different dose levels (0.125, 0.25, 0.50, and 1.0 Gy) at six different time points (0, 2, 4, 8, 24, 48 hours). We categorized the morphology of the monolayer into five general responses and have representative figures to visualize the categories.

Category	Definition
0	Confluent, contact-inhibited monolayer
+1	Intermittent gap formation between 10 and 20% of the cells of the monolayer
+2	Larger gaps were seen between approximately 50 and 60% of cells and a limited number of cells (≤10%) were completely retracted from adjacent cells.
+3	Large regions were occupied by cells that were completely separated from adjacent cells; >90% of all cells exhibited some degree of retraction
RM	Cells had reformed to monolayers

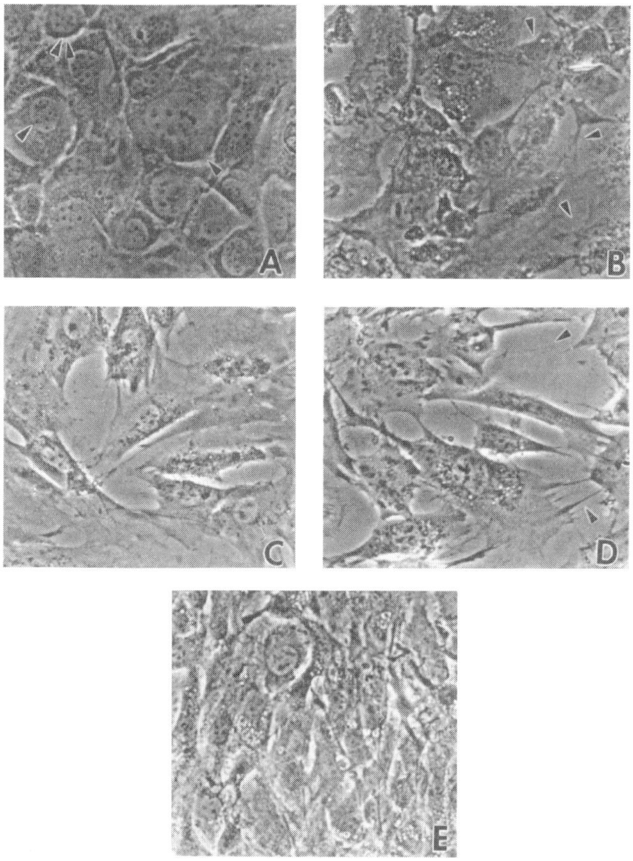

Figure 9.1. Effect of X-irradiation (1.0 Gy) on the morphology of pulmonary microvascular endothelial cell (PMEC) monolayers seeded on Type IV collagen. PMEC were seeded on Type IV collagen and ~24 hr later, confluent monolayers were irradiated and incubated for various time periods, fixed, and observed using phase contrast microscope. (A) Confluent, contact-inhibited PMEC monolayer prior to irradiation (also for sham-irradiated cells). Cells are in tight apposition to one another and nuclei are rounded. (B) At 2 hr postirradiation, there is formation of intermittent gaps in 10 to 20% of cells. (C) At 4 hr postirradiation, 50 to 60% of the cells have retracted. (D) Maximum retraction occurs at 8 hr with 90 to 95% cells retracted. Significant spiking of the cells is observed. (E) Reformation of monolayers 24 hr postirradiation, which is qualitatively identical to pre-irradiated monolayers.

Figures 9.1A-E and 9.2A-E demonstrate the time- and dose-dependent changes in cell morphology after PMEC were exposed to 0.25–1.0 Gy X-radiation, and represent the morphologies corresponding to categories 0, +1, +2, +3, and RM on collagen and Vitrogen, respectively. Individual cells of the monolayer on type IV collagen demonstrated homogeneous cytoplasm with rounded nucleus in the central region of each cell (Figure 9.1A, arrow). Cells within the monolayer were in tight apposition with each other and there was no evidence of gaps between cells. The

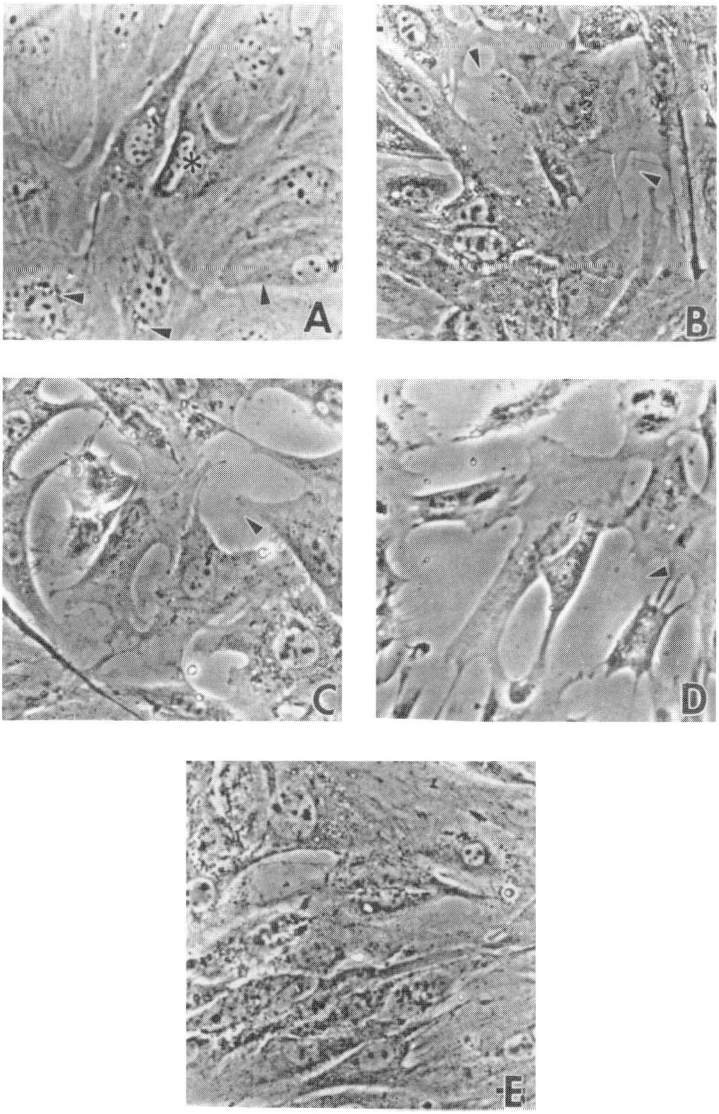

Figure 9.2. Effect of X-irradiation (0.125 Gy) on the morphology of pulmonary microvascular endothelial cell (PMEC) monolayers seeded on Vitrogen (>98% Type I collagen). PMEC were seeded on Vitrogen and ~24 hr later, confluent monolayers were irradiated and incubated for various time periods, fixed, and observed using phase contrast microscope. (A) Confluent, contact-inhibited PMEC monolayer prior to irradiation (also for sham-irradiated cells). Cells are more spindle-shaped with elongated nucleus than cells on Type IV collagen. (B) 2 hr postirradiation, intermittent gaps have formed in 10 to 20% of the cells. (C) By 4 hr postirradiation, 50 to 60% of the cells have retracted. (D) Maximum retraction occurs 8 hr postirradiation, with 90 to 95% of cells retracted. The margins of retracted cells are significantly ruffled. (E) Reformation of monolayers 24 hr postirradiation, which is qualitatively identical to pre-irradiated monolayers.

Table 9.1 Radiation-Induced PMEC Retraction on Type IV Collagen

Time	Dose 0.125 Gy	0.25 Gy	0.5 Gy	1 Gy
0 hours	0	0	0	0
2 hours	0	0	+1	+2
4 hours	+1	+1	+2	+3
8 hours	+2	+2	+3	+3
24 hours	RM	RM	RM	RM
48 hours	RM	RM	RM	RM

Note: 0: No retraction, typical contact-inhibited monolayer; +1: gaps between cells in some regions of the monolayer (10 to 20% of cells retracted); +2: large gaps between majority of the cells in the entire monolayer; loss of cell-cell contact between retracted cells; +3: very large gaps and many areas (~100%) where cells are completely separated from adjacent cells; RM: cells had reformed to monolayers.

Table 9.2 Radiation-Induced PMEC Retraction on Vitrogen

Time	Dose 0.125 Gy	0.25 Gy	0.5 Gy	1 Gy
0 hours	0	0	0	0
2 hours	+1	+1	+2	+2
4 hours	+2	+2	+3	+3
8 hours	+3	+3	+3	+3
24 hours	RM	RM	+2	+2
48 hours	RM	RM	RM	RM

Note: 0: No retraction, typical contact-inhibited monolayer; +1: gaps between cells in some regions of the monolayer (10 to 20% of cells retracted); +2: large gaps between majority of the cells in the entire monolayer; loss of cell-cell contact between retracted cells; +3: very large gaps and many areas (~100%) where cells are completely separated from adjacent cells; RM: cells had reformed to monolayers.

tight apposition between adjacent cells is evident by increased optical density at the cell-cell junctions (Figure 9.1, double arrows). The PMEC plated on Vitrogen often had elongated nuclei (Figure 9.2A, arrow) in contrast to the uniformly rounded nuclei of cells on collagen. Figures 9.1B and 9.2B demonstrate the initial intermittent gap formation that occurred following irradiation of the monolayers. This modest retraction and the gaps seen in the figures are representative of retraction that occurs in 10 to 20% of the cells of the monolayer and have been designated as +1 by the numerical scoring used in Tables 9.1 and 9.2. Figures 9.1C and 9.2C demonstrate the irradiated PMEC monolayer presented as +2. In addition to an increase in the overall number of individual cells that have retracted (relative to a +1), larger gaps between a majority (50 to 60%) of the cells in the monolayer are evident (Figures 9.1C and 9.2C, arrow). Figures 9.1D and 9.2D are representative of the responses categorized as +3. Very large gaps have developed between cells, with 90 to 95% of the cells in the monolayer retracted. Figures 9.1D and 9.2D (arrows) are representative of many samples in which large regions in the monolayer are occupied by cells that are completely separated from each other. Figures 9.1E and 9.2E are representative of the reversal of retraction and the reformation of the monolayer.

B. Analysis of Radiation-Induced Cell Retraction

A dose-dependent increase in the rate and extent of retraction was observed for monolayers on each substrate. However, there were distinct differences in the patterns of responses of PMEC, which we interpreted as substrate-induced differences in PMEC radiation sensitivity.

On Vitrogen, the onset of retraction was clearly visible at even the two lower dose levels (0.125 and 0.25 Gy) at the earliest time point (2 hours). In fact, initial cell retraction was often detectable even 1 hour postradiation (data not presented). In contrast, the onset of retraction of cells on type IV collagen (at these lower dose levels) was only detectable at ≥4 hours postirradiation.

On Vitrogen, maximum retraction (+3) was observed for all four dose levels 8 hours postradiation and as early as 4 hours for the two higher dose levels (0.5 and 1.0 Gy). In contrast, maximum cell retraction (+3) on type IV collagen was only observed at the two higher dose levels. Conversely, all retracted cells on collagen reformed monolayers within 24 hours, whereas cells on Vitrogen exposed to the two higher dose levels did not reform monolayers until 48 hours postirradiation.

C. Analysis of F-actin by Indirect Immunofluorescence Microscopy

We analyzed radiation-induced changes in PMEC monolayer cell morphology at the subcellular level by visualization of a cytoskeletal element, the F-actin microfilaments. Individual cells of PMEC monolayers on type IV collagen and Vitrogen were characterized by distinct arrays of F-actin microfilament stress fibers (Figures 9.3 and 9.4) which reorganized subsequent to radiation exposure.

D. Description of Terms and Tables

We summarize our observations as Tables 9.3 and 9.4, which are analogous to Tables 9.1 and 9.2. Each table is presented in a checkerboard fashion representing the effects of radiation on F-actin organization at the four dose levels and six time intervals. The microfilament organization postradiation was categorized into five general responses:

Category	Definition
0	Microfilament bundles spanning large areas of the cell body
+1	Snow-flake staining pattern of F-actin representative of partial depolymerization
+2	Disorganization (unbundling) of the entire central pool of F- actin stress fibers; the depolymerized F-actin appears as a meshwork in the cell cytoplasmic region; F-actin localized at focal contacts
+3	Complete depolymerization; the microfilament stress fibers totally disappeared; focal contacts were enriched with depolymerized F-actin
R	Reorganization of the F-actin back to stress fibers

Prior to radiation exposure, we observed substrate-dependent differences in the patterns of organization of the F-actin stress fibers (Figures 9.3 and 9.4). Because the patterns of F-actin organization varied with substrate, the '0' designation was subdivided to specify the organization observed on each substrate.

0_C	On collagen, the microfilaments spanning large areas are arranged as thick stress fibers oriented parallel to the cell axis
0_V	On Vitrogen, the microfilament bundles are shorter than those in PMEC on collagen and randomly distributed in a very dense pattern; the short microfilament bundles are arranged perpendicular to the circumference of the cell, terminating at cell borders with a few diffuse filaments observed spanning the ventral plasma membrane

E. Photographic Representation of F-actin Organization

Figures 9.3 and 9.4 are representative of the F-actin organization categorized as $0_C/0_V$, +1, +2, +3, and R. Figures 9.3A and 9.4A represent F-actin organization of sham-irradiated cells or irradiated cells at time "0" postirradiation on type IV collagen and Vitrogen, respectively. On collagen, cells demonstrated microfilament bundles spanning large areas of the cell body. The bundles were oriented in parallel as long thick (F-actin) stress fibers (Figure 9.3A, arrow). Some regions in the cell periphery showed prominent F-actin staining (Figure 9.3A, double arrows). On Vitrogen, cells exhibited short central microfilament bundles which were randomly distributed (Figure 9.4A, arrow). The filaments (due to the random distribution) were intertwined to form a dense mesh-like network of irregularly arranged fibers.

The organization of the F-actin filaments categorized as +1 are represented in Figures 9.3B and 9.4B. Some areas of the central bodies of cells demonstrate initial loss of F-actin fibers or meshwork, although the majority of cells remain virtually unchanged. The initiation of radiation induced changes in microfilament organization was characterized by a "snowflake" staining pattern within the cytoplasm of cells on both substrates (Figure 9.3B and 9.4B, arrow). On type IV collagen, a few residual

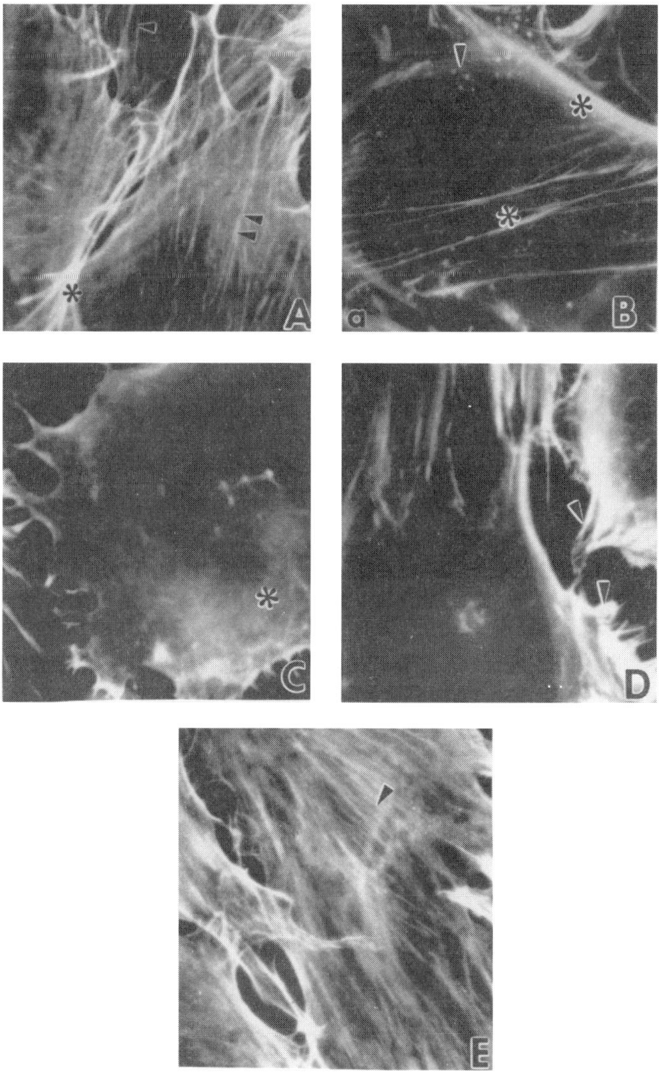

Figure 9.3. Effect of X irradiation (1.0 Gy) on pulmonary microvascular endothelial cell (PMEC) F-actin microfilament organization seeded on Type IV collagen. PMEC were seeded on Type IV collagen and confluent monolayers were irradiated and incubated for various time periods. The cells were permeabilized by lysis squirting, and rhodamine phalloidin was used to stain the microfilaments. (A) In confluent contact-inhibited PMEC monolayers, microfilament bundles were oriented as long thick stress fibers. (B) 2 hr postirradiation, there is partial loss of bundles of F-actin microfilament bundles. (C) 4 hr postirradiation, there is evidence of continued reorganization and the appearance of a prominent "snowflake" pattern of F-actin organization. (D) 8 hr postirradiation, complete reorganization of the central pool of F-actin microfilament bundles, although F-actin staining remains enriched in residual focal contacts. (E) Reorganization of F-actin to centrally located stress fibers occurs by 24 hr postirradiation, concomitant with the reformation of the cell morphology to contact inhibited monolayer.

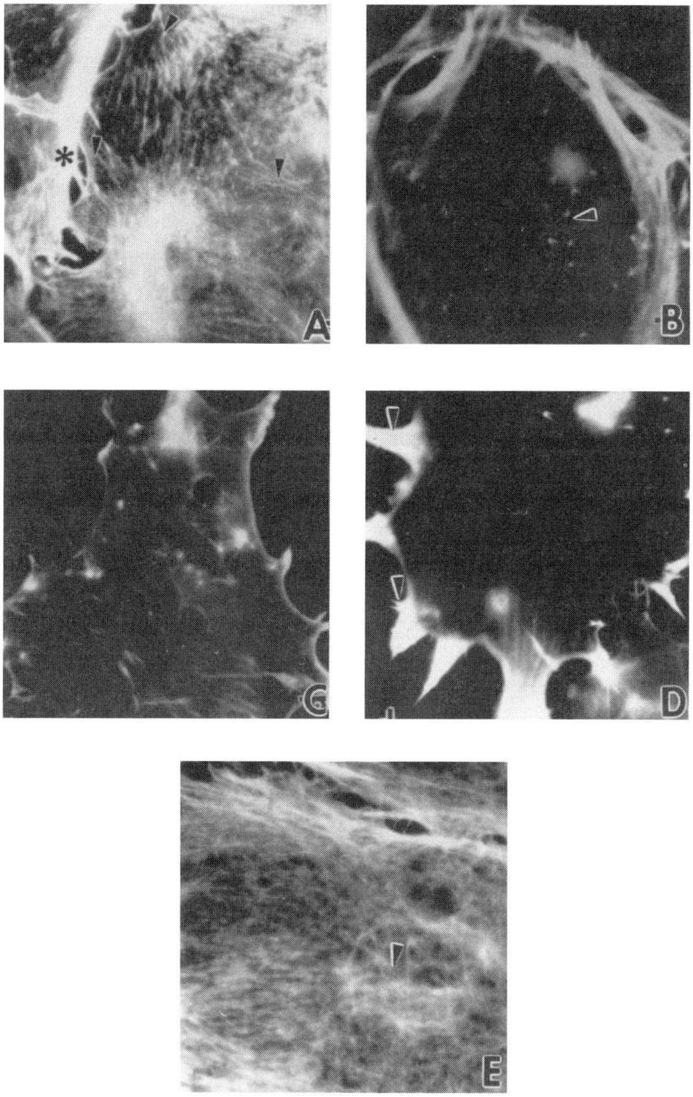

Figure 9.4. Effect of X-irradiation (0.125 Gy) on pulmonary microvascular endothelial cell (PMEC) F-actin microfilament organization plated on Vitrogen. PMEC were seeded on Vitrogen and confluent monolayers were irradiated and incubated for various time periods. The cells were permeabilized by lysis squirting, and rhodamine phalloidin was used to stain the microfilaments. (A) In sham-irradiated cells of confluent contact-inhibited PMEC monolayers, short central microfilament bundles are seen as randomly distributed. (B) 2 hr postirradiation, loss of F-actin microfilament bundles and partial reorganization of F-actin characterized by "snowflake" staining pattern. (C) 4 hr postirradiation, dense peripheral bands of depolymerized actin become visible. (D) 8 hr postirradiation, complete depolymerization of the central pool of actin microfilament bundles has occurred, although F-actin remains enriched in residual focal contacts. (E) 24 hr postirradiation, microfilaments in PMEC reorganized to short actin filaments.

Table 9.3 Radiation-Induced Microfilament Reorganization on Collagen Type IV

Time	Dose 0.125 Gy	0.25 Gy	0.5 Gy	1 Gy
0 hours	0	0	0	0
2 hours	0	0	+1	+1
4 hours	+1	+1	+2	+2
8 hours	+2	+2	+3	+3
24 hours	R	R	R	R
48 hours	R	R	R	R

Note: 0: Bundles of actin microfilaments (stress fibers) located centrally within the cell; +1: partial depolymerization of actin with a few stress fibers still visible in the cell; peripheral staining of F-actin; +2: disorganization (unbundling) of the entire central pool of F-actin with partial depolymerization; F-actin localization at focal contacts; +3: stress fibers disappear; F-actin enriched at focal contacts; R: reorganization of F-actin back to stress fibers.

Table 9.4 Radiation-Induced Microfilament Reorganization on Vitrogen

Time	Dose 0.125 Gy	0.25 Gy	0.5 Gy	1 Gy
0 hours	0	0	0	0
2 hours	+2	+2	+2	+2
4 hours	+2	+2	+3	+3
8 hours	+3	+3	+3	+3
24 hours	R	R	+1	+1
48 hours	R	R	R	R

Note: 0: Bundles of actin microfilaments (stress fibers) located centrally within the cell; +1: partial depolymerization of actin with a few stress fibers still visible in the cell; peripheral staining of F-actin; +2: disorganization (unbundling) of the entire central pool of F-actin with partial depolymerization; F-actin localization at focal contacts; +3: stress fibers disappear; F-actin enriched at focal contacts; R: reorganization of F-actin back to stress fibers.

stress fibers spanning the length of the cell and at the cell periphery parallel to the plasma membrane cell border were still observed (Figure 9.3B, star). Figures 9.3C and 9.4C demonstrate the actin distribution in the irradiated endothelial cells of the PMEC monolayer categorized as +2. The diffuse staining (Figure 9.3C, star) is representative of the continued depolymerization on type IV collagen. On Vitrogen, the depolymerization of the central meshwork of short stress fibers allowed for the visualization of a more scattered and diffuse F-actin staining pattern, with no evidence of a meshwork. Figures 9.3D and 9.4D demonstrate the radiation-induced depolymerization categorized as +3. On both substrates, radiation has induced unbundling and disorganization of the entire central pool of actin stress fibers (i.e., no bundles of stress fibers seen), although there is F-actin enriched at residual focal contacts at the cell periphery in PMEC on both substrates (Figure 9.3D and 9.4D, arrow).

The reorganization of the microfilaments and the reformation of stress fibers (condition "R") is depicted in the Figures 9.3E and 9.4E. During reorganization, the microfilaments start reappearing initially at the cell periphery and a few dispersed stress fiber bundles appear in the cytoplasmic region of the cell (data not shown). After reorganization, PMEC on type IV collagen demonstrated the formation of centrally located stress fibers (Figure 9.3E, arrow) whereas cells on Vitrogen demonstrated short actin filaments as a meshwork dispersed throughout the cell (Figure 9.4E, arrow).

F. Analysis of Dose-Dependent Microfilament Depolymerization of PMEC

Radiation induces dose- and time-dependent changes in F-actin organization in the cells of PMEC monolayers. Increasing levels of radiation increased the extent of F-actin depolymerization/time point. However, the extent of response (i.e., depolymerization) and localization of the F-actin microfilaments post irradiation were substrate dependent.

In PMEC plated on Vitrogen, actin depolymerization reached a level of +2 at 2 hours post radiation at all dose levels (Table 9.4). On type IV collagen, no actin depolymerization was observed at the lower dose levels (0.125 and 0.25 Gy) 2 hours post radiation; and at the two higher levels, only partial depolymerization (+1) was observed. On Vitrogen, complete depolymerization (i.e., level +3) was observed at all four doses within 8 hours of radiation. Moreover, at the two higher doses (i.e., 0.5 and 1.0 Gy), complete depolymerization occurred by 4 hours post radiation. In contrast, cells on type IV collagen never reached the +3 level of depolymerization after exposure to the two lower levels of radiation. However, by 24 hours post irradiation, there was reorganization of the actin (back) to stress fibers within cells on collagen, whereas cells on laminin exposed to the two higher dose levels did not reorganize actin (back) to stress fibers until 48 hours post radiation.

G. Expression of Vitronectin Receptors

Our previous publications demonstrate that radiation is a general agonist for endothelial cell activation and will initiate endothelial responses consonant with the inflammatory response; that is, radiation will stimulate endothelial cell retraction[17] and, in data presented here, the apical (or luminal) expression of the $\alpha v\beta 3$ integrin adhesion receptors. In contrast to the radiation-induced responses of cellular retraction and F-actin reorganization, significant substrate-dependent differences in vitronectin receptor expression were not observed. Radiation (1.0 Gy) significantly stimulated vitronectin ($\alpha v\beta 3$) receptor expression on PMEC of monolayers on either substrate. Increased receptor expression was observed at less than 15 minutes post irradiation and maximum expression was observed between 15 and 30 minutes post irradiation (Figure 9.5). Moreover, the time courses for the down-regulation of receptor expression were also similar, with significant decreases in expression approximately 2 hours post irradiation. We did observe what appeared to be substrate regulation of $\alpha v\beta 3$ integrin expression, but the regulation was not in terms of the time course or magnitude of expression, but rather the spatial arrangement of the expressed receptors. Increased vitronectin receptor expression was more pronounced in the perinuclear regions of cells on type IV collagen, whereas receptor expression was more pronounced in the cytoplasmic extensions of cells on Vitrogen. Moreover, on type IV collagen, a more prominent clustering of receptors was observed; whereas on Vitrogen, the receptors appear as a more diffuse and randomized distribution.

IV. DISCUSSION

The establishment and maintenance of cell-cell contact is a requisite for the macromolecular barrier functions of the vascular endothelium.[3,9,15,16] The integrity

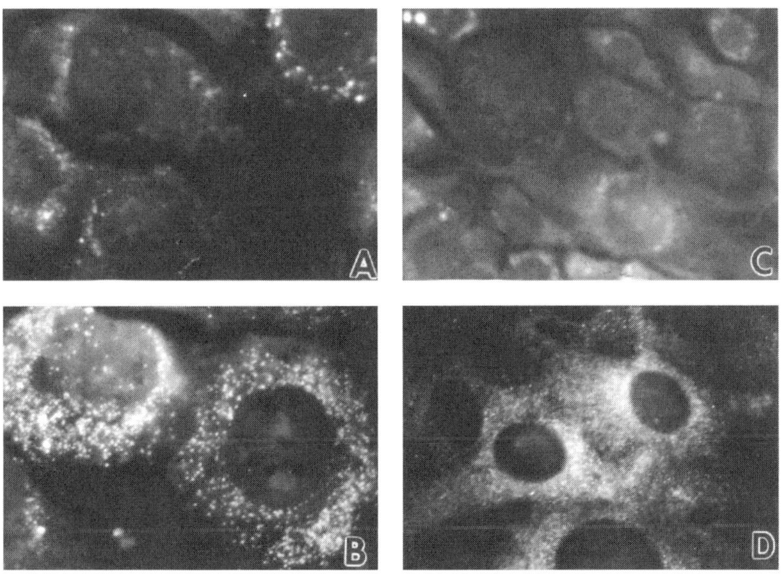

Figure 9.5. Effect of X-radiation (1.0 Gy) on PMEC apical expression of the integrin receptor $\alpha v\beta 3$ (vitronectin receptor). PMEC were seeded on Type IV collagen (A and B) or Vitrogen (C and D) and confluent monolayers were irradiated and incubated for 30 minutes at 37°C. On either substrate, constitutive $\alpha v\beta 3$ (vitronectin receptor) expression is minimal (A and C). Radiation stimulates receptor expression, which reaches a maximum $\leq$30 minutes postirradiation (B and D). The time course and dose dependency of vitronectin receptor expression were similar for PMEC plated on either substrate. The only detectable substrate-dependent differences were in the localization of the vitronectin receptors. Irradiated PMEC on Type IV collagen tended to express receptors in the perinuclear cytoplasmic areas, whereas PMEC on vitrogen tended to express vitronectin receptors in the cell periphery and cytoplasmic extensions.

of the endothelial monolayer is maintained by the tight cell-cell apposition within the monolayer. This requires that the overall morphology of the individual cells within the monolayer be maintained in a fully spread, flattened state. Thus, a change in the morphology of an individual cell (i.e., cellular retraction) can lead to loss of integrity of the monolayer. A disruption of the junctional apposition between the cells can be induced by various mediators, including hormones,[32,33] oxygen metabolites,[16,34–36] and ionizing radiation.[4] The disruption is caused by the reorganization of the cytoskeletal elements and the concomitant contraction of the cell membrane, which results in the formation of gaps between adjacent endothelial cells. A cascade of events follows this loss of endothelial integrity, leading to increased vascular permeability and the development of vascular pathologies.[5,36,37]

We previously reported[12] that low-dose radiation (0.125 to 1.0 Gy) produces significant changes in the morphology and microfilament organization of pulmonary microvascular endothelial cells. However, in that report, we used the traditional method of endothelial monolayer development by seeding cells on plastic wells or glass coverslips. Although the radiation-induced morphological changes described in our study were similar to those reported by others using hormones, drugs, and toxic metabolites, etc., they did not describe a role for the interaction between endothelial cells and subendothelial matrix peptides and the morphological changes that are substrate dependent. The data presented here suggest that extracellular matrix components affect cell morphology, which is regulated by cytoskeleton organization and the interaction of integrin adhesion receptors localized in the focal adhesion plaques on the basal (abluminal) cell surface. Our conclusion is in agreement with other investigators who have demonstrated a regulatory role for extracellular matrix peptides on cellular responses that are regulated by cytoskeletal organization; that is, morphology, differentiation, and migration. That is to say, the extracellular matrix has been proposed to regulate or modulate the overall organization of the cytoskeleton since it has been demonstrated that the cytoskeleton can interact with extracellular matrix via focal adhesion plaques on the basal cell surface.[29,30,38]

Extracellular matrix components are recognized by specific abluminal or basal cell surface receptors that span the membrane and interact with F-actin of the cytoskeleton.[38–40] It is the transmembrane-cytoskeletal linkage that is responsible for the mechanical integration of extracellular signals necessary for altering cell shape and motility.

It is known that upon adhesion to specific extracellular matrix proteins, the basal surface receptors on cells cluster into highly organized cellular structures (i.e., the focal adhesion plaques), which are composed of the basal surface integrin receptors (i.e., $\alpha v\beta 3$) and linker proteins (i.e., vinculin and talin) that connect the extracellular matrix-basal surface adhesion receptor complex to the actin microfilaments.[40] Tranqui et al.[30] have demonstrated that there are two types of adhesion plaques with distinct structural organization. On some substrates, cells attach and flatten using receptors that cluster to form central focal contacts, which are areas of weaker adhesion, but complete spreading is achieved by dynamic peripheral adhesion plaques that support microfilament stress fibers. They also demonstrated that ionophores caused disruption of peripheral adhesion plaques and disappearance of stress fibers, whereas central focal contacts were not disrupted. This description meshes well with our observations of cellular retraction. It may be that these peripheral adhesion plaques that support the microfilament stress fibers are significant regulators of cell adhesion and spreading of endothelial cells on Vitrogen and are more easily disrupted by radiation, whereas PMEC on type IV collagen have more central focal contacts that may be more resistant to the effects of radiation, resulting in greater resistance to radiation injury, but also demonstrate faster recovery when injury occurs.

The goals of the studies presented here were to determine whether subendothelial matrix peptides could modulate radiation-induced changes in PMEC monolayer morphology and cytoskeletal organization and to correlate the changes in morphology with changes in cytoskeletal organization. Furthermore, our goal was to determine whether luminal/apical expression of αvβ3 integrins was predictive, correlative, or causal for substrate-dependent differences in PMEC radiation sensitivity. Previous studies on radiation effects on endothelial cells have not fully considered the parameter of substrate modulation of the radiation-induced effects. To accomplish this goal, pulmonary microvascular endothelial cells were seeded to form monolayers on specific substrate proteins (type IV collagen or Vitrogen). After the development of cell-cell contact and the formation of focal adhesion plaques, the monolayers were irradiated. We observed that the morphological changes (i.e., cell retraction) induced by radiation were substrate dependent. We observed earlier onset of retraction of cells on Vitrogen than that of cells on type IV collagen. In addition, we observed a more pronounced retraction of cells on Vitrogen versus cells on type IV collagen. It is known that cells adhere to different extracellular matrix peptides with different affinities because different basal surface receptors are involved in the formation of specific linkages with the extracellular matrix peptides.[23,24,38] Also, morphological and cytoskeletal changes induced by growth factors (fibroblast growth factor and epidermal growth factor) and cytokines (TGF-β) have been demonstrated to be differentially regulated by the different types of extracellular matrix peptides.[33,39]

Some studies suggest that it is the formation of F-actin microfilament stress fibers that are essential to the development and maintenance of endothelial cell-cell junctional apposition and structural integrity necessary for barrier function. Logically, one might anticipate the breakdown or reorganization of F-actin stress fibers prior to cellular retraction. In fact, this was our observation. The centrally located stress fiber bundles disappear in response to radiation, and it is this radiation-induced reorganization in the microfilament organization that appears to be causal for the morphological change of retraction. The data presented here clearly demonstrate the correlation between radiation-induced actin reorganization and the dose and time effects observed during radiation-induced retraction. The time course for changes in microfilament organization (i.e., F-actin reorganization) correlate precisely with the morphological changes on the individual substrates. What we cannot state with certainty is whether radiation first induces cytoskeletal reorganization, which causes changes in cell morphology (retraction), or whether radiation first induces loss of receptor-matrix interaction, which then leads to cytoskeletal reorganization and retraction. However, since radiation-stimulated αvβ3 expression clearly precedes the onset of detectable F-actin reorganization and cellular retraction, we believe it is not unreasonable to suggest that resolution of the abluminal (basal) focal adhesion plaques, which would free-up αvβ3 integrins for transcytosis and cycling to the luminal (apical) PMEC surface precedes F-actin reorganization. Thus, we anticipated that the expression of αvβ3 integrins would be slower and less numerous on PMEC grown on type IV collagen compared to the rate and magnitude of expression of αvβ3 integrins expressed on PMEC grown on Vitrogen. The presumption being that since PMEC on type IV collagen retract slower and to a lesser extent than PMEC on Vitrogen (for any one given dose and time point), the resolution of focal adhesion plaques on PMEC on type IV collagen would be slower and less numerous than PMEC on Vitrogen, and our presumption being that PMEC on type IV collagen would simply have fewer αvβ3 integrins (released from focal adhesions) to cycle via transcytosis to the apical/luminal surface. Unfortunately, we did not observe substrate-dependent regulation of the rate (time course) or the absolute magnitude of αvβ3 expression on the apical surface. The results of our αvβ3 receptor analyses were equivocal. That is to say, the rate (time-course) and magnitude of αvβ3 expression were similar for PMEC of type IV collagen and Vitrogen. Interestingly, we did observe a substrate-dependent distribution of αvβ3 receptors. We speculate these differences in αvβ3 receptor expression were a consequence of substrate-dependent differences in the location and distribution of focal adhesion plaques.

Retraction of viable endothelial cells is not unidirectional. We observed that radiation-induced endothelial cell retraction and F-actin microfilament reorganization were reversible. The morphologic data demonstrated that on type IV collagen, the retracted endothelial cells respread and were able to reform contact-inhibited monolayers 24 hours post radiation. In parallel, F-actin microfilament reorganization also occurs within 24 hours post radiation. PMEC on Vitrogen exposed to the higher doses (0.5 and 1.0 Gy) fail to achieve contact-inhibited monolayers until 48 hours postirradiation, and, in parallel, the microfilaments failed to completely reorganize until 48 hours postirradiation. These observations demonstrating the reversibility of radiation-induced retraction were significant because the reformation suggested (to us) that monolayer morphology could be restored and microfilament F-actin reorganized. The down-regulation and redistribution of αvβ3 receptors appear to be unique and independent of the reformation of monolayers and the formation of F-actin stress fibers. The loss of surface expression appears to be independent of substrate regulation; αvβ3 receptors are down-regulated and redistributed from the apical surface of PMEC on either substrate by 2 hours post irradiation.

Present studies are directed toward determining the role of specific abluminal endothelial cell substrate adhesion receptors and their role in mediating the radiation-induced changes in cell morphology and cytoskeletal organization. It is now known that the information conveyed from the extracellular matrix to the cytoskeleton involves matrix components and specific abluminal receptors that span the plasma membrane and interact with the actin cytoskeleton and other cytoskeletal proteins. Thus, it would be logical to identify the abluminal receptors and their interaction with the cytoskeleton, and to identify the biochemical mechanism that regulates radiation-induced microfilament depolymerization and redistribution of adhesion plaques and the role of receptor-substrate interactions in modulating cytoskeletal reorganization.

Since the preparation of this manuscript, we have also demonstrated that irradiation upregulates the apical membrane expression of $\alpha_5\beta_1$ in bovine aortic endothelial cells.[41]

REFERENCES

1. Mah, K., Van Dyke, J., Keane, T., and Poon, P. Y., Acute radiation-induced pulmonary damage: a clinical study on the response of fractionated radiation therapy, *Int. J. Radiat. Oncol. Biol. Phys.*, 13 (2), 179, 1987.
2. Awad, H. K., *Radiation Oncology: Radiobiological and Physiological Perspectives.* Kluwer Academic Publishers, 1990, 357.
3. Friedman, M., Ryan, U. S., Davenport, W. C., Chaney, E. L., Strickland, D. L., and Kwock, L., Reversible alterations in cultured pulmonary arterial endothelial cell monolayer morphology and albumin permeability induced by ionizing radiation, *J. Cell. Physiol.*, 129, 237, 1986.
4. Penney, D. P. and Rubin, P., Specific early ultrastructural changes in lung following irradiation, *Int. J. Radiat. Oncol. Biol. Phys.*, 2,1123, 1977.
5. Travis, E. L., Down, J. D., Holmes, S. J., and Hobson, B., Radiation pneumonitis and fibrosis in mouse lung assayed by respiratory frequency and histology, *Radiat. Res.*, 84, 133, 1980.
6. Vergara, J. A., Raymond, U., and Thet, L. A., Changes in lung morphology and cell number in radiation pneumonitis and fibrosis: a quantitative ultrastructural study, *Int. J. Radiat. Oncol. Biol. Phys.*, 13, 723, 1987.
7. Gross, N. J., Pulmonary effects of radiation therapy, *Ann. Intern. Med.*, 86, 81, 1977.
8. Staub, N. C., Pulmonary edema due to increased microvascular permeability to fluid and protein, *Circ. Res.*, 43, 143, 1978.
9. Hurley, V. J., Types of pulmonary microvascular injury, *Ann. N.Y. Acad. Sci.*, 384, 269, 1982.
10. Shankar Giri, P. G., Kimler, B. F., Giri, U. P., Cox, G. G., and Reddy, E. K., Comparison of single fractionated and hyperfractionated irradiation on the development of normal tissue damage in rat lung, *Int. J. Radiat. Oncol. Biol. Phys.*, 11, 527, 1984.
11. Fulkerson, W. J., McLendon, R. E., and Prosnitz, L. R., Adult respiratory distress syndrome after limited thoracic radiotherapy, *Cancer*, 57, 1841, 1986.
12. Kantak, S. S., Diglio, C. A., and Onoda, J. M., Low dose radiation-induced endothelial cell retraction, *Int. J. Radiat. Biol.*, 64, 319, 1993.
13. Aubin, J. E., Alders, E., and Heersche, J. N. M., A primary role for microfilaments, but not microtubules, in hormone-induced cytoplasmic retraction, *Exp. Cell. Res.*, 143, 439, 1983.
14. Wong, M. K. K. and Gotlieb, A. I., Endothelial cell monolayer integrity. I. Characterization of the dense peripheral band of microfilaments, *Arteriosclerosis*, 6, 212, 1986.
15. Shasby, D. M., Shasby, S. S., Sullivan, J. M., and Peach, M. J., Role of endothelial cytoskeleton in the control of endothelial permeability, *Circ. Res.*, 51, 657, 1982.
16. Shasby, D. M., Lind, S. E., Shasby, S. S., Goldsmith, J. C., and Hunninghake, G. W., Reversible oxidant-induced increases in albumin transfer across cultured endothelium: alterations in cell shape and calcium homeostasis, *Blood*, 3, 605, 1985.
17. Gospodarowicz, D. and Ill, C. R., Extracellular matrix and control of proliferation of endothelial cells, *J. Clin. Invest.*, 65, 1351, 1980.

18. Madri, J. A. and Williams, S. K., Capillary endothelial cell cultures: phenotypic modulation by matrix components, *J. Cell Biol.*, 97, 153, 1983.
19. Basson, C. T., Knowles, W. J., Bell, L., Albelda, S. M., Castranova, V., Liotta, L. A., and Madri, J. A., Spatiotemporal segregation of endothelial cell integrin and nonintegrin extracellular matrix-binding proteins during adhesion events, *J. Cell Biol.*, 110, 789, 1990.
20. Young, W. C. and Herman, I. M., Extracellular matrix modulation of endothelial cell shape and motility following injury *in vitro*, *J. Cell Sci.*, 73, 19, 1985.
21. Kubota, Y., Kleinman, H. K., Martin, G. R., and Lawley, T. J., Role of laminin and basement membrane in the morphological differentiation of human endothelial cells into capillary-like structures, *J. Cell Biol.*, 107, 1589, 1988.
22. Semich, R. and Robenek, H., Organization of the cytoskeleton and the focal contacts of bovine aortic endothelial cells cultured on Type I and Type III collagen, *J. Histochem. Cytochem.*, 38 (1), 59, 1990.
23. Burridge, K., Substrate adhesions in normal and transformed fibroblasts: organization and regulation of cytoskeletal, membrane and extracellular matrix components at focal contacts, *Cancer Rev.*, 4, 18, 1986.
24. Kramer, R. H., Chen, Y. F., and Clyman, R., Human microvascular endothelial cells use β_1 and β_3 integrin receptor complexes to attach to laminin, *J. Cell Biol.*, 111, 1233, 1990.
25. Otey, C. A. and Burridge, K., Patterning of the membrane cytoskeleton by the extracellular matrix, *Semin. Cell Biol.*, 1, 391, 1990.
26. Yost, J. C. and Herman, I. M., Substratum-induced stress fiber assembly in vascular endothelial cells during spreading *in vitro*, *J. Cell Sci.*, 95, 507, 1990
27. Akiyama, S. K., Nagata, K., and Yamada, K. M., Cell surface receptors for extracellular matrix components, *Biochem. Biophys. Acta*, 1031, 91, 1990.
28. Zetter, B. R. and Brightman, S. E., Cell motility and the extracellular matrix, *Curr. Opin. Cell. Biol.*, 2, 850, 1990.
29. Gingell, D. and Owens, N., How do cells sense and respond to adhesive contacts? Diffusion-trapping of laterally mobile membrane proteins at maturing adhesions may initiate signals leading to local cytoskeletal assembly response and lamella formation, *J. Cell Sci.*, 101, 255, 1992.
30. Tranqui, L., Usson, Y., Marie, C., and Block, M. R., Adhesion of CHO cells to fibronectin is mediated by functionally and structurally distinct adhesion plaques, *J. Cell. Sci.*, 106, 377, 1993.
31. Diglio, C. A., Grammas, P., Giacomelli, F., and Wiener, J., Primary culture of rat cerebral microvascular endothelial cells. Isolation, growth and characterization, *Lab Invest.*, 46, 554, 1982.
32. Majno, G., Shea, S. J., and Leventhal, M., Endothelial contraction induced by histamine-type mediators, *J. Cell Biol.*, 42,647, 1969.
33. Majno, G., and Palade, G. E., Studies on inflammation. The effect of histamine and serotonin on vascular permeability: an electron microscopic study, *J. Biophys. Biochem. Cytol.*, 11, 571, 1961.
34. Sacks, R., Maldow, C. R., Craddock, P. R., Bowers, T. K., and Jacob, H. S., Oxygen radicals mediate endothelial cell damage by complement-stimulated granulocytes, *J. Clin. Invest.*, 61, 1161, 1978.
35. Tate, R. M., Benthuysen, K. M., Shasby, D. M., McMurty, I. F., and Repine, J. E., Oxygen-radical mediated permeability edema and vasoconstriction in isolated perfused rabbit lungs, *Am. Rev. Respr. Dis.*, 126, 802, 1982.

36. Johnson, K. J., Fantone, J. C., Kaplan, J., and Ward, P. A., *In vivo* damage of rat lungs by oxygen metabolites, *J. Clin. Invest.*, 67, 983, 1981.
37. Karvonen, R. L., Fernandez-Madrid, F., Maughan, R. L., Palmer, K. C., and Fernandez- Madrid, I., An animal model of pulmonary radiation fibrosis with biochemical, physiologic, immunologic, and morphologic observations, *Radiat. Res.*, 111, 68, 1987.
38. Fath, K. R., Edgell, C. S., and Burridge, K., The distribution of distinct integrins in focal contacts is determined by substratum composition, *J. Cell Sci.*, 92, 67, 1989.
39. Madri, J. A., Pratt, B. M., and Tucker, A. M., Phenotypic modulation of endothelial cells by transforming growth factor-β depends upon the composition and organization of the basement membrane, *J. Cell Biol.*, 106, 1375, 1988.
40. Turner, C. E. and Burridge, K., Transmembrane cellular assemblies in cell-extracellular matrix interactions, *Curr. Opin. Cell. Biol.*, 3, 849, 1991.
41. Mooteri, S.N., Podolski, J.L., Drab, E.A., Saclarides, T.J., Onoda, J.M., Kantak, S.S., and Rubin, D.B., WR-1065 and radioprotection of vascular endothelial cells, II. Morphology, *Radiat. Res.*, 145, 217, 1996.

CHAPTER 10

Endothelial-Oriented Strategies to Spare Normal Tissues

William F. Ward, Agostino Molteni, and Chung-hsin Ts'ao

CONTENTS

I. INTRODUCTION

Endothelial-oriented strategies for therapeutic gain in radiation oncology attempt either to increase injury in tumor vasculature or to spare injury in the vasculature of normal tissues, or both. Of the two, attacking tumor vasculature has perhaps received more attention in the recent past. These strategies include among others manipulation of tumor blood flow and pO_2, hyperthermia, photodynamic therapy, immunologic targeting of tumor (or dividing) endothelium, natural and synthetic vascular chemotoxins such as tumor necrosis factor and flavone acetic acid, and

1-8493-4840-4/98/$0.00+$.50

inhibition of angiogenesis. Many of these tumor-based approaches are the subject of individual chapters in this text.

The present chapter will focus on strategies to spare the endothelium of normal tissues from radiation injury. Endothelial cells are one of very few cell types present in every organ (except lens and mature cartilage), and thus are included in every radiotherapy port. The vascular endothelium has important barrier and metabolic functions, both of which are known to be perturbed by ionizing radiation. For decades, a central dogma in radiotherapy held that chronic or late normal tissue injury resulted largely from the gradual compromise of microvascular function.[1] This hypothesis has been challenged, however, in part because the evidence implicating vascular injury in the pathogenesis of late effects was less than rigorous.[2] Much of the original data supporting the vascular basis of late effects was descriptive rather than mechanistic, and invited criticism. As an alternative to the vascular hypothesis, parenchymal cell kill was proposed as the prime cause of both acute and late radiation-induced normal tissue injury.[2] The present text illustrates that the truth may lie between or in a combination of these two competing hypotheses. Functional as well as structural damage does occur in irradiated vasculature, and this damage appears to be important rather than trivial in the development of normal tissue complications. On the other hand, while vascular injury may be a component of, it is not likely to be sufficient for late effects. It is now clear that radiation pathophysiology (as is the case with all insults) involves a complex interrelationship among autocrine, paracrine, and even endocrine signals, superimposed on and influencing the direct radiation cytotoxicity. In this case, attributing tissue response to one or even a few target cells is likely to be simplistic. Similarly, it is important to note Michalowski's observation that "vascular" and "endothelial" are not necessarily synonymous.[3] Vascular radiation responses can include both direct endothelial reactions as well as paracrine interactions among the endothelium, other vascular cell types (pericytes, smooth muscle cells, and fibroblasts), and circulating blood cells, platelets, and hormones. The perivascular and interstitial microenvironments also influence vascular responses to radiation.

The search for modifiers of vascular/endothelial radioresponsiveness is enduring, driven perhaps by the belief that vascular damage contributes to late effects, and by the desire to increase the radiation tolerance of dose-limiting normal tissues. If vascular damage is in fact important, then sparing that damage should reduce the incidence or severity of normal tissue complications. The mechanism of action of the modifier, in turn, should implicate a step in the pathologic process. Effective modifiers, therefore, satisfy both a clinical imperative and elucidate pathways of damage. Interest in modifiers of radiation pathophysiology in general, and of vascular radiopathology in particular, appears to be accelerating. This may reflect in part a disenchantment with the concept that parenchymal cell kill is the only important event in radiation pathology.[2] This hypothesis is a strong disincentive to the search for new approaches in the prophylaxis or treatment of radiation injury. Nevertheless, the idea that radiation pathology is not special, that it may benefit from medical management as does pathology of other etiology, is gaining popularity. Whether this notion can deliver on its promise remains to be determined. Progress to date has been modest at best; and the research effort with any single modifier has been sporadic rather than systematic.

Prophylaxis can assume two forms: (1) a reduction in the initial level of vascular damage using a radioprotector such as WR-2721 present at the time of radiation, or (2) the postexposure administration of modifying agents prior to the development of normal tissue complications. Alternatively, radiation reactions can be treated pharmacologically once they appear. This chapter will describe all three approaches. A review chapter such as this can adopt any of several organizational schema. Modifiers can be grouped by their presumed mechanism of action; organs can be considered individually; or the several vascular/endothelial functions can be evaluated in turn. The latter approach has been selected as perhaps the most coherent biologically. The chapter will describe attempts to ameliorate radiation-induced vascular histopathology *in vivo* and endothelial cytotoxicity *in vitro*, beginning with those studies that primarily evaluated structure or survival rather than function. Then, the vascular/endothelial functions of perfusion, permeability, coagulation, fibrinolysis, eicosanoid metabolism, and other enzyme activities will be reviewed. Both effective and ineffective modifiers have been included, in the belief that lessons can be learned from failure as well as success. This chapter is not intended as a comprehensive review of all modifiers of radiation pathophysiology, but will focus on those whose target, real or presumed, is the vascular endothelial cell. References in the tables are listed in chronological order.

II. STRATEGIES TO SPARE ENDOTHELIUM OF NORMAL TISSUES

A. Vascular Histopathology and Endothelial Cytotoxicity

Vascular lesions figure prominently in the histopathology of radiation injury in many normal tissues, including heart,[4] lung,[5,6] liver,[7] kidney,[8] and skin.[9] These reactions have been summarized by Fajardo and colleagues,[10] and are discussed in detail elsewhere in this text. Morphologic data such as these constituted the primary basis of the widely held hypothesis that late effects of radiation were caused by vascular damage.[1] Some endothelial reactions such as focal detachment from the basement membrane (blebbing) occur almost immediately after irradiation (Figure 10.1A), and are essentially permanent after high doses (Figure 10.1B).

Effective modifiers of radiation-induced vascular histopathology *in vivo* and endothelial cell cytotoxicity *in vitro* are summarized in Table 10.1A. Agents effective *in vivo* include trihydroxyethyl-rutosidum, a synthetic derivative of the flavonoid rutin,[11] the collagen antagonist D-penicillamine,[12] the steroidal anti-inflammatory agent dexamethasone,[13,14] and the platelet activating factor (PAF) antagonist BN52021.[15] Basic fibroblast growth factor (bFGF) also prevents radiation-induced apoptosis in the pulmonary endothelium of mice,[16] although this finding was not confirmed in a similar study.[17]

The *in vivo* work with bFGF grew out of very interesting *in vitro* studies demonstrating that bFGF protected bovine aortic endothelial cells from the lethal effects

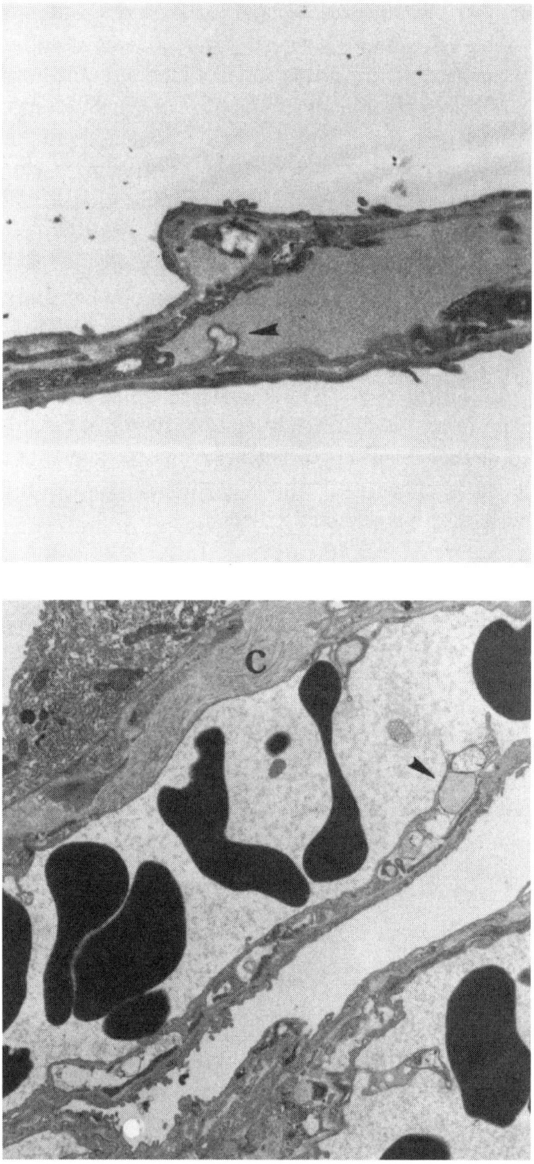

Figure 10.1. (A) Electron micrograph of rat lung 24 hr after exposure to a single dose of 25-Gy X-rays. Note focal detachment of endothelium from basement membrane (arrow). Uranyl acetate and lead citrate; original magnification 4000×. (B) Electron micrograph of rat lung 6 months after exposure to a single dose of 25-Gy X-rays. Note focal detachment of endothelium from basement membrane (arrow) and interstitial collagen (C). Uranyl acetate and lead citrate; original magnification 3500×.

of radiation, using both colony formation[18] and apoptosis[19] as endpoints. The ability of bFGF to ameliorate radiation-induced endothelial apoptosis appears to be mediated by protein kinase C (PKC), since PKC activation by the phorbol ester 12-O-tetradecanoylphorbol-13-acetate (TPA) mimicks bFGF.[19] Of interest, radiation itself stimulates endothelial cells to synthesize and release bFGF, resulting in an autocrine repair process that improves survival.[18] Neutralizing antibodies to bFGF block this autorepair pathway.[18] Endotoxin-mediated enhancement of radiation-induced apoptosis in endothelial cells also can be inhibited by interleukin-10 (IL-10) and by neutralizing antibodies to tumor necrosis factor-alpha (TNFα).[20]

Other indices of nonreproductive toxicity in irradiated endothelial cells *in vitro*, such as detachment from the monolayer and release of lactate dehydrogenase (LDH) and ^{51}Cr from prelabeled cells, are spared by cyclic adenosine monophosphate (cAMP)[21,22] and retinol.[23] It is curious that some agents such as dexamethasone,[24] the synthetic prostaglandin misoprostol,[25] and the angiotensin converting enzyme (ACE) inhibitor captopril,[22] all of which spare radiation reactions *in vivo*, are ineffective as modifiers of endothelial radiation cytotoxicity *in vitro* (Table 10.1B). An unfortunate conclusion from these data is that *in vitro* screening assays for potential radiation modifiers may yield false negatives.

Table 10.1A Effective Modifiers of Radiation-Induced Vascular Histopathology and Endothelial Cytotoxicity

Modifier	Response	Organ	Species	Reference
Trihydroxyethyl-rutosidum	Histopathology *in vivo*	Ear chamber	Rabbit	Klemm, 1967[11]
Penicillamine	EM changes *in vivo*	Lung	Rat	Port & Ward, 1982[12]
SOD	Histopathology *in vivo*	Lung	Rat	Malaker & Das, 1988[109]
cAMP	LDH & ^{51}Cr release *in vitro*	Aorta	Bovine	Ward et al., 1991[21]
bFGF	Colony formation *in vitro*	Aorta	Bovine	Haimovitz-Friedman et al., 1991[18]
Dexamethasone	Histopathology & function *in vivo*	Kidney	Rat	Geraci et al., 1993[13] Geraci et al,. 1995[14]
bFGF	Apoptosis *in vitro*	Aorta	Bovine	Haimovitz-Friedman et al., 1994[19] Fuks et al., 1994[16]
TPA	Apoptosis *in vitro*	Aorta	Bovine	Haimovitz-Friedman et al., 1994[19,117]
bFGF	Apoptosis & survival *in vivo*	Lung	Mouse	Fuks et al., 1994[16]
Retinol	Detachment *in vitro*	Adrenal	Bovine	Bump et al., 1994[23]
Antibodies to virus hemagglutinin	Leukocyte adhesion *in vitro*	Umbilical vein	Human	Colden-Stanfield et al., 1994[81]
IL-10 and antibodies to TNF-α	LPS-mediated apoptosis *in vitro*	Microvascular	Human	Eissner et al., 1995[20]
PAF antagonist (BN52021)	Leukocyte adhesion *in vivo*	Skin chamber	Rat	Kimura et al., 1995[15]

Table 10.1B Ineffective Modifiers of Radiation-Induced Vascular Histopathology and Endothelial Cytotoxicity

Modifier	Response	Organ	Species	Reference
Dexamethasone	Detachment & LDH release *in vitro*	Aorta	Bovine	Lam et al., 1985[24]
Vitamin C	Detachment *in vitro*	Aorta	Bovine	Eldor et al., 1987[97]
Eicosanoids	^{3}H-thymidine incorporation *in vitro*	Aorta	Bovine	Rubin et al., 1991[25]
Captopril	Detachment, LDH & ^{51}Cr release *in vitro*	Aorta	Bovine	Ward et al., 1993[22]
bFGF	Apoptosis, survival & histopathology *in vivo*	Lung	Mouse	Tee & Travis, 1995[17]

B. Defects in Perfusion and Hemodynamics

Hypoperfusion is one of the most universal responses of normal tissues to radiation, although it may take weeks to months to develop. Many agents have been reported to ameliorate radiation-induced defects in vascular perfusion and hemodynamics (Table 10.2A). These include the radioprotector WR-2721,[26] the vasodilator pentoxifylline,[27–29] dexamethasone,[14,30,31] the ACE inhibitors captopril and enalapril,[32–38] and essential fatty acids such as γ-linoleic acid.[39–41] Ischemia resulting in ataxia in rats irradiated to the spinal cord exhibits elements of reperfusion injury in that it responds positively to a low iron diet and the iron chelator desferrioxamine.[42] There is no indication of a tissue- or species-specificity to these modifiers of radiation-induced perfusion defects (Table 10.2A). Neither does a single unifying mechanism of action emerge from this diverse group of modifiers. ACE inhibitors may act by blocking production of the vasoconstrictor angiotensin II by the vascular endothelium. Supporting this hypothesis is the ability of angiotensin II receptor antagonists to prevent radiation nephropathy in rats.[43] The effectiveness of dexamethasone, pentoxifylline, and essential fatty acids in modifying radiation-induced perfusion defects might implicate inflammatory reactions or hyperproduction of arachidonate products in this response. However, both the thromboxane synthetase inhibitor CGS-13080[44] and the thromboxane receptor blocker daltroban[45] are ineffective in reducing radiation injury in normal tissues.

Some agents such as penicillamine,[46] captopril,[32–37] and essential fatty acids[39–41] are effective under some circumstances, but not under others.[30,32,41] Whether these apparently contradictory findings are due to differences in drug dosage or timing, radiation regimen, or observation time is not clear at present.

In an interesting reversal of the normal clinical strategy, there have been several attempts to exacerbate hypoxia in normal tissues and thereby decrease their radiosensitivity. These are summarized in Table 10.2C. Included among these approaches are very early attempts to reduce the severity of radiation reactions in the skin of patients by simple compression.[47] Later efforts to reduce blood flow to normal tissues prior to radiotherapy involved mechanical blockade of vessels with catheters[48,49] or starch microspheres,[50] as well as pharmacological manipulation of blood flow with epinephrine[51–53] or with inhibitors of nitric oxide (NO) synthetase.[54]

Table 10.2A Effective Modifiers of Radiation-Induced Defects in Vascular Perfusion and Hemodynamics

Modifier	Response	Organ	Species	Reference
Penicillamine	Perfusion	Lung	Rat	Ward, 1981[45]
Captopril	GFR	Kidney	Pig	Robbins & Hopewell, 1986[32]
WR-2721	Microangiography	Brain	Rat	Plotnikova et al., 1988[26]
Pentoxifylline	Gross pathology	Skin	Mouse	Dion et al., 1989[27]
Pentoxifylline	Necrosis	Soft tissue	Human	Dion et al., 1990[28]
Dipyridamole, low iron diet, & desferrioxamine	Ataxia	Spinal cord	Rat	Hornsey et al., 1990[42]
Dexamethasone	Perfusion	Lung	Rat	Peterson et al., 1992[30] Peterson et al., 1992[31]
ACE inhibitors	Function	Kidney	Rat	Cohen et al., 1992[33] Moulder et al., 1993[34] Moulder et al., 1993[35] Juncos et al., 1993[36] Cohen et al., 1994[37] Geraci et al., 1995[14]
Captopril	Hypertension	Lung	Rat	Ward et al., 1993[38]
Essential fatty acids	Ischemic necrosis	Skin	Pig	Hopewell et al., 1993[39] Hopewell et al., 1994[40]
Essential fatty acids	Paralysis	Spinal cord	Pig	Hopewell et al., 1994[41]
Aspirin	Function	Kidney	Mouse	Verheij et al., 1995[45]
Pentoxifylline	Perfusion	Lung	Rat	Koh et al., 1995[29]
Dexamethasone and/or captopril	Function	Kidney	Rat	Geraci et al., 1995[14]
Angiotensin II receptor blocker	Function	Kidney	Rat	Moulder et al., 1996[43]

Table 10.2B Ineffective Modifiers of Radiation-Induced Defects in Vascular Perfusion and Hemodynamics

Modifier	Response	Organ	Species	Reference
Indoramin, captopril	GFR, ERPF ERPF	Kidney	Pig	Robbins & Hopewell, 1986[32]
Verapamil	Ataxia	Spinal cord	Rat	Hornsey et al., 1990[42]
TXA_2 synthetase inhibitor (CGS-13080)	Perfusion	Intestine	Dog	Barter et al., 1991[44]
Captopril, cyproheptadine, diethylcarbamazine, penicillamine, theophylline	Perfusion	Lung	Rat	Peterson et al., 1992[30]
Essential fatty acids	GFR	Kidney	Pig	Hopewell et al., 1994[41]
TXA_2 receptor blocker (daltroban)	EDTA clearance	Kidney	Mouse	Verheij et al., 1995[45]

Table 10.2C Approaches to Normal Tissue Hypoxia as a Radioprotective Strategy

Approach	Organ	Species	Reference
Compression	Skin	Human	Schwarz, 1909[47]
Catheter	Small intestine	Dog	Kligerman et al., 1962[48]
Epinephrine	Kidney	Dog	Johnson et al., 1968[51]
Epinephrine	Kidney	Dog, human	Steckel et al., 1969[52]
Catheter	Kidney	Dog, human	Steckel et al., 1969[49]
Epinephrine	Kidney	Dog, human	Steckel et al., 1974[53]
Starch microspheres	Gut	Rat	Forsberg & Jung, 1978[50]
NG-nitro-L-arginine DEA/NO	Bone marrow	Mouse	Liebmann et al., 1994[54]

C. Vascular Hyperpermeability

Increased permeability is probably the most common as well as the earliest response of the vascular endothelium to radiation. Vascular hyperpermeability essentially occurs immediately after clinically relevant radiation doses,[55] and has been observed in almost every tissue. This response is monitored with relative ease by the extravasation of dyes or radiolabeled serum proteins, and reflects a compromise of normal endothelial barrier function.

Those agents reported to ameliorate radiation-induced vascular hyperpermeability are listed in Table 10.3A. Particularly effective are steroidal and nonsteroidal anti-inflammatory compounds.[31,56–63] For reasons that are not clear, however, positive results are not always obtained with steroids.[64,65] The involvement of proteases in this radiation reaction is suggested by the significant sparing provided by protease inhibitors such as soybean trypsin inhibitor[66] and ε-amino-*n*-caproic acid.[66,67] Essential fatty acids reduce the severity of erythema in irradiated pig skin, but only when given after and not for 4 weeks before irradiation.[39,40] Depletion of complement with cobra venom factor,[68] inhibition of polyamine synthesis with difluoromethylornithine,[69,70] and blocking antibodies to CD18 and ICAM-1[71] also ameliorate radiation-induced vascular hyperpermeability, underscoring the biological complexity of this seemingly simple radiation reaction.

D. Coagulopathy

Coagulopathy is a well-documented manifestation of radiation-induced vascular injury, particularly in liver and lung. Radiation hepatitis has long been described as a form of venocclusive disease,[7,72] in which the lumen of central veins becomes blocked by fibrillar material. Obstruction of irradiated vessels with platelet aggregates has also been observed in lung[73] and heart.[10] The anticoagulants heparin,[74–77] dicumarol,[78] and warfarin[77] spare radiation reactions in lung, liver, and peripheral nerves in several species including humans (Table 10.4), although Moss et al.[79] were unable to prevent the radiation-induced decrease in compliance in rat lung with heparin. Steroids have also been reported to ameliorate radiation injury in liver.[62,72]

Using human umbilical vein endothelial cells (HUVEC) *in vitro*, Verheij et al.[80] demonstrated increased release of von Willebrand factor and increased platelet

Table 10.3A Effective Modifiers of Radiation-Induced Vascular Hyperpermeability

Modifier	Organ	Species	Reference
Soybean trypsin inhibitor, ε-amino-*n*-caproic acid, *meso*-inositol hexanicotinate, hydroxy-chloroquine sulfate, trihydroxyethyl-rutoside	Skin	Rabbit, guinea pig	Jolles & Harrison, 1966[66]
Indomethacin, triamcinolone	Skin	Guinea pig	Song et al., 1968[56]
ε-amino-*n*-caproic acid	Skin	Rabbit	Eassa & Casarett, 1973[67]
Dexamethasone	Brain	Rat	Blomstrand et al., 1975[57]
Cobra venom factor	Skin	Rat	Ullrich & Casarett, 1977[68]
Dexamethasone	Lung	Rat	Evans et al., 1987[58]
Dexamethasone	Spinal cord	Rat	Delattre et al., 1988[59]
Methylprednisolone	Lung	Sheep	Loyd et al., 1988[60]
Dexamethasone, indomethacin, cromolyn, cyproheptadine, vitamin D_3, theophylline, diethylcarbamazine	Lung	Rat	Graham et al., 1990[61]
Dexamethasone	Lung	Rat	Geraci et al., 1992[62]
Dexamethasone	Lung	Rat	Peterson et al., 1992[31]
Methylprednisolone	Lung	Rat	Ward et al., 1993[63]
γ-Interferon	Lung	Rat	Rosiello et al., 1993[118]
Essential fatty acids	Skin	Pig	Hopewell et al., 1993[39] Hopewell et al., 1994[40]
Difluoromethylornithine	Brain	Dog	Fike et al., 1994[69] Fike et al., 1995[70]
Antibodies to CD18 and ICAM-1	Small intestine	Rat	Panes et al., 1995[71]

Table 10.3B Ineffective Modifiers of Radiation-Induced Vascular Hyperpermeability

Modifier	Organ	Species	Reference
Hydrocortisone	Intestine	Rat	Willoughby, 1960[64]
Mepyramine maleate, triprolidine hydrochloride, lysergic acid diethylamide, ovomucoid trypsin inhibitor	Skin	Rabbit, guinea pig	Jolles & Harrison, 1966[66]
Methylprednisolone	Lung	Mouse	Gross, 1980[65]
Captopril, cobra venom factor, penicillamine, trapidil, ε-amino-*n*-caproic acid, dapsone	Lung	Rat	Graham et al., 1990[61]
Antibodies to P-selectin	Small intestine	Rat	Panes et al., 1995[71]

adherence to irradiated monolayers. Interestingly, the radiation-induced platelet adherence could be blocked by antibodies to von Willebrand factor.[80] This study represents one of very few[18,20,71,81] that has attempted to block radiation reactions with neutralizing antibodies to putative mediators of the injury.

Table 10.4 Effective Modifiers of Radiation-Induced Coagulopathy

Modifier	Response	Organ	Species	Reference
Heparin	Histopathology	Lung	Rabbit	Boys & Harris, 1943[74]
Dicumarol	Histopathology	Lung	Human	Macht & Perlberg, 1950[78]
Heparin	Histopathology	Lung	Dog	Fleming et al., 1962[75]
Diuretics, steroids	Symptoms, function, survival	Liver	Human	Ingold et al., 1965[72]
Heparin	Histopathology, function	Liver	Rat	Kinzie et al., 1972[76]
Dexamethasone	Function	Liver	Rat	Geraci et al., 1992[62]
Heparin, warfarin	Symptoms, function	Nervous system	Human	Glanz et al., 1994[77]
Antibodies to von Willebrand Factor	Platelet adhesion *in vitro*	HUVEC	Human	Verheij et al., 1994[80]

E. Defective Fibrinolysis

Ionizing radiation induces defects in fibrinolysis in most tissues, but the nature of the defect is tissue specific. The activity of plasminogen activator, a major component of tissue fibrinolytic activity, decreases after irradiation in blood vessels,[82,83] liver,[84] and lung,[85–88] but increases in the gastrointestinal tract[89,90] and the central nervous system.[91,92] The explanation for this tissue specificity is not clear.

Diverse agents such as actinomycin D,[75] the fibrinolytic stimulator dextran sulfate,[93] and ACE inhibitors[86–88] prevent the radiation-induced suppression in plasminogen activator activity in lung of several species including humans (Table 10.5A). Dextran sulfate reduces radiation histopathology in the lung of humans, but not of mice.[93] Radiation also suppresses plasminogen activator release from bovine aortic endothelial cells *in vitro*, and phorbol ester partially prevents this response.[94] The phorbol ester TPA also protects irradiated endothelial cells from apoptosis.[19] In contrast, several agents including cortisone,[75] penicillamine,[95] and pentoxifylline[96] fail to ameliorate the radiation-induced suppression in lung plasminogen activator activity (Table 10.5B). However, these agents all spare the development of perfusion defects in irradiated lung (Table 10.2A).

Inhibitors of the elevated fibrinolytic activity in irradiated gastrointestinal (GI) tract also have been identified (Table 10.5A). The antiplasmin agents *trans*-4-aminomethylcyclohexane carboxylic acid[84] and tranexamic acid[90] ameliorate GI dysfunction and histopathology in irradiated rats. The fact that modifiers of radiation-induced defects in fibrinolytic activity — whether that activity is decreased or increased — reduce the severity of the radiation reaction suggests that this vascular dysfunction is not trivial in the pathogenesis of radiation injury in normal tissues.

F. Altered Eicosanoid Metabolism

Perturbed eicosanoid metabolism is a common response of the vascular endothelium to insult, including that induced by ionizing radiation, although there appears

Table 10.5A Effective Modifiers of Radiation-Induced Defects in Fibrinolysis

Modifier	Response	Organ	Species	Reference
	Stimulators of Depressed Fibrinolysis			
Actinomycin D	Histopathology	Lung	Dog	Fleming et al., 1962[75]
Dextran sulfate	Radiographic	Lung	Human	Ohno et al., 1978[93]
ACE inhibitors	Function	Lung	Rat	Ward et al., 1988[86] Ward et al., 1989[87] Ward et al., 1989[88]
Phorbol ester	Suppressed PA release *in vitro*	Aorta	Bovine	Raymond et al., 1990[94]
	Inhibitors of Enhanced Fibrinolysis			
Trans-4-aminomethyl-cyclohexane, carboxylic acid	Excretion	Intestine	Rat	Kondo et al., 1976[89]
Tranexamic acid	Histopathology	Stomach	Rat	Stenberg et al., 1980[90]

Table 10.5B Ineffective Modifiers of Radiation-Induced Defects in Fibrinolysis

Modifier	Response	Organ	Species	Reference
Cortisone	Histopathology	Lung	Dog	Fleming et al., 1962[75]
Dextran sulfate	Histopathology	Lung	Mouse	Ohno et al., 1978[93]
Aspirin	Histopathology	Liver	Dog	Henderson et al., 1983[84]
Penicillamine	Function	Lung	Rat	Ward et al., 1987[95]
Pentoxifylline	Function	Lung	Rat	Ward et al., 1992[96]

to be no universal pattern of response. In some studies, the initial endothelial reaction to radiation *in vitro*[97,98] and *in vivo*[99] is a suppression in the production of arachidonate products even after radiation doses as low as 5 cGy.[98] When present *in vivo*, this suppression is transient, however, and is followed by a sustained hyperproduction of eicosanoids in several tissues including lung,[86,87,95] intestine,[100] brain[101] and liver.[101] In other studies,[102] the hyperproduction of arachidonate products occurs immediately after irradiation. The effect of irradiation on arachidonate metabolism, and the use of anti-inflammatory drugs in the treatment and prevention of this radiation response, has been reviewed in detail by Michalowski.[3] The obvious and most frequently cited explanation for radiation-induced hyperproduction of eicosanoids is the activation of phospholipase, the enzyme primarily responsible for cleaving the substrate arachidonic acid from plasma membrane phospholipids. To our knowledge, however, phospholipase activity has been measured only once in irradiated tissue, and curiously was found to be decreased in irradiated rat lung.[103] The same lung tissue was hyperproducing thromboxane A_2 and contained a subnormal amount of the putative phospholipase inhibitor annexin I, yet phospholipase activity also was suppressed.[103] These data suggest that phospholipase activity should be measured before it is assumed to be elevated, and that alternative pathways to arachidonate release may operate after irradiation.

Effective modifiers of radiation-induced changes in endothelial eicosanoid metabolism include the radioprotector WR-2721,[102] steroidal[60] and nonsteroidal[100] anti-inflammatory compounds, Vitamins C[97] and A,[98] and interleukin-2 (IL-2)[104] *in vitro* (Table 10.6A). A structure-function analysis of thiol and nonthiol ACE inhibitors revealed that only the former were effective in sparing the radiation-induced hyperproduction of PGI_2 and TXA_2 by rat lung,[105] suggesting that modifier activity in this case was independent of the renin-angiotensin system. Confirming this hypothesis, penicillamine, a thiol compound with virtually no ACE inhibitory activity, also ameliorated the radiation-induced hyperproduction of PGI_2 (Figure 10.2) and TXA_2 by rat lung.[95] Vitamin C has yielded both positive[97] and negative[98] results as a modifier of suppressed PGI_2 production by endothelial cells irradiated *in vitro*, although this discrepancy may be due to marked differences in radiation dose.

G. Endothelial Enzyme Activity

In addition to perturbing the activity of key enzymes in arachidonate metabolism and fibrinolysis, ionizing radiation also suppresses angiotensin converting enzyme (ACE),[86–88] 5′-nucleotidase,[106] and alkaline phosphatase[107] activities in endothelial cells irradiated *in vitro* or *in vivo*. Steroidal[24] and nonsteroidal[106] anti-inflammatory agents ameliorate the radiation-induced suppression in ACE activity (Table 10.7A).

Table 10.6A Effective Modifiers of Radiation-Induced Changes in Eicosanoid Metabolism

Modifier	Response	Organ	Species	Reference
WR-2721	Excretion of PGE, $PGF_{2\alpha}$, & TXB_2	Whole body	Rat	Donlon et al., 1985[102]
Penicillamine	Hyperproduction of PGI_2 & TXA_2	Lung	Rat	Ward et al., 1987[95]
Vitamin C	Impaired PGI_2 production *in vitro*	Aorta	Bovine	Eldor et al., 1987[97]
Methylprednisolone	Increased TXA_2 in lymph	Lung	Sheep	Loyd et al., 1988[60]
ACE inhibitors (thiol)	Hyperproduction of PGI_2 & TXA_2	Lung	Rat	Ward et al., 1988[86] Ward et al., 1989[87]
Indomethacin	Hyperproduction of PGE_2, $PGF_{2\alpha}$, & PGI_2	Jejunum	Dog	Summers et al., 1991[100]
α-Tocopherol	Suppressed PGI_2 production *in vitro*	Aorta	Bovine	Hosoi et al., 1993[98]
Interleukin-2	Suppressed PGI_2 production *in vitro*	Umbilical vein	Human	Verheij et al., 1994[104]

Table 10.6B Ineffective Modifiers of Radiation-Induced Changes in Eicosanoid Metabolism

Modifier	Response	Organ	Species	Reference
ACE inhibitors (nonthiol)	Hyperproduction of PGI_2 & TXA_2	Lung	Rat	Ward et al., 1989[88]
Pentoxifylline	Hyperproduction of PGI_2 & TXA_2	Lung	Rat	Ward et al., 1992[96]
Vitamin C	Suppressed PGI_2 production *in vitro*	Aorta	Bovine	Hosoi et al., 1993[98]

Curiously, ACE inhibitors themselves administered after irradiation also spare the reduction in ACE activity observed in rat lung.[86–88] Both thiol and nonthiol ACE inhibitors are effective in this case, whereas the thiol compound penicillamine is not (Figure 10.3). Pentoxifylline does not spare the acute (8 weeks) radiation-induced suppression in lung ACE activity,[96] but has a beneficial effect on this response when administered continuously for 40 weeks.[29] In irradiated mouse lung, ACE activity increases in bronchoalveolar lavage fluid (BAL), possibly as a result of transudation of serum enzyme.[108] This response can be prevented by the radioprotector WR-2721.[108] Indomethacin also ameliorates the radiation-induced suppression in 5′′-nucleotidase activity in rabbit lung.[106]

III. SUMMARY

The literature on endothelial-oriented strategies to spare normal tissue complications after irradiation is large but fragmented. From this complex and sometimes contradictory body of information, certain events emerge in the pathogenesis of radiation-induced normal tissue injury. Steroids, particularly dexamethasone, are the most consistently effective radiation modifier in the laboratory, and the most extensively employed in the clinic. The widespread success with steroids and NSAIDS strongly implicates an inflammatory component in acute vascular/endothelial radiation response, consistent with the descriptive histopathology. The effectiveness of antioxidants such as SOD,[109] Vitamins A and C,[97,98] and retinol[23] suggests that oxygen radical species may contribute to the inflammatory reaction. These are predictable conclusions.

Perhaps less predictable is the efficacy of ACE inhibitors, of which captopril is the prototype, as modifiers of experimental radiation pathophysiology in kidney and lung. The mechanism of therapeutic action of captopril is not clear at present. The drug might be expected to improve perfusion and hemodynamic defects in irradiated tissues (Table 10.2A), as a result of inhibition of production of the vasoconstrictor angiotensin II. On the other hand, the ability of captopril to spare both the hyperproduction of PGI_2 and TXA_2 and the accumulation of collagen in irradiated rat lung is not shared by all ACE inhibitors, only those containing a thiol group.[105] Thus, captopril's therapeutic efficacy appears to consist of at least two components: ACE inhibition and a nonspecific thiol effect. The latter component is shared by thiols such as penicillamine, which lack ACE inhibitory activity.[105] While captopril spares radiation reactions in the vasculature of both kidney and lung, there may be differences in the responsiveness of the two organs to the drug. In kidney, thiol (captopril) and nonthiol (enalapril) ACE inhibitors are equally effective in sparing radiation nephropathy,[34] although blood pressure reduction by other drugs is ineffective.[34,36] In irradiated lung, all ACE inhibitors are not equal.[105]

Effective modifiers, where studied in sufficient detail, do not modify every radiation defect in any single organ, nor do they modify the same defect (e.g., hyperpermeability) in every organ. However, all of these modifiers with the exception of WR-2721 can be effective when administered after irradiation, i.e., in the prophylaxis or treatment of the radiation response. This would argue against direct

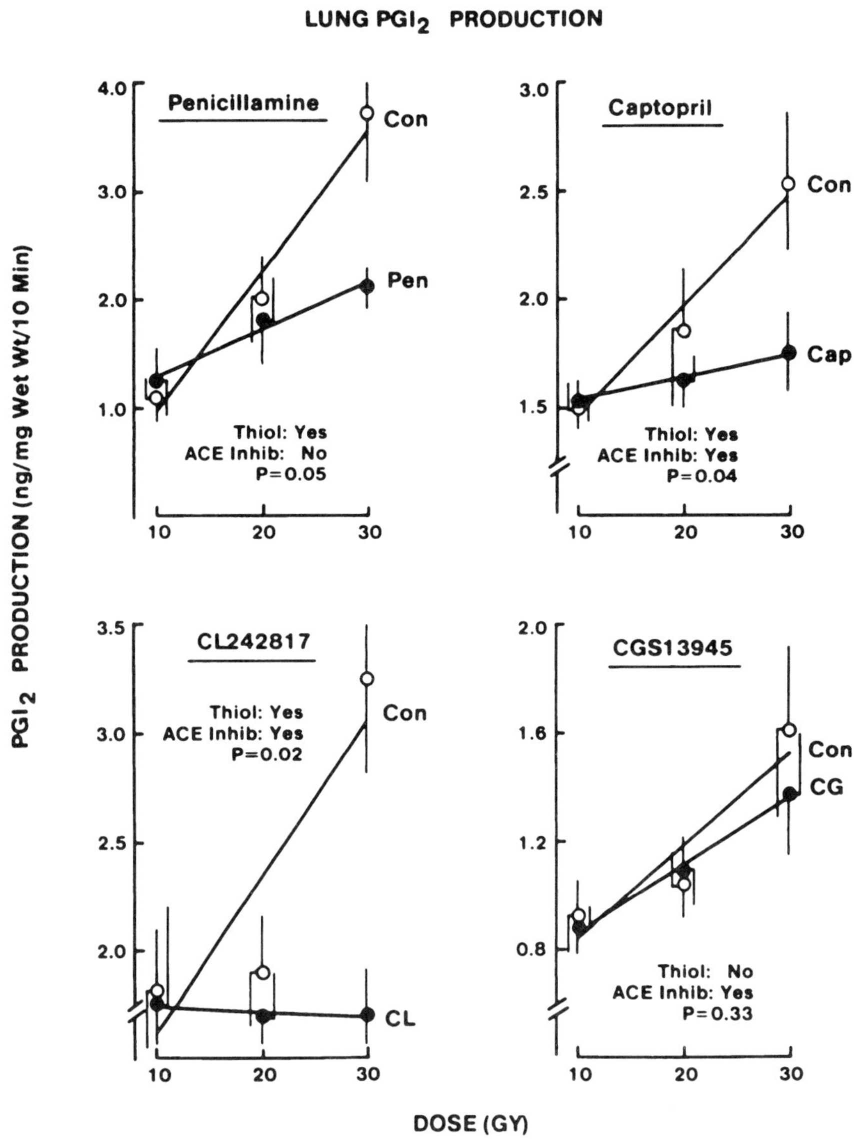

Figure 10.2. Prostacyclin (PGI_2) production by right lung of rats 2 months after right hemithoracic irradiation. Animals consumed either control diet (Con, open circles) or diet containing one of four modifiers (solid cirlces) continuously after irradiation. Mean ± SEM; n=6-8 rats per group. Lines fitted by linear regression analysis. Upper left: penicillamine, a thiol compound with no ACE inhibitory activity; Upper right: captopril, a thiol-containing ACE inhibitor; Lower left: CL242817, a thio-acetate ACE inhibitor, activated to free thiol; Lower right: CGS13945, a nonthiol ACE inhibitor.

Table 10.7A Effective Modifiers of Radiation-Induced Changes in Endothelial Enzyme Activity

Modifier	Response	Organ	Species	Reference
Dexamethasone	Suppressed ACE activity *in vitro*	Aorta	Bovine	Lam et al., 1985[24]
WR-2721	Elevated ACE activity in BAL	Lung	Mouse	Travis et al., 1987[108]
ACE inhibitors	Suppressed ACE & PA activity	Lung	Rat	Ward et al., 1988[86] Ward et al., 1989[87] Ward et al., 1989[88]
Indomethacin	Suppressed ACE & 5′-nucleotidase activity	Lung	Rabbit	Organos et al., 1994[106]
Pentoxifylline	Suppressed ACE activity	Lung	Rat	Koh et al., 1995[29]

Table 10.7B Ineffective Modifiers of Radiation-Induced Changes in Endothelial Enzyme Activity

Modifier	Response	Organ	Species	Reference
Penicillamine	Suppressed ACE & PA activity	Lung	Rat	Ward et al., 1987[95]
Pentoxifylline	Suppressed ACE & PA activity	Lung	Rat	Ward et al., 1992[96]

radiation-induced cell kill as the sole mechanism of injury. Not every class of modifier found effective in the laboratory has undergone clinical trial in radiotherapy patients. ACE inhibitors are perhaps the best example of this, despite the fact that they are taken by more than 10 million people worldwide in the management of hypertension and congestive heart failure.

The radiation modifiers described in this chapter are subject to the general criticism that they have not received systematic testing. At least two additional criticisms seem warranted. First, very few studies have examined the efficacy of combinations of modifiers — either simultaneous or sequential — as opposed to single agents. Yet the probability that any single modifier will represent the optimum therapeutic approach is small. To our knowledge, only the studies of Peterson et al.,[30] in which second agents were administered during dexamethasone taper after lung irradiation, and of Geraci et al.,[14] in which a regimen of dexamethasone plus captopril proved superior to either agent alone in sparing radiation nephropathy, have combinations of modifiers been tested.

A second criticism of the existing literature on radiation modifiers is that very few studies were designed to evaluate the effect of the agent on both acute vascular/endothelial responses and late effects such as fibrosis, thus missing the opportunity to test the hypothesis of a causal relationship between the two.[110] Ullrich and Casarett[68] conducted such a study with cobra venom factor (complement depletion) in irradiated rat skin, and Ward et al. have tested this relationship with captopril in

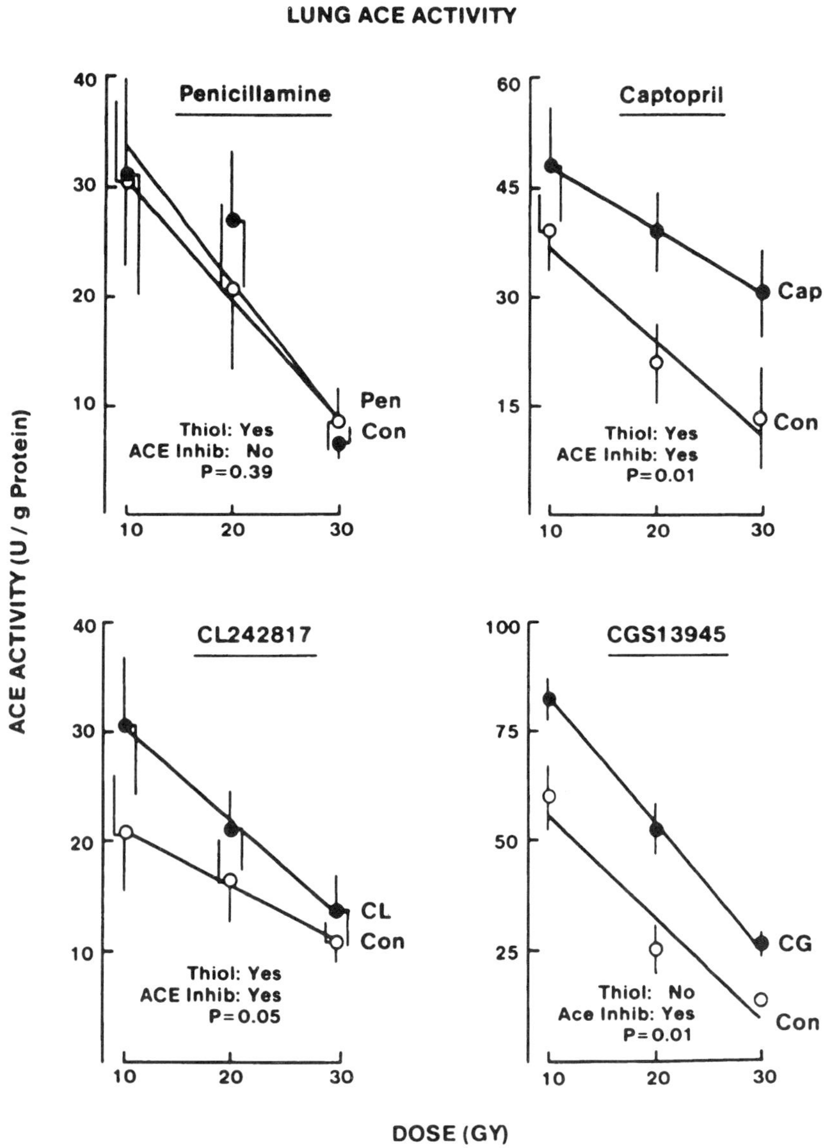

Figure 10.3. Angiotensin converting enzyme (ACE) activity in right lung of rats 2 months after right hemithoracic irradiation. Animals consumed either control diet (Con, open circles) or diet containing one of four modifiers (solid circles) continuously after irradiation (see Figure 10.2 caption).

irradiated rat lung.[38,86,111] In both cases, a reduction in the severity of the acute vascular reaction was followed by an amelioration in fibrosis. These data are necessary support but not sufficient proof of the hypothesis.

IV. FUTURE RESEARCH

The classical pharmacologic approach to modifiers of radiation-induced vascular/endothelial injury is likely in the near future to identify promising new agents and combinations of agents, particularly if these are studied on a systematic basis. A firmer rational basis for this research will emerge as the molecular and cellular basis of both radiation injury and modifier action becomes clearer. These studies will identify the autocrine, paracrine, and endocrine pathways in vascular radiopathology, and will permit the testing of appropriate receptor antagonists, blocking antibodies, or antisense oligonucleotides to these growth factors and mediators. To our knowledge, these agents have been used only rarely as modifiers of radiation damage in endothelial cells *in vitro* and *in vivo*. More work is needed in this area as these paracrine interactions are clarified.

The possibility of genetically engineering endothelial cells to combat specific radiation-induced defects in normal tissues is an attractive prospect. For many years it has been possible to stably transfect aortic and pulmonary artery endothelial cells *in vitro*, and to demonstrate expression of the transfected genes.[112,113] This approach now is feasible *in vivo*, possibly on a site-specific basis, and endothelial and vascular smooth muscle cells expressing transfected ACE and growth hormone genes will colonize arterial walls and lungs of mice.[114–116] Given the data summarized in the present chapter, for example, it might be useful to implant in an irradiated organ endothelial cells engineered to overexpress plasminogen activator or PGI_2 production. While many obstacles remain in the clinical application of this technology, progress has been impressive in the fields of vascular and transplant biology, and these lessons may someday be applied to radiation pathology. The vascular endothelial cell is a logical target in this regard.

REFERENCES

1. Rubin, P. and Casarett, G. W., *Clinical Radiation Pathology*, Saunders, Philadelphia, 1968.
2. Withers, H. R., Peters, L. J., and Kogelnik, H. D., The pathobiology of late effects of irradiation, in *Radiation Biology in Cancer Research*, Meyn, R. E. and Withers, H. R., Eds., Raven Press, New York, 1980, 439.
3. Michalowski, A. S., On radiation damage to normal tissues and its treatment. II. Anti-inflammatory drugs, *Acta Oncol.*, 33, 139, 1994.

4. Yeung, T. L., Lauk, S., Simmons, R. H., Hopewell, J. W., and Trott, K.-R., Morphological and functional changes in the rat heart after X irradiation: strain differences, *Radiat. Res.*, 119, 489, 1989.
5. Slauson, D. O., Hahn, F. F., and Chiffelle, T. L., The pulmonary vascular pathology of experimental radiation pneumonitis, *Am. J. Pathol.*, 88, 635, 1977.
6. Adamson, I. Y. R. and Bowden, D. H., Endothelial injury and repair in radiation-induced pulmonary fibrosis, *Am. J. Pathol.*, 112, 224, 1983.
7. Reed, G. B. Jr. and Cox, A. J., The human liver after radiation injury: a form of veno-occlusive disease, *Am. J. Pathol.*, 48, 597, 1966.
8. Jaenke, R. S., Robbins, M. E. C., Bywaters, T., Whitehouse, E., Rezvani, M., and Hopewell, J. W., Capillary endothelium: target site of renal radiation injury, *Lab. Invest.*, 68, 396, 1993.
9. Hopewell, J. W., Foster, J. L., Gunn, Y., Moustafa, H. F., Patterson, T. J. S., Wiernik, G., and Young, C. M. A., Role of vascular damage in the development of late radiation effects in the skin, in *Late Biological Effects of Ionizing Radiation*, Vol. 1, International Atomic Energy Agency, Vienna, 1978, 483.
10. Fajardo, L. F. and Berthrong, M., Vascular lesions following radiation, *Pathol. Ann.*, 23, 297, 1988.
11. Klemm, J., Die Veranderungen der Kapillarendothelzelle unter dem Einfluss ionisierender Strahlen und mogliche Schutzwirkung von Trihydroxyaethylrutosidum, *Strahlentherapie*, 134, 31,1967.
12. Port, C. D. and Ward, W. F., The ultrastructure of radiation injury in rat lung: modification by D-penicillamine, *Radiat. Res.*, 92, 61, 1982.
13. Geraci, J. P., Mariano, M. S., and Jackson, K. L., Amelioration of radiation nephropathy in rats by dexamethasone treatment after irradiation, *Radiat. Res.*, 134, 86, 1993.
14. Geraci, J. P., Sun, M.-C., and Mariano, M. S., Amelioration of radiation nephropathy in rats by postirradiation treatment with dexamethasone and/or captopril, *Radiat. Res.*, 143, 58, 1995.
15. Kimura, H., Wu, N. Z., Dodge, R., Spencer, D. P., Klitzman, B. M., McIntyre, T. M., and Dewhirst, M. W., Inhibition of radiation-induced up-regulation of leukocyte adhesion to endothelial cells with the platelet-activating factor inhibitor, BN52021, *Int. J. Radiat. Oncol. Biol. Phys.*, 33, 627, 1995.
16. Fuks, Z., Persaud, R. S., Alfieri, A., McLoughlin, M., Ehleiter, D., Schwartz, J. L., Seddon, A. P., Cordon-Cardo, C., and Haimovitz-Friedman, A., Basic fibroblast growth factor protects endothelial cells against radiation-induced programmed cell death *in vitro* and *in vivo*, *Cancer Res.*, 54, 2582, 1994.
17. Tee, P. G. and Travis, E. L., Basic fibroblast growth factor does not protect against classical radiation pneumonitis in two strains of mice, *Cancer Res.*, 55, 298, 1995.
18. Haimovitz-Friedman, A., Vlodavsky, I., Chaudhuri, A., Witte, L., and Fuks, Z., Autocrine effects of fibroblast growth factor in repair of radiation damage in endothelial cells, *Cancer Res.*, 51, 2552, 1991.
19. Haimovitz-Friedman, A., Balaban, N., McLoughlin, M., Ehleiter, D., Michaeli, J., Vlodavsky, I., and Fuks, Z., Protein kinase C mediates basic fibroblast growth factor protection of endothelial cells against radiation-induced apoptosis, *Cancer Res.*, 54, 2591, 1994.
20. Eissner, G., Kohlhuber, F., Grell, M., Ueffing, M., Scheurich, P., Hieke, A., Multhoff, G., Bornkamm, G. W., and Holler, E., Critical involvement of transmembrane tumor necrosis factor-alpha in endothelial programmed cell death mediated by ionizing radiation and bacterial endotoxin, *Blood*, 86, 4184, 1995.

21. Ward, W. F., Ts'ao, C., Molteni, A., and Hinz, J. M., Non-reproductive dysfunction in irradiated endothelial cells in culture: modification by dibutyryl cyclic AMP, *Int. J. Radiat. Biol.*, 60, 39, 1991.
22. Ward, W. F., Ts'ao, C., Molteni, A., Yoon, S., and Taylor, J., Captopril does not spare endothelial cell radiation sensitivity in vitro. Presented at the 41st Annual Meeting of the Radiation Research Society, Dallas, TX, March 20–25, 1993.
23. Bump, E. A., Braunhut, S. J., Palayoor, S. T., Medeiros, D., Lai, L. L., Cerce, B. A., Langley, R. E., and Coleman, C. N., Novel concepts in modification of radiation sensitivity, *Int. J. Radiat. Oncol. Biol. Phys.*, 29, 249, 1994.
24. Lam, T. T., Rubin, D., and Drab, E. A., The effect of dexamethasone on the cytotoxic and enzymatic response of cultured endothelial cells to radiation, *Radiat. Res.*, 103, 240, 1985.
25. Rubin, D. B., Drab, E. A., Stone, A. M., Walden, T. L. Jr., and Hanson, W. R., The influence of exogenous eicosanoids on the radiation response of cultured bovine aortic endothelial cells, *Radiat. Res.*, 125, 41, 1991.
26. Plotnikova, E. D., Levitman, M. K., Shaposhnikova, V. V., Koshevoj, J. V., and Eidus, L. K., Protection of microvasculature in rat brain against late radiation injury by gammaphos, *Int. J. Radiat. Oncol. Biol. Phys.*, 15, 1197, 1988.
27. Dion, M. W., Hussey, D. H., and Osborne, J. W., The effect of pentoxifylline on early and late radiation injury following fractionated irradiation in C3H mice, *Int. J. Radiat. Oncol. Biol. Phys.*, 17, 101, 1989.
28. Dion, M. W., Hussey, D. H., Doornbos, J. F., Vigliotti, A. P., Wen, B. C., and Anderson, B., Preliminary results of a pilot study of pentoxifylline in the treatment of late radiation soft tissue necrosis, *Int. J. Radiat. Oncol. Biol. Phys.*, 19, 401, 1990.
29. Koh, W.-J., Stelzer, K. J., Peterson, L. M., Staker, B. L., Ward, W. F., Russell, K. J., and Griffin, T. W., Effect of pentoxifylline on radiation-induced lung and skin toxicity in rats, *Int. J. Radiat. Oncol. Biol. Phys.*, 31, 71, 1995.
30. Peterson, L. M., Evans, M. L., Graham, M. M., Eary, J. F., and Dahlen, D. D., Vascular response to radiation injury in the rat lung, *Radiat. Res.*, 129, 139, 1992.
31. Peterson, L. M., Evans, M. L., Thomas, K. L., and Graham, M. M., Vascular response to fractionated irradiation in the rat lung, *Radiat. Res.*, 131, 224, 1992.
32. Robbins, M. E. C. and Hopewell, J. W., Physiological factors effecting renal radiation tolerance: a guide to the treatment of late effects, *Br. J. Cancer*, 53 (Suppl. VII), 265, 1986.
33. Cohen, E. P., Fish, B. L., and Moulder, J. E., Treatment of radiation nephropathy with captopril, *Radiat. Res.*, 132, 346, 1992.
34. Moulder, J. E., Fish, B. L., and Cohen, E. P., Treatment of radiation nephropathy with ACE inhibitors, *Int. J. Radiat. Oncol. Biol. Phys.*, 27, 93, 1993.
35. Moulder, J. E., Cohen, E. P., Fish, B. L., and Hill, P., Prophylaxis of bone marrow transplant nephropathy with captopril, an angiotensin converting enzyme inhibitor, *Radiat. Res.*, 136, 404, 1993.
36. Juncos, L. I., Duenas, S. C., Cornejo, J. C., Broglia, C. A., and Cejas, H., Long-term enalapril and hydrochlorothiazide in radiation nephritis, *Nephron*, 64, 249, 1993.
37. Cohen, E. P., Moulder, J. E., Fish, B. L., and Hill, P., Prophylaxis of experimental bone marrow transplant nephropathy, *J. Lab. Clin. Med.*, 124, 371, 1994.
38. Ward, W. F., Lin, P.-J. P., Wong, P. S., Behnia, R., and Jalali, N., Radiation pneumonitis in rats and its modification by the angiotensin-converting enzyme inhibitor captopril evaluated by high-resolution computed tomography, *Radiat. Res.*, 135, 81, 1993.

39. Hopewell, J. W., Robbins, M. E. C., van den Aardweg, G. J. M. J., Morris, G. M., Ross, G. A., Whitehouse, E., Horrobin, D. F., and Scott, C. A., The modulation of radiation-induced damage to pig skin by essential fatty acids, *Br. J. Cancer*, 68, 1, 1993.
40. Hopewell, J. W., van den Aardweg, G. J. M. J., Morris, G. M., Rezvani, M., Robbins, M. E. C., Ross, G. A., Whitehouse, E. M., Scott, C. A., and Horrobin, D. F., Amelioration of both early and late radiation-induced damage to pig skin by essential fatty acids, *Int. J. Radiat. Oncol. Biol. Phys.*, 30, 1119, 1994.
41. Hopewell, J. W., van den Aardweg, G. J. M. J., Morris, G. M., Rezvani, M., Robbins, M. E. C., Ross, G. A., and Whitehouse, E. M., Unsaturated lipids as modulators of radiation damage in normal tissues, in *New Approaches to Cancer Treatment*, Horrobin, D. F., Ed., Churchill Commun. Europe, London, 1994, 88.
42. Hornsey, S., Myers, R., and Jenkinson, T., The reduction of radiation damage to the spinal cord by post-irradiation administration of vasoactive drugs, *Int. J. Radiat. Oncol. Biol. Phys.*, 18,1437, 1990.
43. Moulder, J. E., Fish, B. L., Cohen, E. P., and Bonsib, S. M., Angiotensin II receptor antagonists in the prevention of radiation nephropathy, *Radiat. Res.*, 146, 106, 1996.
44. Barter, J. F., Marlow, D., Kamath, R. K., Harbert, J., Torrisi, J. R., Barnes, W. A., Potkul, R. K., Newsome, J. T., and Delgado, G., Lack of protective effect of thromboxane synthetase inhibitor (CGS-13080) on single dose radiated canine intestine, *Gynecol. Oncol.*, 40, 218, 1991.
45. Verheij, M., Stewart, F. A., Oussoren, Y., Weening, J. J., and Dewit, L., Amelioration of radiation nephropathy by acetylsalicylic acid, *Int. J. Radiat. Biol.*, 67, 587, 1995.
46. Ward, W. F., Radiation-induced pulmonary arterial perfusion defects: modification by D-penicillamine, *Radiology*, 139, 201, 1981.
47. Schwarz, G., Uber Desensibiliserung gegen Rontgen und Radiumstrahlen, *Munchener Medizinische Wochenschrift*, 24, 1, 1909.
48. Kligerman, M. M., Bond, V. P., Held, B. T., Mather, G., and Nielsen, N., Anoxic protection of small intestine of dogs by intra-arterial catheterization, *Radiology*, 78, 284, 1962.
49. Steckel, R. J., Tobin, P. L., Ross, G., Stein, J. J., and Stevens, G. H., Radiation protection of vital organs, using a selective arterial catheter — Experimental and clinical aspects, *Am. J. Roentgenol.*, 106, 841, 1969.
50. Forsberg, J. O. and Jung, B., Abdominal radiation response modified by hypoxia after intra-aortal injection of starch microspheres, *Acta Radiol.*, 17, 353, 1978.
51. Johnson, R. E., Doppman, J. L., Harbert, J. C., Steckel, R. J., and MacLowry, J. D., Prevention of radiation nephritis with renal artery infusion of vasoconstrictors, *Radiology*, 91, 103, 1968.
52. Steckel, R. J., Tobin, P. L., Stein, J. J., and Bennett, R. L., Intra-arterial epinephrine protection against radiation nephritis, *Radiology*, 92, 1341, 1969.
53. Steckel, R. J., Collins, J. D., Snow, H. D., Lagasse, L. D., Barenfus, M., Anderson, D. P., Weisenburger, Hauskins, L. A., and Ross, N. A., Radiation protection of the normal kidney by selective arterial infusions, *Cancer*, 34, 1046, 1974.
54. Liebmann, J., DeLuca, A. M., Coffin, D., Keefer, L. K., Venzon, D., Wink, D. A., and Mitchell, J. B., *In vivo* radiation protection by nitric oxide modulation, *Cancer Res.*, 54, 3365, 1994.
55. Krishnan, L., Krishnan, E. C., and Jewell, W. R., Immediate effect of irradiation on microvasculature, *Int. J. Radiat. Oncol. Biol. Phys.*, 15, 147, 1988.

56. Song, C. W., Drescher, J. J., and Tabachnick, J., Effect of anti-inflammatory compounds on beta-irradiation-induced increases in vascular permeability, *Radiat. Res.*, 34, 616, 1968.
57. Blomstrand, C., Johansson, B., and Rosengren, B., Dexamethasone effect on blood-brain barrier damage caused by acute hypertension in X-irradiated rats, *Acta Neurol. Scand.*, 52, 331, 1975.
58. Evans, M. L., Graham, M. M., Mahler, P. A., and Rasey, J. S., Use of steroids to suppress vascular response to radiation, *Int. J. Radiat. Oncol. Biol. Phys.*, 13, 563, 1987.
59. Delattre, J. Y., Rosenblum, M. K., Thaler, H. T., Mandell, L., Shapiro, W. R., and Posner, J. B., A model of radiation myelopathy in the rat. Pathology, regional capillary permeability changes and treatment with dexamethasone, *Brain*, 111, 1319, 1988.
60. Loyd, J. E., Bolds, J. M., Wickersham, N., Malcolm, A. W., and Brigham, K. L., Corticosteroids prevent acute lung dysfunction caused by thoracic irradiation in unanesthetized sheep, *Am. Rev. Respir. Dis.*, 138, 1227, 1988.
61. Graham, M. M., Evans, M. L., Dahlen, D. D., Mahler, P. A., and Rasey, J. S., Pharmacological alteration of the lung vascular response to radiation, *Int. J. Radiat. Oncol. Biol. Phys.*, 19, 329, 1990.
62. Geraci, J. P., Mariano, M. S., Jackson, K. L., Taylor, D. A., and Still, E. R., Effects of dexamethasone on late radiation injury following partial-body and local organ exposures, *Radiat. Res.*, 129, 61, 1992.
63. Ward, H. E., Kemsley, L., Davies, L., Holecek, M., and Berend, N., The effect of steroids on radiation-induced lung disease in the rat, *Radiat. Res.*, 136, 22, 1993.
64. Willoughby, D. A., Pharmacological aspects of the vascular permeability changes in the rat's intestine following abdominal radiation, *Br. J. Radiol.*, 33, 515, 1960.
65. Gross, N. J., Radiation pneumonitis in mice. Some effects of corticosteroids on mortality and pulmonary physiology, *J. Clin. Invest.*, 66, 504, 1980.
66. Jolles, B. and Harrison, R. G., Enzymic processes and vascular changes in the skin radiation reaction, *Br. J. Radiol.*, 39, 12, 1966.
67. Eassa, E.-H. M. and Casarett, G. W., Effect of epsilon-amino-n-caproic acid (EACA) on radiation-induced increase in capillary permeability, *Radiology*, 106, 679, 1973.
68. Ullrich, R. L. and Casarett, G. W., Interrelationship between the early inflammatory response and subsequent fibrosis after radiation exposure, *Radiat. Res.*, 72, 107, 1977.
69. Fike, J. R., Gobbel, G. T., Marton, L. J., and Seilhan, T. M., Radiation brain injury is reduced by the polyamine inhibitor alpha-difluoromethylornithine, *Radiat. Res.*, 138, 99, 1994.
70. Fike, J. R., Gobbel, G. T., Chou, D., Wijnhoven, B. P., Bellinzona, M., Nakagawa, M., and Seilhan, T. M., Cellular proliferation and infiltration following interstitial irradiation of normal dog brain is altered by an inhibitor of polyamine synthesis, *Int. J. Radiat. Oncol. Biol. Phys.*, 32, 1035, 1995.
71. Panes, J., Anderson, D. C., Miyasaka, M., and Granger, D. N., Role of leukocyte-endothelial cell adhesion in radiation-induced microvascular dysfunction in rats, *Gastroenterology*, 108, 1761, 1995.
72. Ingold, J. A., Reed, G. B., Kaplan, H. S., and Bagshaw, M. A., Radiation hepatitis, *Am J. Roentgenol. Radium Ther. Nucl. Med.*, 93, 200, 1965.
73. Maisin, J. R., The ultrastructure of the lung of mice exposed to a supra-lethal dose of ionizing radiation on the thorax, *Radiat. Res.*, 44, 545, 1970.
74. Boys, F. and Harris, I. D., The effect of heparinization on experimental post-irradiation tissue changes in the lung, *Am. J. Roentgenol. Radium Ther.*, 50, 1, 1943.

75. Fleming, W. H., Szakacs, J. E., and King, E. R., The effect of gamma radiation on the fibrinolytic system of dog lung and its modification by certain drugs: relationship to radiation pneumonitis and hyaline membrane formation in lung, *J Nucl. Med.*, 3, 341, 1962.
76. Kinzie, J., Studer, R. K., Perez, B., and Potchen, E. J., Noncytokinetic radiation injury: anticoagulants as radioprotective agents in experimental radiation hepatitis, *Science*, 175, 1481, 1972.
77. Glanz, M. J., Burger, P. C., Friedman, A. H., Radtke, R. A., Massey, E. W., and Schold, S. C. Jr., Treatment of radiation-induced nervous system injury with heparin and warfarin, *Neurology*, 44, 2020, 1994.
78. Macht, S. H. and Perlberg, H., Use of anticoagulant (dicoumarol) in preventing post-irradiation tissue changes in the human lung, *Am. J. Roentgenol.*, 63, 335, 1950.
79. Moss, W. T., Haddy, F. J., and Sweany, S. K., Some factors altering the severity of acute radiation pneumonitis: variation with cortisone, heparin, and antibiotics, *Radiology*, 75, 50, 1960.
80. Verheij, M., Dewit, L. G. H., Boomgaard, M. N., Brinkman, H.-J. M., and van Mourik, J. A., Ionizing radiation enhances platelet adhesion to the extracellular matrix of human endothelial cells by an increase in the release of von Willebrand factor, *Radiat. Res.*, 137, 202, 1994.
81. Colden-Stanfield, M., Kalinich, J. F., and Gallin, E. K., Ionizing radiation increases endothelial and epithelial cell production of influenza virus and leukocyte adherence, *J. Immunol.*, 153, 5222, 1994.
82. Astedt, B. Bergentz, S., and Svanberg, L., Effect of irradiation on the plasminogen activator content in rat vessels, *Experientia*, 30, 1466, 1974.
83. Svanberg, L., Astedt, B., and Kullander, S., On radiation-decreased fibrinolytic activity of vessel walls, *Acta Obstet. Gynecol. Scand.*, 55, 49, 1976.
84. Henderson, B. W., Bicher, H. I., and Johnson, R. J., Loss of vascular fibrinolytic activity following irradiation of the liver — an aspect of late radiation damage, *Radiat. Res.*, 95, 646, 1983.
85. Dancewicz, A. M., Mazanowska, A., and Gerber, G. B., Late biochemical changes in the rat lung after hemithoracic irradiation, *Radiat. Res.*, 67, 482, 1976.
86. Ward, W. F., Kim, Y. T., Molteni, A., and Solliday, N. H., Radiation-induced pulmonary endothelial dysfunction in rats: modification by an inhibitor of angiotensin converting enzyme, *Int. J. Radiat. Oncol. Biol. Phys.*, 15, 135, 1988.
87. Ward, W. F., Molteni, A., and Ts'ao, C., Radiation-induced endothelial dysfunction and fibrosis in rat lung: modification by the angiotensin converting enzyme inhibitor CL242817, *Radiat. Res.*, 117, 342, 1989.
88. Ward, W. F., Molteni, A., Kim, Y. T., and Ts'ao, C., Structure-function analysis of angiotensin-converting enzyme inhibitors as modifiers of radiation-induced pulmonary endothelial dysfunction in rats, *Br. J. Radiol.*, 62, 348, 1989.
89. Kondo, M., Nakanishi, K., Bamba, T., Hosokawa, K., and Masuda, M., Experimental protein-losing gastroenteropathy: role of tissue plasminogen activator, *Gastroenterology*, 71, 631, 1976.
90. Stenberg, B., Risberg, B., and Peterson, H.-I., Irradiation and gastro-intestinal fibrinolysis: an experimental study in the rat, *Eur. J. Clin. Invest.*, 10, 139, 1980.
91. Soreq, H. and Miskin, R., Plasminogen activator in the rodent brain, *Brain Res.*, 216, 361, 1981.
92. Sawaya, R., Rayford, A., Kono, S., Ang, K. K., Feng, Y., Stephens, L. C., and Rao, J. S., Elevated levels of plasminogen activators in the pathogenesis of delayed radiation damage in rat cervical spinal cord *in vivo*, *Radiat. Res.*, 138, 386, 1994.

93. Ohno, M., Matsuura, K., Nakagawa, E., Ohba, T., and Nakata, H., Effects of dextran sulfate on pulmonary radiation injury: clinical and experimental studies, *Nippon Kyobu Shikkan Gakki Zasshi*, 16, 756, 1978.
94. Raymond, J., Yoon, S. C., and Ts'ao, C., Stimulation of radiation-impaired plasminogen activator release by phorbol ester in aortic endothelial cells, *Proc. Soc. Exp. Biol. Med.*, 195, 213, 1990.
95. Ward, W. F., Molteni, A., Ts'ao, C., and Solliday, N. H., Functional responses of the pulmonary endothelium to thoracic irradiation in rats: differential modification by D-penicillamine, *Int. J. Radiat. Oncol. Biol. Phys.*, 13, 1505, 1987.
96. Ward, W. F., Kim, Y. T., Molteni, A., Ts'ao, C., and Hinz, J. M., Pentoxifylline does not spare acute radiation reactions in rat lung and skin, *Radiat. Res.*, 129, 107, 1992.
97. Eldor, A., Vlodavsky, I., Riklis, E., and Fuks, Z., Recovery of prostacyclin capacity of irradiated endothelial cells and the protective effect of vitamin C, *Prostaglandins*, 34, 241, 1987.
98. Hosoi, Y., Yamamoto, M., Ono, T., and Sakamoto, K., Prostacyclin production in cultured endothelial cells is highly sensitive to low doses of ionizing radiation, *Int. J. Radiat. Biol.*, 63, 631, 1993.
99. Ts'ao, C., Ward, W. F., and Port, C. D., Radiation injury in rat lung. I. Prostacyclin (PGI_2) production, arterial perfusion, and ultrastructure, *Radiat. Res.*, 96, 284, 1983.
100. Summers, R. W., Glenn, C. E., Flatt, A. J., and Elahmady, A., Radiation and indomethacin effects on morphology, prostaglandins, and motility in dog jejunum, *Am. J. Physiol.*, 261, G145, 1991.
101. Pausescu, E., Chirvasie, R., Teodosiu, T., and Paun, C., Effects of ^{60}Co γ-radiation on the hepatic and cerebral levels of some prostaglandins, *Radiat. Res.*, 65, 163, 1976.
102. Donlon, M., Steel, L., Helgeson, E. A., Wolfe, W. W., and Catravas, G. N., WR-2721 inhibition of radiation-induced prostaglandin excretion in rats, *Int. J. Radiat. Biol.*, 47, 205, 1985.
103. Ts'ao, C., Ts'ao, F. H. C., Taylor, J. M., Ward, W. F., and Molteni, A., Annexin I concentration, phospholipase activity and thromboxane synthesis in irradiated rat lung, *Radiat. Res.*, 142, 85, 1995.
104. Verheij, M., Koomen, G. C., van Mourik, J. A., and Dewit, L., Radiation reduces cyclooxygenase activity in cultured human endothelial cells at low doses, *Prostaglandins*, 48, 351, 1994.
105. Ward, W. F., Molteni, A., Ts'ao, C., Kim, Y. T., and Hinz, J. M., Radiation pneumotoxicity in rats: modification by inhibitors of angiotensin converting enzyme, *Int. J. Radiat. Oncol. Biol. Phys.*, 22, 623, 1992.
106. Orfanos, S. E., Chen, X. L., Burch, S. E., Ryan, J. W., Chung, A. Y., and Catravas, J. D., Radiation-induced early pulmonary endothelial ectoenzyme dysfunction *in vivo*: effect of indomethacin, *Toxicol. Appl. Pharmacol.*, 124, 112, 1994.
107. Schultz-Hector, S. and Balz, K., Radiation-induced loss of endothelial alkaline phosphatase activity and development of myocardial degeneration. An ultrastructural study, *Lab. Invest.*, 71, 252, 1994.
108. Travis, E. L., Newman, R. A., and Helbing, S. J., WR 2721 modification of type II cell and endothelial cell function in mouse lung after single doses of radiation, *Int. J. Radiat. Oncol. Biol. Phys.*, 13, 1355, 1987.
109. Malaker, K. and Das, R. M., Effect of superoxide dismutase on early radiation injury of lungs in the rat, *Mol. Cell. Biochem.*, 84, 141, 1988.
110. Law, M. P., Radiation-induced vascular injury and its relation to late effects in normal tissues, *Adv. Radiat. Biol.*, 9, 37, 1981.

111. Ward, W. F., Molteni, A., Ts'ao, C., and Hinz, J. M., Captopril reduces collagen and mast cell accumulation in irradiated rat lung, *Int. J. Radiat. Oncol. Biol. Phys.*, 19, 1405, 1990.
112. Zweibel, J. A., Freeman, S. M., Kantoff, P. W., Cornetta, K., Ryan, U. S., and Anderson, W. F., High-level recombinant gene expression in rabbit endothellial cells transduced by retroviral vectors, *Science*, 243, 220, 1989.
113. Brigham, K. L., Meyrick, B., Christman, B., Berry, L. C. Jr., and King, G., Expression of a prokaryotic gene in cultured lung endothelial cells after lipofection with a plasmid vector, *Am. J. Respir. Cell Mol. Biol.*, 1, 95, 1989.
114. Nabel, E. G., Plautz, G., and Nabel, G. J., Site-specific gene expression *in vivo* by direct gene transfer into the arterial wall, *Science*, 249, 1285, 1990.
115. Morishita, R., Gibbons, G. H., Kaneda, Y., Ogihara, T. and Dzau, V. J., Novel *in vitro* gene transfer method for study of local modulators in vascular smooth muscle cells, *Hypertension*, 21, 894, 1993.
116. Brigham, K. L., Meyrick, B., Christman, B., Conary, J. T., King, G., Berry, L. C. Jr., and Magnuson, M. A., Expression of human growth hormone fusion genes in cultured lung endothelial cells and in the lungs of mice, *Am. J. Respir. Cell Mol. Biol.*, 8, 209, 1993.
117. Haimovitz-Friedman, A., Kan, C. C., Ehleiter, D., Persaud, R. S., McLoughlin, M., Fuks, Z., and Kolesnick, R. N., Ionizing radiation acts on cellular membranes to generate ceramide and initiate apoptosis, *J. Exp. Med.*, 180, 525, 1994.
118. Rosiello, R. A., Merrill, W. W., Rockwell, S., Carter, D., Cooper, Jr., J. A. D., Care, S., and Amento, E. P., Radiation pneumonitis. Bronchoalveolar lavage assessment and modulation by a recombinant cytokine, *Am. Rev. Respir. Dis.*, 148, 1671, 1993.

CHAPTER 11

Tumor Bed Phenomenon: Effect on Tumor Progression and Therapy

Luka Milas and Mira Milas

CONTENTS

I. INTRODUCTION

The process of tumor establishment and growth, both at the primary and metastatic sites, takes place in an environment of nonmalignant tissue. In the process, there is a rapid formation of tumor stroma, a loosely defined structural framework that surrounds and infiltrates the tumor. The stroma consists of connective tissue elements, blood vessels, and lymphatic channels, as well as of lymphocytes, macrophages, granulocytes, and other cell infiltrates. The composition of normal tissue elements within the stroma varies greatly among different tumor types. More important than the structural relationship, however, is the role of tumor stroma as an essential regulator of tumor growth, dissemination, and response to therapy, including radiation. This regulation is achieved through direct cell-to-cell interaction between normal and tumor cells, and also indirectly by soluble factors, like growth factors and cytokines or similar signals, produced by both normal and tumor cells.

1-8493-4840-4/98/$0.00+$.50

The vascular component of tumor stroma in particular is a key determinant of the unrestricted growth of tumor cells. In the early phase of tumor growth, small avascular clusters of tumor cells, usually containing fewer than 10^6 cells, are able to survive adequately on nutrients and oxygen that diffuse from the surrounding microenvironment.[1–3] Similarly, leukemic cells and malignant cells in ascites grow unanchored, with nutritional needs met by their immediate surroundings. Sustained growth and metastatic spread of solid tumors, however, require formation of vasculature within the tumor itself.[4] Tumor angiogenesis is a neovascularization process in which the stroma of the microenvironment, specifically the endothelial cells of the existing capillaries and venules, provide the source for the formation of new tumor blood vessels. New blood vessel growth is otherwise rare under normal physiologic conditions, and angiogenesis in the human body occurs in select circumstances such as cycles of uterine endometrium and in response to wounding. Preventing new blood vessels from forming within an avascular tumor leads to a dormant state during which small tumor clusters persist in viable form for long periods of time without evidence of further development or invasion. In some instances the tumor completely regresses.[5,6] Folkman et al.[7] showed that progression of paraneoplastic lesions to malignant tumors requires oncogene expression and abnormal cell proliferation, but that these events alone, without angiogenesis, were not sufficient for this progression.

Tumor angiogenesis has been characterized in enough detail to appreciate the complexity of regulatory factors involved in the overall sequence of events.[1,4,8,9] At some stage in their malignant transformation, tumor cells start producing substances that ultimately induce the vessels of the surrounding tumor stroma to form vascular sprouts directed toward the tumor. These substances include angiogenic factors that diffuse into the stroma and directly activate endothelial cells to form sprouts. In addition, tumor cells elaborate cytokines or other factors that attract macrophages, mast cells, and neutrophils, which are also capable of producing angiogenic factors. Tumor cells also secrete enzymes that release angiogenic factors stored within tumor stroma and stop production of molecules that inhibit angiogenesis.[10] These various signals initiate a sequence of overlapping events that lead to neovascularization of the tumor.[1,9] Within the tumor stroma, capillaries and small venules, which lack an investing smooth muscle layer, become dilated and permeable. Endothelial cells retract from each other and secrete proteases and collagenase that dissolve their own basement membrane and allow migration of endothelial cells through the extracellular matrix and tumor stroma toward the source of the angiogenic stimulus. These migratory strands, the sprouts of new vessels, consist of actively dividing endothelial cells that eventually arrange to form lumina, anastomose with neighboring sprouts to form loops, and lay down new basement membrane. Pericytes and fibroblasts migrate to the region and become associated with the maturing mesh of capillary loops, which eventually grow into the tumor mass itself. Depending on the persistence of angiogenic stimuli and the nutritional needs of the growing tumor, there is considerable remodeling of the vascular stroma within the tumor; repair and regeneration of normal tissue components occurs in the surrounding tumor stroma. The final arrangement of tumor microvasculature following angiogenesis does not follow a consistent pattern, but varies from tumor to tumor and even within different regions of the same tumor. Most of the newly formed vessels in the tumor never acquire a smooth muscle sheath or innervation.[1] They become dilated and tortuous and may even contain tumor cells within the endothelial layer of the vessels.[11] Compared to blood flow in normal tissue, blood flow through the new vasculature in the tumor is actually poorer and decreases with increasing tumor size.

Formation of blood vessels precedes measurable tumor growth (Figure 11.1). Once it begins, the growth of blood vessels is exponential for a few days, but then it slows. The beginning of the vascularization process is accompanied by a rapid increase in tumor mass. Although the extent of angiogenesis is heavily dependent on tumor type, in the majority of solid tumors, the formation of blood vessels eventually lags behind the proliferation rate of tumor cells, leading to inadequate blood supply to certain tumor parts, and thus to necrosis. This early inadequacy further results in microenvironments deficient in oxygen, abnormal in their metabolic activity, and deprived of adequate nutrients. The lack of oxygen and nutrients becomes greater as the distance between cells and blood vessels increases. Hypoxia frequently appears as close as 100 μm from the blood vessel. Naturally, these microenvironments influence cell growth behavior and their response to cytotoxic agents. For example, hypoxic cells become quiescent (nonproliferative) and highly resistant to radiotherapy.[12]

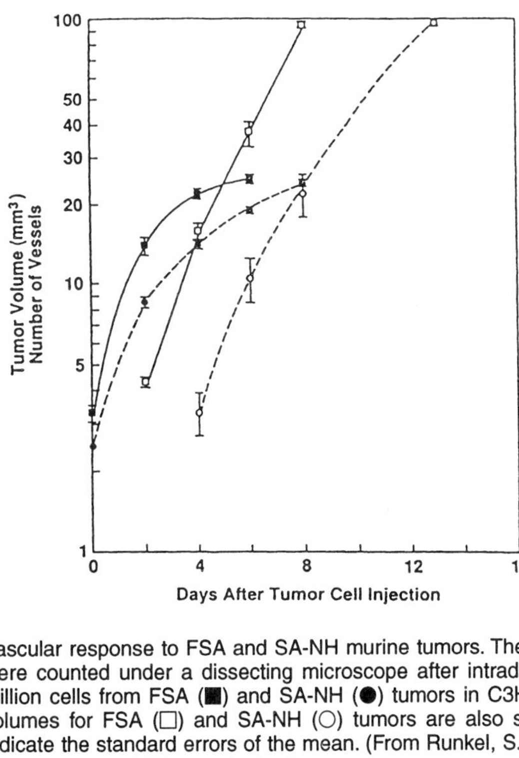

Figure 11.1. Vascular response to FSA and SA-NH murine tumors. The number of vessels were counted under a dissecting microscope after intradermal injection of 1 million cells from FSA (■) and SA-NH (●) tumors in C3Hf/Kam mice. Tumor volumes for FSA (□) and SA-NH (○) tumors are also shown. Vertical bars indicate the standard errors of the mean. (From Runkel, S. et al., *Radiat. Res.*, 126, 237, 1991. With permission.)

Thus, it is evident that endothelial cell interaction with malignant cells is crucial if a tumor is to develop beyond its initial phase of growth to a stage where proliferation is uncontrolled and metastatic spread more likely. The tumor stroma directly and indirectly contributes to the final dynamics of a tumor mass. The multiple components of the tumor stroma are arranged in various combinations and vary from tumor to tumor. The neovasculature created in the growing tumor mass has abnormal features of vessel morphology and lacks a shared pattern of organization that can be recognized among different tumor types. These conditions in general create a microenvironment conducive to diverse cell-to-cell or cell-to-stromal-matrix interactions that can generate virtually limitless heterogeneity in tumor cells' growth behavior and their response to hormones, growth factors, or other stimuli, and therapies. Perturbation of tumor stroma by certain treatments, such as radiation, can modulate these responses and be reflected in specific changes within the tumor stroma, the process of vascularization, and the biological characteristics of the growing tumor mass.

II. TUMOR BED EFFECT

The tumor bed phenomenon refers to observations about tumor growth derived from application of ionizing radiation to tumor stroma. When tumor cells are transplanted into subcutaneous tissue exposed to high doses of ionizing radiation, they require more time to establish tumors than cells transplanted into nonirradiated tissues.[13–16] Further growth of tumors in the irradiated host tissue has also been shown to be retarded,[13–16] and this decrease in tumor growth rate in pre-irradiated tissue has been referred to as the tumor bed effect (TBE). Figure 11.2 provides an example of the TBE effect in a sarcoma tumor cell line. The phenomenon of TBE was first recognized in 1914 by Frankl and Kimbell[13]; early work has been reviewed by Stenstrom et al. (1955),[17] Summers et al. (1964),[14] and Hewitt and Blake (1968).[15]

Irradiation of normal connective tissues produces TBE, which manifests as the growth inhibition of subsequently transplanted tumor cells. Exposing the tumor bed stroma to radiation alters the normal stromal response to the growing tumor and interferes with the natural progression of tumor growth. The inhibition of tumor growth is thought to result from a reduced ability of radiation-injured normal tissue to provide new blood vessels to tumors.[14–16,18] The process of tumor angiogenesis is impaired and blood flow in such tumors is compromised, being as low as 50% of that in tumors growing in nonirradiated tissue.[18] The tumors transplanted into pre-irradiated stroma are also more necrotic than the same size tumors growing in nonirradiated tissue.[19] Furthermore, tumor recurrences after treatment with radiation frequently grow more slowly than corresponding untreated tumors, a finding that is also attributed to the TBE. Using an intradermal assay[20] we recently developed to quantify tumor angiogenesis, we demonstrated that the radiation dose of 25 Gy given 1 day before tumor cell inoculation significantly reduced the number of new vessels formed at the tumor cell injection site. This inhibition of angiogenesis manifested in a reduced growth rate of tumors (Figure 11.3).

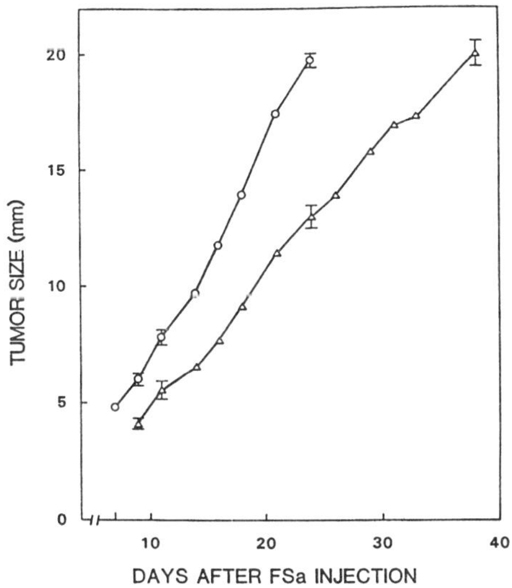

Figure 11.2. Growth of FSa in pre-irradiated tissue. Mice were given injections of 2×10^5 FSA cells into their right thighs, which were either untreated or had been irradiated with a 30-Gy single dose of γ-rays 1 day earlier. The curves trace tumor growth in unirradiated mice (○) and pre-irradiated mice (△). Bars indicate the standard error of the mean. Groups consisted of eight mice each. (From Ito, H. et al., *Int. J. Radiat. Oncol. Biol. Phys.*, 11, 547, 1985. With permission.)

Both the development and magnitude of the TBE depend on the radiation dose delivered to the tumor bed. For most tumors, it has been observed that the TBE begins to appear at radiation doses of approximately 10 Gy, steeply increases as the dose increases, and then plateaus when the dose of irradiation is between 20 and 30 Gy.[21,22] This typical pattern is demonstrated in the example of murine mammary carcinoma MCA-4 (Figure 11.4). The development of TBE in the MCA-4 tumor is evident in both diminished tumor growth rate and prolonged tumor latency. Growth rate was, in this case, affected by lower doses of irradiation than was tumor latency.

The time between irradiation and tumor grafting is also an important factor in the development of TBE.[14,15,23] Studies of TBE are commonly performed by transplanting tumor cells into subcutaneous tissue irradiated 1 day earlier. However, the TBE can be evident when irradiation of the tumor bed takes place months or even more than 1 year before tumor cell transplantation.[15,23] In our own studies using MCA-4 tumor, the time interval between irradiation and tumor cell inoculation was 1 to 200 days.[24] The greatest TBE was observed when cells were transplanted 1 day after irradiation, and the degree of TBE significantly decreased when the cells were transplanted 200 days after irradiation. Interestingly, the degree of TBE greatly depended on the number of tumor cells injected so that it was greater as the number of tumor cells injected was smaller. This suggests that larger number of tumor cells provided stronger angiogenic stimulus that reduced development of TBE. This aspect of TBE will be considered in more detail later in the text.

Although large, single doses of irradiation are usually applied to induce and study TBE in an experimental setting, the doses and schedules of irradiation commonly

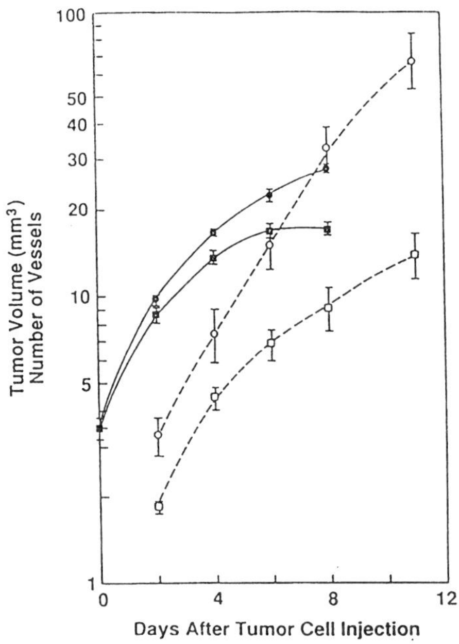

Figure 11.3. Influence of pre-irradiation of tumor bed on tumor neoangiogenesis. The number of vessels (closed symbols) were recorded after the injection of one million cells of SA-NH into the normal skin flap (circles) and into the skin flap that had been irradiated with 25 Gy, 1 day previously (squares). Open symbols represent corresponding tumor volumes. All values are mean ± SEM. (From Runkel, S. et al., *Radiat. Res.*, 126, 237, 1991. With permission.)

used in the clinic can also produce TBE.[22] Daily fractions of 2 Gy, delivered five times per week to a total dose of 60 Gy, was almost as effective in inducing TBE as a single dose of 16.7 Gy.[22] The latter dose was estimated to be equivalent to 60 Gy delivered in 27 fractions on the basis of an α/β model for late-responding tissues that assumes an α/β value of 3.7.[25] These observations imply that TBE may commonly be expressed in the clinical radiotherapy setting as well.

As already noted, the inhibition of tumor growth has been attributed to radiation-induced damage of stromal vessels and other tissue required to support tumor growth. The TBE is generally considered to be the expression of the late-responding tissue, which basically is tissue that proliferates slowly and includes the endothelium of blood vessels. However, a recent analysis of the TBE phenomenon using the linear quadratic model for radiation damage showed that the α/β ratio is 6.2, a value that is on the border of α/β values for acutely responding tissues.[26] This ratio implies that the TBE may not be entirely the expression of late-responding tissues, and that stromal vascular damage alone cannot be the sole explanation for TBE. Recent observations by others[27] and by us[19,21,22] have shown that the development and severity of TBE are strongly dependent on tumor type. These findings imply that the interaction between tumor cells and the irradiated stroma is a major component in the development of the TBE, and that the pathogenesis of TBE is more complex than previously considered.

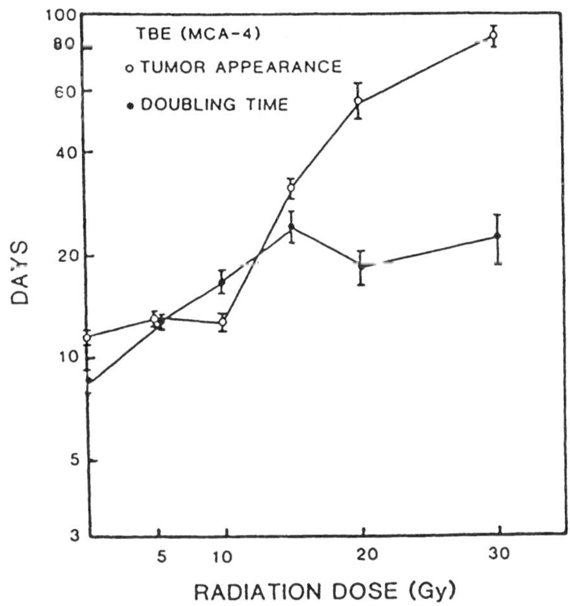

Figure 11.4. Dependency of the tumor-bed effect (TBE) on radiation dose. The right hind thighs of mice were exposed to graded doses of irradiation, and 1 day later mice were given subcutaneous injections of 4×10^5 MCA-4 cells. Tumor appearance is expressed as the time in days from tumor cell injection to a tumor size of 5 mm in diameter. Tumor diameter doubling time was expressed as the time in days needed for tumors to grow from 6 to 12 mm in diameter. Bars indicate standard errors. (From Milas, L. et al., *Cancer Res.*, 46, 723, 1986. With permission.)

The dependence of TBE on tumor type is exemplified by our studies of ten murine tumors, five carcinomas, and five sarcomas (Table 11.1). The tumors were assayed for their TBE expression when transplanted into subcutaneous tissue exposed to 20 Gy γ -rays, a radiation dose that causes maximal or near-maximal TBE. The tumors varied greatly in their manifestation of TBE. However, all five carcinomas showed significant TBE, while only two of the five sarcomas did. The magnitude of TBE was evaluated on the basis of tumor growth rates. The diameter doubling times of tumors growing in irradiated tissue were divided by the diameter doubling times of tumors growing in nonirradiated tissue. The resulting ratios ranged from 1.0 to 2.36, demonstrating the wide range in ability of different tumors to exhibit TBE. These results led to the conclusion that TBE is greatly dependent on tumor type, with carcinomas exhibiting a greater TBE than sarcomas. Therefore, even when extensive stromal radiation damage is present, certain tumors can circumvent the restrictive effect of TBE on tumor growth.

The processes that account for variable expression of TBE among different tumor types are not known. There is substantial evidence that tumors vary in their ability to produce angiogenesis and that the extent of neovascularization depends on the amount of angiogenic factors secreted by tumor cells.[4,8,28] Therefore, a possible reason for the variation in TBE that deserves consideration is the angiogenic capacity

TABLE 11.1 Tumor-Bed Effect in Murine Sarcomas and Carcinomas

	Tumor Diameter Doubling Time[a]		
Tumor Type	**Nonirradiated Tumor Bed**	**Irradiated Tumor Bed (20 Gy)**	**Modification Factor[b]**
Sarcomas			
SA-NH	9.8 ± 1.5[c]	15.4 ± 0.4	1.57
FSA-II	5.3 ± 0.4	6.6 ± 0.4	1.25 (NS)
SA-4020	6.4 ± 0.5	7.7 ± 0.4	1.2 (NS)
FSA	8.3 ± 0.3	19.6 ± 0.5	2.36
NFSA	7.5 ± 0.3	7.4 ± 0.3	1.0 (NS)
Carcinomas			
MCA-35	30.5 ± 2.0	43.6 ± 3.0	1.43
HCA-I	8.0 ± 0.4	10.7 ± 0.5	1.34
OCA-I	27.3 ± 3.5	46.0 ± 5.4	1.69
MCA-4	7.4 ± 0.5	15.3 ± 2.0	2.1
MCA-K	9.4 ± 0.7	16.1 ± 1.1	1.71

[a] The right hind thighs of mice were exposed to single 20-Gy doses of γ-rays, and 1 day later they were given subcutaneous inoculations of 1.5×10^5 (FSA, NFSA) or 3×10^5 (all other tumors) tumor cells. The tumor diameter doubling time was taken as days needed for tumors to grow from 6 to 12 mm in diameter. Groups consisted of 7 to 8 mice each.

[b] Obtained by dividing the tumor diameter doubling time in the irradiated group by that in the nonirradiated group. **NS** indicates not significant.

[c] Mean in days ± standard error.

From Milas, L. et al., *Cancer Res.*, 48, 2116, 1988. With permission.

of a given tumor. There is some indirect evidence from our own studies to support this relationship. First, a highly significant negative correlation exists between the TBE and the number of clonogenic cells in the tumor (Figure 11.5):[29] TBE decreases as the number of clonogenic tumor cells increases. Tumors that contain more clonogenic cells also establish tumors upon transplantation much more easily than those containing fewer clonogenic cells. Since published evidence shows that development of tumor vascularization is a prerequisite for tumor establishment and subsequent progressive growth,[8,23,30] it can be inferred that tumors with more clonogenic cells are more angiogenic, and, for example, better secretors of angiogenic factors. Additional supportive evidence is the already-mentioned dependence of TBE development on the number of tumor cells injected: fewer cells injected result in a stronger TBE.[29]

Second, angiogenic factors are indirectly implicated in TBE from a study with irradiated tumor cells.[22] After irradiation, tumor cells themselves do not lose the property of angiogenesis induction.[28] We found that when irradiated cells from a tumor not exhibiting TBE were admixed with viable cells of a tumor exhibiting TBE and injected into irradiated tissue, the resulting tumors did not exhibit TBE. In explaining this finding, we hypothesized that tumors that do not exhibit TBE could be strong stimulators of vascularization. More direct evidence that irradiated tumor cells stimulate neovascularization comes from our recent study in which irradiated

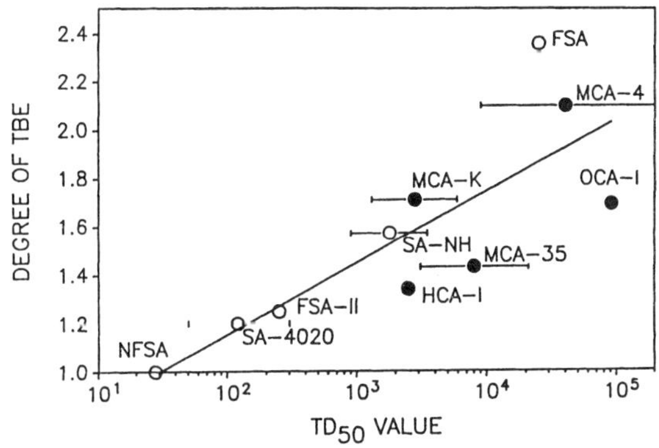

Figure 11.5. Correlation between degree of tumor-bed effect (TBE) and tumor dose (TD_{50}) values of murine tumors. Individual TBE values and individual TD_{50} values ± standard errors have been reported by Milas et al.[19] and Milas,[29] respectively. The correlation coefficient is 0.81 ($p < 0.005$). The ○ indicates sarcoma; ●, carcinoma; horizontal bars represent 95% confidence limits of TD_{50} values. (From Milas, L., *Radiat. Res.*, 123, 232, 1990. With permission.)

tumor cells were injected intradermally and newly formed blood vessels counted.[20] Tiny "tumors" appeared at the injection site as a result of a few postirradiation tumor cell divisions, but these "tumors" soon disappeared. Neovascularization, however, developed at the tumor cell injection site and it persisted for several days after the tumors had disappeared.

Tumor cells are not the only secretors of angiogenic factors. Other cellular components of the tumor microenvironment, such as lymphoid cells that migrate to the region and infiltrate tumors, can secrete angiogenic factors and other soluble substances that stimulate angiogenesis. The variety of substances identified as stimulating endothelial cell proliferation is considerable:[31] vascular endothelial growth factor (VEGF), heparin-binding endothelial cell growth factors like acidic and basic fibroblast growth factors,[32] the transforming growth factors TGF-α and TGF-β,[33] angiogenin,[34] tumor necrosis factor-α (TNF-α),[35] interleukin-1α (IL-1α),[36] platelet-derived endothelial cell growth factor (PD-ECGF),[37] and heparin. Many of these angiogenic factors, like TGF-β, TNF-α, and IL-1α, are known to be secreted by leukocytes. TGF-α can act directly on endothelial cells to induce DNA synthesis and cell proliferation.[33] TGF-β induces a granulation response that draws macrophages to the tumor area, which then release substances (basic fibroblast growth factor, TNF) that stimulate the vascular bed. It is thus conceivable that infiltrates of these cells may contribute to the angiogenesis of tumors in a way that influences expression of the TBE. The participation of macrophages, especially, has been suggested by studies that note a negative correlation between macrophage content of tumors and the TBE:[29] high macrophage content in tumors could be conducive to tumor cell proliferation. Since the tumors presented in Figure 11.5 contain 9% to 83% macrophages,[38] some of this variability may be reflected in the differential expression of the TBE by the tumors.

The TBE thus appears to reflect a similar complexity of interactions that is, in general, present among the components of the tumor stroma and tumor cells. The

TBE varies with radiation dose, tumor type, and macrophage content of tumors, and these factors in turn can be correlated with other important aspects of angiogenesis like endothelial cell-stimulating products of normal and tumor cells. We hypothesize from these various observations that, in the presence of a potent angiogenic stimulus, radiation damage to the vascular system of a tumor bed could be circumvented or minimized. This minimization of damage could be accomplished by eliciting more vigorous capillary sprouting from the irradiated vessels or by attracting endothelial cells to the irradiated bed from nearby nonirradiated sites. Related to this is an interesting observation by Baumann et al.[39] on RIH rat rhabdomyosarcoma. These investigators implanted into 15-Gy pre-irradiated subcutaneous tissue — either single-cell suspensions of tumor cells from which stromal cells were removed by prior *in vitro* culturing of tumor cell suspensions, or tumor fragments that contained both tumor cells and stromal elements. Tumors resulting from the inoculated single-cell suspensions exhibited TBE, whereas those that grew from the inoculated fragments did not. The authors attributed the lack of TBE in the latter situation to the growth of new vessels from the existing vascular vessels in the transplanted fragments.

III. EFFECT OF CHEMOTHERAPEUTIC DRUGS AND HYPERTHERMIA ON TBE

Compared to the TBE induced by radiation, less is known about the effect of other cytotoxic agents, such as chemotherapeutic drugs and hyperthermia, on the tumor bed. With respect to chemotherapy, we observed no TBE after treatment of mice with cyclophosphamide, melphalan, cisplatin, carmustine, or bleomycin (unpublished observation). The effect of chemotherapy agents on irradiation-caused TBE also remains to be explored.

Although there have been more studies on TBE induced by hyperthermia, they report that hyperthermia, similar to chemotherapy, causes no significant TBE.[40–43] The nature of thermal damage to stroma is different from that caused by irradiation; heating stimulated endothelial cells to proliferate rapidly to replace dead cells so that repair with complete recovery of the tumor bed was evident in about 10 days.[42] In experiments where TBE was observed following hyperthermia, the timing of tumor implantation into the heated tumor bed proved important. In contrast to radiation-induced TBE, here the TBE rapidly decreased with increasing time interval from heat application.

Hyperthermia can, however, modify radiation-induced TBE. Depending on experimental conditions, both thermal enhancement and thermal reduction of radiation damage have been demonstrated.[41–43] The TBE expressed after combined heat-radiation treatment appears to vary indirectly with heat dose: the smaller the heat dose applied immediately after irradiation, the larger the TBE compared to irradiation alone.[42] When heat was applied to tumor stroma prior to irradiation, the subsequently observed TBE was found to depend on the time interval between the hyperthermia treatment and irradiation. Short intervals between the two treatment modalities were associated with lower TBE, and when irradiation occurred more than 21 days after heating, the TBE was comparable to the TBE induced by radiation alone.

IV. TBE AND METASTASIS

The role of tumor stroma in metastasis formation has, in general, been acknowledged as important, although it is still poorly understood. The various components of the tumor stroma can function as a physical barrier to tumor cell entry into lymphatic or blood circulation, interact with tumor cells to change their metastatic proclivity, or form microenvironments within a tumor conducive or hostile to tumor cell survival and cell exit into the circulation. By disrupting the anatomic and functional integrity of tumor bed stroma, ionizing radiation could affect the propensity of tumors to metastasize.

There have been only a few studies concerning the effect of pre-irradiated tumor bed on the ability of tumors to spread.[44–46] These studies were limited to only a few tumors, but suggested that radiation-injured tumor bed stroma can facilitate metastatic spread. The mechanism hypothesized from these studies to explain enhanced metastasis formation was that ionizing radiation made the tumor bed, particularly the vasculature but also the lymphatics, more permeable to tumor cell escape. Ionizing radiation increases capillary permeability to fluids, an effect that can already be observed within 1 hour, usually lasts for several weeks, and can persist for several months.[47] In fact, radiation-induced capillary damage has been invoked as a major cause for the local thoracic enhancement of tumor nodule formation in the irradiated thoracic tissue by IV-injected tumor cells.[48] Evidence that the irradiated capillaries are analogously more permeable to tumor cells is, however, scarce. It was recently shown that endothelial cells cultured *in vitro* and exposed to ionizing radiation retract and release enzymes such as heparinase that degrade subendothelial matrix.[49] The resulting gaps between endothelial cells facilitate the attachment of tumor cells to subcellular matrix, which the released enzymes damage and make more permeable, thereby increasing the probability that tumor cells will pass through the capillary wall. Similarly, Heisel et al.[50] observed that irradiated capillary wall components became more susceptible to degradation by tumor cells. Thus, these findings imply that irradiated capillaries are likely to be more permeable than nonirradiated capillaries.

Further studies in our laboratory have contributed to a better understanding of metastasis from tumors growing in the irradiated tumor bed.[19,46] In particular, these studies addressed whether the metastasis-enhancing effect of radiation was dependent on TBE expression. Of the nine different murine tumors used, all six that exhibited TBE were more metastatic when they were grown in pre-irradiated tissue (Table 11.2). This enhanced metastatic capability was manifested as an increase in both the proportion of mice with metastases and the number of metastases per lung. Of the three tumors that did not exhibit TBE, only one showed increased metastatic spread when it grew in irradiated tissue, and this was at a considerably lower level than tumors exhibiting TBE. We also established that pre-irradiation of the tumor bed had no influence on either the incidence or number of lung metastases if tumor size did not exceed 5 mm in diameter. Only tumors with diameters of 7 mm or larger showed enhanced metastatic spread if they grew in pre-irradiated tissue.[19] It is important to note that this dependence on tumor size implies that metastatic events occur later, after the tumor cells have had some time to proliferate and the tumor

TABLE 11.2 Effect of Pre-irradiated Tumor Bed on Metastatic Spread

Tumor	Size at Amputation (mm)	Lung Metastasis[a]			
		Metastasis Incidence[b]		No. Lung Metastases per Mouse[c]	
		Nonirradiated Tumor Bed	Irradiated Tumor Bed	Nonirradiated Tumor Bed	Irradiated Tumor Bed
Tumors exhibiting TBE					
FSA	14	4/15 (27%)	15/16 (94%)	1.5 ± 0.8	15.0 ± 3.7
SA-NH	11	12/31 (39%)	23/26 (88%)	< 1	12.6 ± 3.8
MCA-4	12	0/12 (0%)	7/9 (78%)	0	13.9 ± 5.7
MCA-K	10	7/12 (58%)	9/10 (90%)	5.6 ± 1.0	60.0 ± 21.8
HCA-I	8	5/15 (33%)	11/14 (79%)	< 1	5.0 ± 2.0
OCA-I	12	8/15 (53%)	8/9 (89%)	< 1	16.8 ± 10.7
Tumors not exhibiting TBE					
FSA-II	12	0/10 (0%)	5/10 (50%)	0	2.5 ± 1.3
SA-4020	8	9/14 (64%)	7/15 (47%)	4.2 ± 3.5	1.5 ± 0.8
NFSA	12	2/10 (20%)	2/10 (20%)	< 1	<1

[a] Tumors were generated by injecting 3×10^5 viable tumor cells subcutaneously into the right hind thighs that were normal or had been irradiated with 20 Gy of γ-rays 1 day earlier. When tumors reached a designated size, they were surgically removed; and 18 to 50 days later, depending on tumor type, lungs were checked for the presence of metastases.

[b] Number of mice with metastases over total number of mice; numbers in parentheses are percentages of mice with metastases.

[c] Mean ± standard error.

From Milas, L. et al., *Cancer Res.*, 48, 2116, 1988. With permission.

enlarges. For instance, the tumors in our study reached a diameter of 7 mm approximately 25 days after irradiation. In contrast, vascular permeability to fluids increases virtually from the time of irradiation, and the increase usually lasts for several weeks.[47] Thus, our experimental findings do not support the hypothesis[44,45] that radiation-induced vascular permeability is a major mechanism for the enhanced metastasis of tumors in the irradiated tumor bed.

There are, however, a few alternative explanations. One could argue that during the first few weeks of growth, the tumors were so small that, even in the presence of increased vascular permeability, they could not release into the circulation a sufficient number of tumor cells to result in greater metastasis formation. The argument is supported by our recent observation that even the most metastatic tumors used in the present study rarely metastasized when smaller than 6 to 8 mm in diameter in nonirradiated legs.[51] However, the fact that two of three tumors lacking significant TBE did not show metastasis enhancement even when they were large enough, is strong evidence that the increased permeability is not predominantly responsible for the increased metastatic spread of tumors growing in pre-irradiated tissue. Tumors exhibiting the TBE show a significant reduction in their vascular space and blood flow when growing in pre-irradiated tissues.[18,52] It is plausible that this reduction in total vascular space could minimize the "prometastatic" effect of radiation-induced increase in vascular permeability discussed above.

The second mechanism to be considered, and the one we favor, takes into account the retardation in tumor growth rate exhibited by tumors growing in the pre-irradiated bed. This mechanism assumes a similar rate of tumor cell release into the circulation by tumors growing in the pre-irradiated bed as by those growing in the nonirradiated bed. Since the former tumors take more time to reach a given size, they have more time to release cells into the circulation and start establishing overt metastatic foci. That all six tumors that exhibited the TBE also showed greater metastatic activity conforms to this reasoning. The regression analysis curves in Figure 11.6 show that the percentage increase in metastatic spread per unit time was similar for tumors growing in nonirradiated or pre-irradiated tissue. This further supports the argument that the extra time available for tumors growing in pre-irradiated tissues to metastasize is the main reason for increased metastasis of tumors of a given size. Variability in the time available for tumor growth could also be responsible for the marked differences in the number of metastatic lung nodules disseminated from the same-sized tumors of certain types, such as MCA-K (Table 11.2) and SA-NH,[19] growing in the pre-irradiated tissue. This is because the primary tumors were not all removed at the same time after tumor transplantation, but over a period of several days.

A final consideration is that of necrosis, a common feature of solid tumors, which is also known to be a promoting factor in tumor dissemination.[53] Tumors growing in pre-irradiated tissue were more necrotic than the same-size tumors growing in nonirradiated tissue. The amount of necrosis increased as the tumor size increased, and this finding correlates well with the increase in metastatic spread, whether or not the tumor bed was pre-irradiated. At first glance, these data could be taken to mean that the increase in necrosis with tumor bed pre-irradiation was the underlying process in the TBE-related metastasis enhancement. However, this interpretation is not supported by our further observations that tumors that exhibited neither TBE

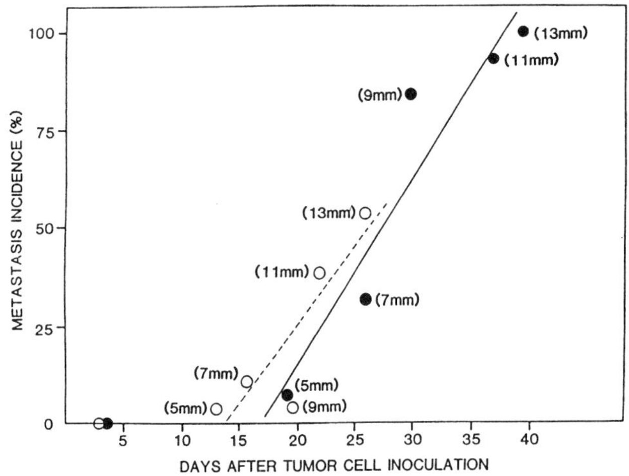

Figure 11.6. Relationship between metastasis incidence and time of tumor amputation after tumor cell inoculation. SA-NH tumors were generated by 3×10^5 tumor cells injected subcutaneously into mice's right hind thighs that were normal (○) or that had been irradiated with 20 Gy of γ-rays (●) 1 day earlier. The tumors were surgically removed at the indicated time (in parentheses are tumor diameter sizes at amputation). (From Milas, L. et al., *Cancer Res.*, 48, 2116, 1988. With permission.)

nor metastases enhancement showed similar increases in the amount of necrosis when growing in the pre-irradiated or nonirradiated beds.[19]

Recurrences after local tumor irradiation also grow more slowly than the corresponding untreated tumors, which has been attributed to the TBE.[54] Based on the data presented above, these recurrences can therefore also be anticipated to be more metastatic at a given size than the same-size untreated tumors. Information supporting this expected behavior of the irradiated tumor comes from studies on the incidence of metastases after failure of radiation therapy.[55] The practical significance of these various findings about tumor and tumor bed irradiation and metastases is that one must attempt to control postirradiation recurrences as early as possible after their diagnosis to prevent or reduce the probability of their metastatic spread.

V. TBE AND RESPONSE TO TUMOR THERAPY

The ramifications of radiation-induced vascular damage are not limited to the direct effects, vis-a-vis the tumor; for instance, the reduced tumor blood supply and slower growth rate. Disruption of the vascular integrity of the irradiated tumor bed can compromise drug delivery to tumors or make tumors more resistant to radiation therapy because of a fractional increase in radioresistant hypoxic cells.

Studies using two murine sarcomas demonstrate the effects of pre-irradiation on tumor curability with chemotherapy or radiotherapy.[23,56] The tumors growing in pre-irradiated tissue responded better to cyclophosphamide than did tumors growing in

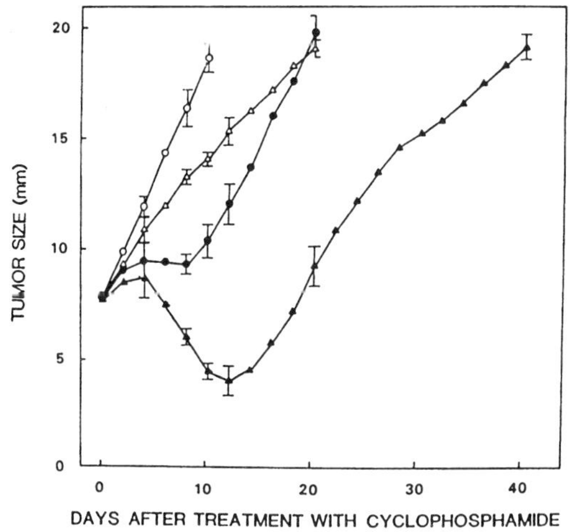

Figure 11.7. Effect of cyclophosphamide on FSA growing in pre-irradiated tissue. Mice were given injections of 2×10^5 FSA cells into their right thighs, which had either been left untreated or been irradiated with single 30-Gy doses of γ-rays 1 day earlier. When tumors grew to 8 mm in diameter, mice were given intraperitoneal injections of 100 mg/kg of cyclophosphamide. The curves trace tumor growth in unirradiated mice (○), unirradiated mice treated with cyclophosphamide (●), pre-irradiated mice (△), and pre-irradiated mice treated with cyclophosphamide (▲). Bars indicate the standard error of the mean. (From Ito, H. et al., *Int. J. Radiat. Oncol. Biol. Phys.,* 11, 547, 1985. With permission.)

nonirradiated tissue (Figure 11.7). A similar effect was noted when tumors were treated with irradiation instead of cyclophosphamide. In both cases, tumor growth delay was the designated endpoint of tumor response. When tumor cure was considered the endpoint of treatment, however, pre-irradiation of the tumor bed significantly lowered the effects achieved by both cyclophosphamide and irradiation.[23,56]

Specifically, the TBE-caused reduction in tumor radiocurability was evidenced by the increase of the tumor control dose (TCD_{50}) values, i.e., the radiation dose yielding 50% tumor cure (Figure 11.8). Tumor irradiation under hypoxic conditions yielded increased TCD_{50} values for tumors that grew in pre-irradiated legs, and the hypoxic cell radiosensitizer misonidazole reduced these TCD_{50} values more than it did the TCD_{50} values of tumors that grew in unirradiated legs.[56] This disparity observed with misonidazole treatment implies that the TBE-induced increase in TCD_{50} was due to an increase in the hypoxic fraction of cells in the tumor.

The hypoxic cell fraction in tumors following irradiation has been estimated to increase by as much as fourfold. For instance, the hypoxic cell fraction increased from 3 to 12% in a 6-mm SA-NH sarcoma. However, there was a wide variability among individual tumors in the extent of hypoxic fraction increase. This is implied by the shallow slope of the radiocurability curve in the TBE group (Figure 11.8). The same assortment of complex factors, including angiogenic stimuli, that contribute to the differential expression of TBE probably also participated in the tumor response in these experiments. The wide heterogeneity in tumor radioresponse is most likely related to uneven neovascularization by the irradiated tumor stroma.

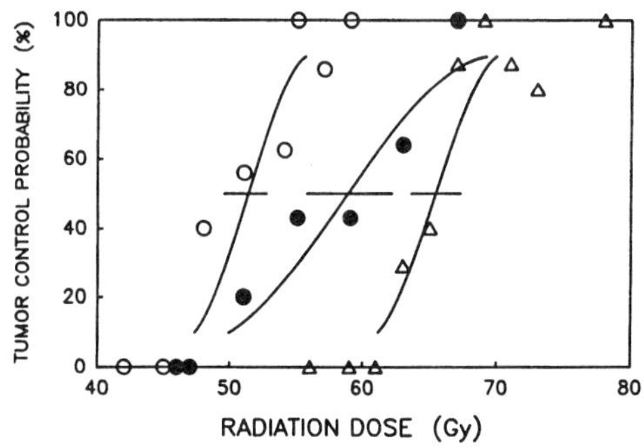

Figure 11.8. Radiation dose-response curves for local control of SA-NH tumors growing in normal mouse legs and irradiated under normal conditions (○) or in 30-Gy pre-irradiated mouse legs irradiated under normal (●) or hypoxic (△) conditions. Tumors were 6 mm in diameter at the time of radiotherapy. (From Milas, L. et al., *Int. J. Radiat. Oncol. Biol. Phys.*, 16, 139, 1989. With permission.)

It can be concluded that the TBE influence on tumor response to irradiation or cyclophosphamide depends on the endpoint chosen to mark tumor therapy response. TBE did indeed delay tumor regrowth, but it reduced tumor curability. We have proposed a model (Figure 11.9) that explains these apparently conflicting results as different phases on a continuum of tumor response to therapy that is affected by TBE impairment of tumor blood supply, TBE influence on hypoxia within the tumor, and tumor cell growth kinetics. Initially, the blood supply impairment might be expected to reduce tumor radiocurability through an increase in the proportion of hypoxic cells and to reduce tumor curability by cyclophosphamide, and likely by other chemotherapeutic agents, through poorer drug access to tumor cells. The reduced curability of tumors growing in pre-irradiated tissue indicates that the net cell kill achieved by a given dose of cytotoxic agents is reduced. However, even a moderate reduction in cell kill can still translate into longer tumor regrowth delay because of the still-present TBE effect of retarding growth of surviving tumor cells. Thus, depending on the endpoint of tumor treatment, the TBE can be concluded to be either beneficial or deleterious. Since the TBE is also caused by radiation doses and fractionation schedules commonly used in the clinic,[22] it is important to recognize that the TBE requires a closer assessment of the final outcome of clinical radiotherapy.

VI. CONCLUSION

Since it was first observed as the slowed tumor growth after pre-irradiation of the tumor environment, the tumor bed effect (TBE) has been the subject of many experimental studies. These have characterized the TBE phenomenon in considerable detail and affirmed the important implications for clinical radiotherapy. Major discoveries concerning TBE are summarized in Table 11.3. The TBE results from a radiation-induced insult to the tumor-bed stroma, which impairs ability of the stroma to provide the tumor with new blood vessels via the angiogenic mechanism. By

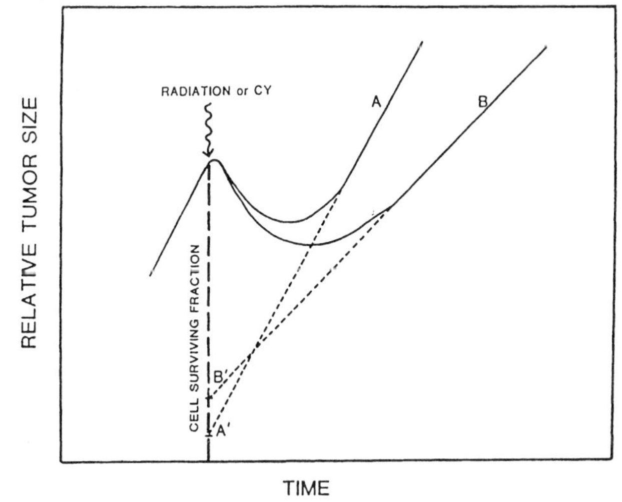

Figure 11.9. Model demonstrates prolonged growth delay but reduced curability of tumors growing in pre-irradiated tissue. Pre-irradiation of the tumor bed reduces the cell kill (increases the fraction of cells surviving) resulting from subsequent radiation or cyclophosphamide (CY) treatment from A′ to B′, thus reducing the probability of tumor cure, but retards the growth rate of surviving cells such that the therapeutic effect measured by growth delay is enhanced (B vs. A). (From Ito, H. et al., *Int. J. Radiat. Oncol. Biol. Phys.*, 11, 547, 1985. With permission.)

disrupting the complex interaction of angiogenesis, where endothelial cells of the stroma and angiogenic factors produced by tumor and other cells are crucial components, radiation leads to decreased tumor blood supply. Tumors growing in such pre-irradiated environments are deprived of oxygen and other nutrients, they have an increased fraction of hypoxic cells, and they become more necrotic. The measurable expressions of the TBE include reductions in growth of the initially transplanted tumor and tumor recurrences. These have been shown to vary in magnitude, depending on tumor type, radiation dose, macrophage content of tumors, and kind of therapy (hyperthermia vs. chemotherapy vs. radiation). More important, the TBE enhances metastasis formation and in this respect makes tumors appear more malignant, likely, as we hypothesized, because the slowed tumor growth rate allows more time for seeding to distant sites.

Table 11.3 Tumor Bed Effect

Cause	Pathophysiologic Changes	Growth Behavior	Response to Therapy
Reduced neovascularization	Reduced tumor blood flow Increased hypoxia Metabolic deprivation Increased necrosis	Prolonged latency Tumor growth retardation Increased metastatic spread	Reduced radiocurability Reduced chemocurability

All of these changes caused by the TBE ultimately influence tumor response to therapy, in that both radiotherapy and chemotherapy become less curative. Thus, the TBE may have a significant impact on the clinical management of cancer patients. Continued studies of the TBE phenomenon have every reason to be fruitful because TBE, as an experimental model, addresses many important factors: tumor stroma, tumor cell growth and behavior, and local changes associated with various therapies. An increased knowledge of this poorly understood field of tumor biology may culminate in enabling us to regulate tumor stroma influences on tumor growth to the benefit of the tumor host, and also to utilize this knowledge for developing more effective therapeutic strategies.

REFERENCES

1. Paweletz, G. L. and Kierim, M., Tumor-related angiogenesis, *Critical Rev. Oncol./Hematol.*, 9, 197, 1989.
2. Thomlinson, R. H. and Gray, L. H., The histological structure of some human lung cancers and the possible implications for radiotherapy, *Br. J. Cancer*, 9, 539, 1955.
3. Sutherland, R. M. and Durand, R. E., Radiation response of multicell spheroids — An *in vitro* tumor model, *Curr. Concepts Radiat. Res.,* 11, 87, 1976.
4. Folkman, J., Tumor Angiogenesis, *Adv. Cancer Res.*, 19, 331 1974.
5. Ausprunck, D. H., Falterman, K., and Folkman, J., The sequence of events in the regression of corneal capillaries, *Lab. Invest.*, 38, 284, 1978.
6. Folkman, J., Langer, R., Linhardt, R. J., Haudenschild, C., and Taylor, S., Angiogenesis inhibition and tumor regression caused by heparin or a heparin fragment in the presence of cortisone, *Science*, 221, 719, 1983.
7. Folkman, J., Watson, K., Ingber, D., and Hanshan, D., Induction of angiogenesis during the transition from hyperplasia to neoplasia, *Nature*, 339, 58, 1989.
8. Folkman, J., Tumor angiogenesis, *Adv. Cancer Res.*, 43, 175, 1985.
9. Furcht, L. T., Critical factors controlling angiogenesis: cell products, cell matrix, and growth factors, *Lab. Invest.*, 55, 505, 1986.
10. Bouck, N., Angiogenesis: a mechanism by which oncogenes and tumor suppressor genes regulate tumorigenesis, in *Oncogenes and Tumor Suppressor Genes in Human Malignancies*, Benz, C. C. and Liu, E. T., Eds., Kluwer Academic, Boston, 1993, chap. 19.
11. Jain, R. K., Delivery of novel therapeutic agents in tumors. Physiological barriers and strategies, *J. Natl. Cancer Inst.,* 81, 570, 1989.
12. Withers, H. R. and Peters, L. J., Biologic aspects of radiation therapy, in *Textbook of Radiotherapy,* 3rd ed., Fletcher, G. H., Ed., Lea and Febiger, Philadelphia, 1980, 103.
13. Frankl, O. and Kimball C. P., Uber die Beeinflussung von Mause-Tumoren durch Rontgenstrahlen, *Wien Klin Wochenschr*, 27, 1448, 1914.
14. Summers, W. C., Clifton, K. H., and Vermund, H., X-irradiation of the tumor bed. I. A study of the indirect actions of radiation on transplantable tumors, *Radiology*, 82, 691, 1964.
15. Hewitt, H. B. and Blake, E. R., The growth of transplanted murine tumors in pre-irradiated sites, *Br. J. Cancer*, 22, 808, 1968.

16. Urano, M. and Suit, H. D., Experimental evaluation of tumor bed effect of C3H mouse mammary carcinoma and of C3H mouse fibrosarcoma, *Radiat. Res.*, 45, 41, 1970.
17. Stenstrom, K. W., Vermund, H., Mosser, D. G., and Marrin, J. F., Effects of roentgen irradiation on the tumor bed, I. The inhibiting action of local pretransplantation roentgen irradiation (1500 r) on the growth of mouse mammary carcinoma, *Radiat. Res.*, 2, 180, 1955.
18. Jirtle, R., Rankin, J. H. G., Clifton, K. H., et al., Effect of X-irradiation of tumour bed on tumour blood flow and vascular response to drugs, *Br. J. Cancer*, 37, 103, 1978.
19. Milas, L., Hirata, H., Hunter, N., and Peters, L. J., Effect of radiation-induced injury of tumor bed stroma on metastatic spread of murine sarcomas and carcinomas, *Cancer Res.*, 48, 2116, 1988.
20. Runkel, S., Hunter, N., and Milas, L., An intradermal assay for quantification and kinetics studies of tumor angiogenesis in mice, *Radiat. Res.*, 126, 237, 1991.
21. Begg, A. C. and Denekamp, J., Stromal damage as a complication in the interpretation of tumor growth delay, *Eur. J. Cancer Clin. Oncol.*, 19, 1639, 1983.
22. Milas, L., Ito, H., Hunter, N., et al., Retardation of tumor growth in mice caused by radiation-induced injury of tumor bed stroma: dependency on tumor type, *Cancer Res.*, 46, 723, 1986.
23. Ito, H., Barkley, T. Jr., Peters, L. J., and Milas, L., Modification of tumor response to cyclophosphamide and irradiation by preirradiation of the tumor bed: prolonged growth delay but reduced curability, *Int. J. Radiat. Oncol. Biol. Phys.*, 11, 547, 1985.
24. Milas, L., Hunter, N., and Peters, L. J., The tumor bed effect: dependence of tumor take, growth rate, and metastasis on the time interval between irradiation and tumor cell transplantation, *Int. J. Radiat. Oncol. Biol. Phys.*, 13, 379, 1987.
25. Withers, H. R., Thames, H. D., Jr., and Peters, L. J., A new isoeffect curve for change in dose per fraction, *Radiother. Oncol.*, 1, 187, 1983.
26. Begg, A. C. and Terry, N. H. A., The sensitivity of normal stroma to fractionated radiotherapy measured by a tumor growth rate assay, *Radiother. Oncol.*, 2, 333, 1984.
27. Begg, A. C. and Terry, N. H. A., Stromal radiosensitivity: influence of tumor type on the tumor bed effect assay, *Br. J. Radiol.*, 58, 93, 1985.
28. Auerbach, R., Arensman, R., Kubai, L., and Folkman, J., Tumor-induced angiogenesis: lack of inhibition by irradiation, *Int. J. Cancer*, 15, 241, 1975.
29. Milas, L., Tumor bed effect in murine tumors: relationship to tumor take and murine macrophage content, *Radiat. Res.*, 123, 232, 1990.
30. Shor, A. M. and Shor, S. L., Tumor angiogenesis, *J. Pathol.*, 141, 385, 1983.
31. Mahadevan, V. and Hart, I. R., Metastasis and angiogenesis, *Acta Oncologica*, 29, 97, 1990.
32. Gospodarewicz, D., Neufeld, G. M., and Schweigerer, L., Fibroblast growth factor: structural and biological properties, *J. Cell Physiol.*, 5(Suppl.), 15, 1987.
33. Schreiber, A. B., Winker, M. E., and Derynck, R., Transforming growth factor α: a more potent angiogenic mediator than epidermal growth factor, *Science*, 232, 1250, 1986.
34. Folkman, J. and Klagsbrun, M., Angiogenic factors, *Science,* 235, 442, 1987.
35. Leibovich, S. J., Polverini, P. J., Shepard, H. M., et al., Macrophage-induced angiogenesis is mediated by tumor necrosis factor α, *Nature,* 329, 630, 1987.
36. Mahadevan, V., Hart I. R., and Lewis, G. P., Factors influencing blood supply in wound granuloma quantitated by a new *in vivo* technique, *Cancer Res.*, 49, 415, 1989.

37. Ishikawa, F., Miyazono, K., Hellman, U., et al., Identification of angiogenic activity and the cloning and expression of platelet-derived endothelial cell growth factor, *Nature*, 338, 557, 1989.
38. Milas, L., Wike, J. Hunter, N., et al., Macrophage content of murine sarcomas and carcinomas: associations with tumor growth parameters and tumor radiocurability, *Cancer Res.*, 47, 1069, 1987.
39. Baumann, M., Wurschmidt, F., Twardy, A., and Beck-Bornholdt, H. P., Impact of tumor stroma on expression of the tumor bed effect in R1H rat rhabdomyosarcoma, *Radiat Res.,* 140, 432, 1994.
40. Urano, M., and Cunningham, M., Insignificant tumor bed effect after pretransplantation hyperthermia, *Cancer Res.*, 40, 26, 1980.
41. Wheldon, T. E. and Hingston, E. C., Differential effect of hyperthermia and X-irradiation on regrowth rate and tumor-bed effect for a rat sarcoma, *Br. J. Cancer*, 45, 265, 1982.
42. Wondergem, J., Begg, A. C., and Haveman, J., Effects of hyperthermia and X-irradiation on mouse stromal tissue, *Int. J. Radiat. Biol.*, 50, 65, 1986.
43. Hill, S., Smith, K. A., and Denekamp, J., Stromal sensitivity to radiation and hyperthermia, *Cancer*, 56, 383, 1987.
44. Van den Brenk, H. A. S., Crowe, M. C., and Stone, M. G., Reactions of the tumor bed to lethally irradiated tumor cells and Revesz effect, *Br. J. Cancer*, 36, 94, 1977.
45. Rappaport, D. S. and Brown, J. M., Influence of localized preinoculation irradiation of the foot on lymphatic metastasis from a primary tumor growing in the foot, *Radiat. Res.*, 78, 108, 1979.
46. Milas, L., Hunter, N., and Peters, L. J., The tumor bed effect: dependence of tumor take, growth rate, and metastasis on the time interval between irradiation and tumor cell transplantation, *Int. J. Radiat. Oncol. Biol. Phys.,* 13, 379, 1987.
47. Casarett, G. W., *Radiation Histopathology,* CRC Press LLC, Boca Raton, FL, 1980, Vol. I, 41.
48. Milas, L. and Peters, L. J., Conditioning of tissues for metastasis formation by radiation and cytotoxic drugs, in *Cancer Invasion and Metastasis: Biological and Therapeutic Aspects*, Nicolson, G. L. and Milas, L., Eds., Raven Press, New York, 1984, 321.
49. Nicolson, G. L., Custead, S. W., Dulski, K. M., and Milas, L., Effects of gamma irradiation on cultured rat and mouse microvessel endothelial cells: metastatic tumor cell adhesion, subendothelial matrix degradation, and secretion of tumor cell growth factors. *Clin. Exp. Metastasis*, 9, 457, 1991.
50. Heisel, M. A., Long, W. E., Stowe, S. M., and Jones, P. A., Effects of X-irradiation on artificial blood vessel wall degradation by invasive tumor cells, *Cancer Res.*, 44, 2441, 1984.
51. Milas, L., Hunter, N., Basic, I., Volpe, J. P., and Tofilon, P. J., Effect of the radiosensitizer misonidazole and the radioprotector diethyldithiocarbamate on spontaneous metastasis formation of murine tumors, *Int. J. Radiat. Oncol. Biol. Phys.*, 12, 1071, 1986.
52. Jirtle, R. and Clifton, K. H., Effect of preirradiation of the tumor bed on the relative vascular space of mouse gastric adenocarcinoma 328 and mammary adenocarcinoma CA 755, *Cancer Res.*, 33, 764, 1973.
53. Weiss, L., *Principles of Metastasis,* Academic Press, New York, 1985, 50.
54. Suit, H. D., Sedlacek, R., Fagundes, L., et al., Time distributions of recurrences of immunogenic and nonimmunogenic tumors following local irradiation, *Radiat. Res.*, 73, 251, 1978.

55. Ramsay, J., Suit, H. D., and Sedlacek, R., Experimental studies on the incidence of metastases after failure of radiation treatment and the effect of salvage surgery, *Int. J. Radiat. Oncol. Biol. Phys.*, 14, 1165, 1988.
56. Milas, L., Hunter, N., and Peters, L. J., Tumor bed effect-induced reduction of tumor radiocurability through the increase in hypoxic cell fractions, *Int. J. Radiat. Oncol. Biol. Phys.*, 16, 139, 1989.

CHAPTER 12

Photodynamic Therapy and Endothelial Cells

David C. West and James V. Moore

CONTENTS

I. INTRODUCTION

The most famous face in photodynamic therapy is undoubtedly that of Friedrich Meyer-Betz, who in 1912 injected himself with a massive dose of hematoporphyrin and then exposed various parts of his person to sunlight.[1] In 1913, by which time he had presumably recovered somewhat, he published the results of this self-experiment, including "before and after" photographs that show his face and hands with the classic signs of a severe inflammatory response. Eighty-five years later, numerous studies reveal that the observed short- and long-term effects of photodynamic therapy (PDT) on normal tissues and tumors are strongly mediated by known inflammatory regulators and modifiers of blood flow, vessel tone, and endothelial biochemistry. Increased understanding of the detailed mechanisms of vascular-mediated injury leads to the possibility of manipulating effect, either to enhance the destruction of tumors or to reduce acute or chronic damage to normal tissue. A point we will

1-8493-4840-4/98/$0.00+$.50

emphasize is that the outcome of PDT often seems to arise from more subtle effects than mere release of all cell contents after the death of, in the present context, an endothelial cell. The precise biochemical/physiological outcome is conditioned by the doses of both photosensitizer and light, in complex interaction with various aspects of the target cell population(s).

II. INFLAMMATION-LIKE RESPONSES AND PHOTOTOXICITY

In rodent skin, PDT induces a prompt (30 to 60 min) and delayed (>6 hours) "erythema," the early time associated with large changes in vascular permeability leading to edema,[2,3] the latter with leukocyte adherence to endothelium followed by infiltration.[2] As will become apparent throughout this chapter, many of the events occurring after PDT are both drug-and light-dose dependent; for example, the extent of skin edema and the probability of subsequent necrosis of the tissue[3] (Figure 12.1). At the cellular level, the number of leukocytes adherent to venular endothelium in rat cremaster muscle was shown to rise as a function of light dose in the PDT treatment.[4] Mechanistically, Lim and colleagues employed mutant strains of mice to demonstrate that the phototoxic vascular response was determined in a complex manner by activation of the complement system, degranulation of mast cells, and local leukocyte concentration.[5] Animals deficient in any one of these parameters exhibited a greatly reduced (but not totally abolished) phototoxic response.

III. VASCULAR CONFORMATION AND PERMEABILITY: BEFORE AND AFTER PDT

Vascular permeability might seem a potentially important parameter for PDT, in that some commonly used injected photosensitizers circulate in association with macromolecules, such as low-density lipoprotein and albumin. The probability of drug uptake into the tumor mass might then be a product of the "leakiness" of the tumor neovasculature. In recent experiments with mouse skin, Roberts and Hasan[6] found that making vessels artificially "leaky" (tumor-like) did indeed increase uptake of a variety of photosensitizers. Of additional interest however, there was no evidence of increased rates of *retention*, a parameter important for the therapeutic differential between tumor and normal tissue believed to be obtained by allowing a 2- to 3-day gap between sensitizer injection and light delivery. Subsequently, a positive correlation was found between (1) the amount of vascular permeability factor (VPF), or vascular endothelial growth factor (VEGF) produced by a series of preclinical tumors, (2) the amount of extravasation induced by VPF, and (3) the initial uptake of AlPcS. However there was again little indication of differential rates of retention.[7]

PDT itself induces marked changes in conformation and permeability of vessels and water balance, within a few hours. Vasoconstrictions have been observed in both tumor and normal tissue models, as a result of (1) mural thrombus, in which case red blood cell column diameter may be reduced without changes in the vessel wall diameter[8-10] (Table 12.1); but also, (2) genuine vasoconstrictions appear to occur in

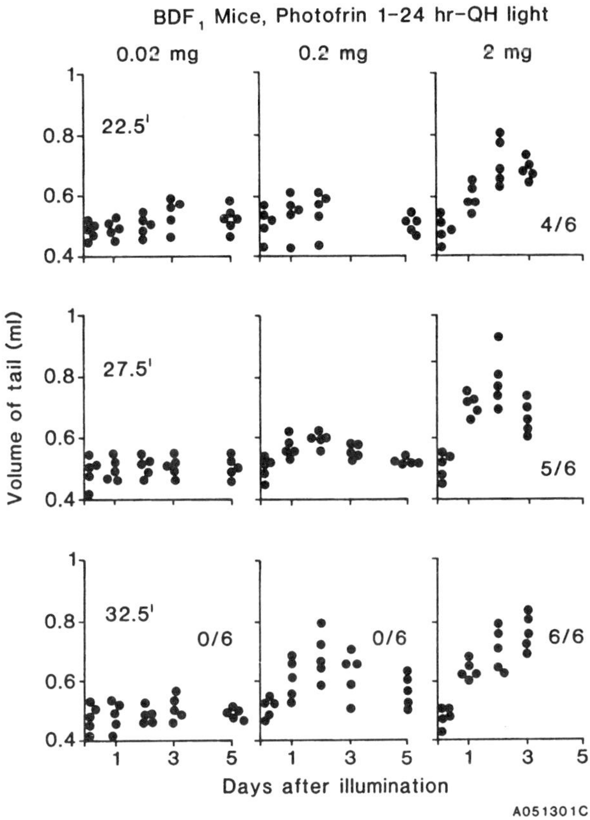

Figure 12.1. Degree of edema in murine tail skin as a function of: dose of hematoporphyrin derivative (0.02, 0.2, and 2.0 mg kg^{-1}), duration of illumination (22.5, 27.5, and 32.5 min), and time after illumination (6 hr to 5 days). Data points are for individual animals. Probability of subsequent necrosis of the tissue between 5 and 30 days is given as 0/6, 4/6, etc. (From Moore, J. V. et al., *Br. J. Radiol.*, 59, 257, 1986. With permission.)

arterioles and to a lesser extent in venules.[4] Effects on the endothelium itself may participate in these vasoconstrictions, in the impairment of production or release of the nitric oxide-related, endothelium-derived relaxing factor (EDRF) (e.g., in isolated rat aortas or *in situ* in rat cremaster muscle, in arterioles of the mouse brain).[11-14] This is the opposite effect to that invoked by UV irradiation, which induces a vasodilatory effect mediated by increased endothelial EDRF release.[15,16] Manipulating EDRF levels in PDT-damaged brain endothelium (e.g., reducing synthesis by the application of arginase or increasing it with L-arginine), had pronounced effects on the capture of platelets by the damaged endothelium (respectively increasing and decreasing capture).[17] Fingar et al.[4] made the further interesting observation of *discrete* areas of constriction and dilation along the length of individual arterioles and speculated on a central role of local endothelial damage leading to high local concentrations of vasoconstrictory compounds. Whether these hypothesized endothelial "hot spots" would be expected to be random or morphologically/anatomically related, is an intriguing question. An overall effect of these vascular events is commonly an increase in tumor fluid pressure, an outcome again susceptible to manipulation.[18,19] In a chondrosarcoma, gross tumor interstitial pressure fluctuated in a complex way, with an initial drop that could be abrogated by the serotonin antagonist ketanserin, and a subsequent rise that was in turn prevented by the cyclooxygenase inhibitor indomethacin (Table 12.1).[18]

The effect of indomethacin suggests that a component of PDT outcome may be mediated by the release of eicosanoids. This is seemingly more subtle than simply a nonspecific consequence of cell death. *In vitro*, cultured human umbilical vein and omental microvascular endothelial cells, loaded with uroporphyrin (UP) or protoporphyrin (PP), selectively released $PGF_{2\alpha}$ (a vasoconstrictor *in vivo*) on illumination.[20] In contrast, Henderson et al.[21] exposed human umbilical vein endothelial cell (HUVEC) and bovine aortic endothelium (BAE) cultures to $ZnPcS_{1,2}$ (zinc phthalocyanine mono-, disulfonate) and light, and detected the release of primarily PGF_{1a} (a vasodilator *in vivo*) and only some $PGF_{2\alpha}$ (PGF = prostaglandin F). The difference in these findings may reflect selectivity due to the different sensitizers, but may also

Table 12.1 Effect of PDT on Vascular-Related Parameters in Tumors and Normal Tissues

Tumor/tissue (& host)	Treatment	Endpoint	Time* and Size of Maximum Change	Ref.
Cremaster muscle (rat)	PhII-24hr-135 J/cm²:	Venule diameter	% of control, at 60′	
	5 mg/kg		100	4
	10		92	
	25		60	
	25 + indomethacin		100	
Cremaster muscle (rat)	FITC-BSA-30′-360J/cm²	RBC column diameter	% of control, at 60′	11
	w/o indomethacin		92	
	w indomethacin		49	
Cremaster muscle (rat)	PhII-24hr-135 J/cm²:	Venule permeability to FSCN-tagged albumin	Relative extravascular fluorescence:	
	5 mg/kg		0	4
	10		70′ +250	
	25		>90′ +280	
	25 + indomethacin		0	
Chondrosarcoma (rat)	PhII-24hr-135 J/cm²:	Change in tumor interstitial pressure	(change, mmHg)	
	5 mg/kg		0	
	10		30′ −0.3 120′ +1.4	18
	25		30′ −0.7 100′ +2.1	
	25 mg/kg-24hr-hυ:			
	135 J/cm²		30′ −0.5 120′ +2.2	
	135 + indomethacin		0	
	270		60′ −0.8 160′ +1.8	
	540		60′ −0.8 170′ +1.5	
Melanoma (hamster)	5 mg/kg PhII-24hr-100 J/cm²	Change in tumor interstitial pressure	(change, mmHg) 60–180′ +4 24hr −8	19

be an aspect of the endothelia: the $PGF_{1\alpha}$ levels were much lower in the human cell medium than the bovine.[21]

In vivo, PDT treatment of rodent skin using hematoporphyrin (HPD) raised plasma levels of leukotriene C_4.[20] Further evidence of selectivity comes from Fingar and colleagues[22] who found four- to fivefold increases in thromboxane B_2 (TXB_2, a spontaneous hydrolysis product of thromboxane A_2) and leukotriene B_4 immediately after treatment of a chondrosarcoma in rats, and similar increases for TXB_2 in the plasma of PDT-treated human patients. In both species, levels of 5-HETE (5-hydroxyeicosatetraenoate) remained unchanged. These raised levels of vasoactive eicosanoids were associated with good control of tumor growth and, more significantly, tumor responsiveness to therapy could be largely or entirely blocked by eicosanoid inhibitors such as imidazole or indomethacin.[22] The degree of induction of eicosanoids for a given class of sensitizer can vary markedly with the precise pharmacokinetic properties of the molecule; thus, the hydrophobic monosulfonated- or tetrabutylated zinc phthalocyanines caused large increases in plasma TXB_2 and LTB_4, with lesser increases after the more hydrophilic $ZnPcS_2$ and $ZnPcS_4$.[23]

Thus, some *in vitro* results appear to contradict the *in vivo* studies in that the endothelial cells release mainly PGI_2, a potent vasodilator, while serum shows increases in TXA_2, LTB_4, and LTC_4, which would induce vasoconstriction and leukocyte accumulation. The use of TXA_2 inhibitors, *in vivo,* inhibits PDT-precipitated thrombus formation and reduces both vasoconstriction and vascular permeability, suggesting that TXA_2 may mediate many of the vascular effects observed with PDT. The increased serum levels of TXA_2, LTB_4, and LTC_4 do not apparently originate from the endothelium and are probably released from activated platelets and leukocytes, either directly by PDT or through the PDT "activation" of the endothelium.

IV. VESSEL OCCLUSION

The very rapid action of PDT on bringing about vessel occlusion via thrombi can be directly visualized in arterioles of the hamster cheek-pouch preparation.[24] Using a vessel-confined "photosensitizer" — FITC-dextran (FITC = fluorescein isothiocyanate) — the time to complete occlusion in these vessels could be as little as 45 seconds, with the occlusion persisting for hours. Because of the relative dwell time of platelets and endothelium in the illumination field, the authors felt that the occlusive effects were more likely to result from primary endothelial injury, with secondary responses of the platelets. One potential mechanism for this is suggested in the *in vitro* work of Sporn and Foster,[8] who found a rapid microtubule depolymerization in cultured HUVECs treated with Photofrin™ and light. At sublethal treatment doses, this effect was transient, and they postulated that the depolymerization resulted from increased intracellular calcium concentrations. Disruption of endothelial cell microtubules would disrupt the endothelial permeability barrier and potentially expose the thrombogenic subendothelial matrix, thus contributing to local thrombus formation. Using the same experimental model, Foster et al.[9] also demonstrated a rapid release of von Willebrand Factor (see below). This release was also dependent on increased intracellular calcium levels.

In the hamster cheek-pouch study cited earlier,[24] acetylsalicylic acid (ASA) greatly extended the time to clotting or even abrogated the effect. Experimental interventions by compounds such as ASA lead to the important conclusion that the time-scale and/or nature of vascular injury induced by different photosensitizers may differ: Stern et al.[25] observed that whereas DHE (dihematoporphyrin) induced very rapid hemorrhage in a rat observation chamber, ClAlPcS (chloraluminum phthalocyanine tetrasulfonate) did not cause maximal vascular response until after 4 hours. With ClAlPcS-mediated PDT, ASA had little or no effect if given immediately after PDT, but the vascular effect was totally abolished with ASA given 6 hr after treatment.[26] Where such vessel disruption/aggregation occurs first and its extent have been the subject of investigation. Rosenblum and El-Sabban,[27] using FSCN (fluorescein) as a photosensitizer, found that aggregation invariably occurred first in venules of normal murine brain, one possible reason advanced being the discouragement of aggregation by rapid blood flow in arterioles. Similarly, in the rat cremaster muscle model, embolization after DHE- or FITC-BSA-mediated PDT was more complete in smaller venules than in arterioles.[11,28] Biochemical as well as mechanical differences in vessels may mediate these differential effects; thus, in the study of mouse brain just cited,[27] cyclooxygenase inhibitors blocked thrombus formation in arterioles but not venules.

What processes underlie these observations? We have already noted the release from endothelia of eicosanoids that might have direct or indirect effects on thrombus formation. Given the *in vivo* evidence for eicosanoids being associated with aggregation after PDT, attention should again be drawn to the *in vitro* results of Henderson et al.,[21] using HUVEC and bovine pulmonary artery EC (endothelial cells). Monolayers were illuminated following incubation with either DHE or $ZnPcS_{1,2}$. The supernatant was harvested and tested for its effect on platelet aggregation in the presence of inducers such as ADP or collagen. Both sensitizer-light combinations profoundly inhibited aggregation *in vitro*. Additionally, and seemingly perversely given what has gone before, indomethacin inhibited the inhibition. It is striking that among the particular eicosanoids released from bovine pulmonary aortic EC into the supernatant, 6-keto-$PGF_{1\alpha}$ (hydrolysis product of PGI_2) was particularly prominent (six times that of $PGF_{2\alpha}$). PGI_2 (prostacyclin) is known to inhibit platelet aggregation. However, other endothelia may release different proportions of differently acting eicosanoids; for HUVECs PGI_2 constitutes a much smaller proportion.[21,29] Differences in prostaglandin production are known for different types of vessels: microvessels prepared from bovine brain, heart, and retina produce similar amounts of PGI_2, PGE_2, and $PGF_{2\alpha}$, while larger vessels produce mainly PGI_2. Species differences may also confuse the picture as rabbit coronary microvessels and endothelial cells produce mainly PGE_2. It is apparent also that the cocktail released, and thereby outcome, may be therapy dose-dependent: moderate doses (0.1 μg/ml) of $ZnPcS_{1,2}$ caused inhibition of aggregation as described above, but high doses (>5 μg/ml) caused no anti-aggregation. Such differences may well depend on the degree of damage to endothelia; as Henderson et al.[21] speculate, very high therapeutic doses may inactivate the very enzymes necessary for the production of the eicosanoids measured in their assay. Additionally, the *in vitro* studies examine a pure population of endothelial cells; *in vivo*, as we have observed above, a number of cell types may well contribute to final outcome.

Of other factors that might be involved in inducing vessel occlusion after PDT, the kinetics of release of von Willebrand Factor (vWF) are very rapid *in vitro*, with 50% of the depot of bovine adrenal medulla endothelium entering the medium within 15 min of a PDT treatment with ClAlPcS.[30] Importantly, these rapid kinetics are quite different from those of ^{51}Cr release from the same cells as they die, and appears to be an active process in that it is inhibited at temperatures lower than 37°C. In HUVECs, onset of release was somewhat slower[9] (1 hr, still much more rapid than after, say, ionizing radiation, which peaks at 24 hr[31]). This study further emphasized the nonrandom nature of the process in that light dose-dependent release from viable cells was of the large multimeric form of vWF, while the distribution of the dimer pool remained unchanged. The trigger for release was an influx of Ca^{2+} and the whole process could be abrogated by chelating agents.

In concert with this rapid release of von Willebrand Factor, there will be a rapid translocation of P-selectin, a neutrophil adhesion molecule, from the Weibel-Palade bodies to the endothelial cell surface.[33,34] Thrombin, presumably activated as part of the PDT-induced local coagulation, and various stimuli increasing endothelial intracellular calcium, also induce the rapid translocation of P-selectin and release of platelet-activating factor (PAF). PAF on the endothelial surface activates adherent neutrophils facilitating their extravasation and neutrophil release of larger amounts of PAF.[35] PAF also acts directly on the endothelium, platelets and leukocytes, leading to the adherence of platelets and inflammatory cells to the endothelial surface, followed by platelet aggregation, vessel leakage, leukocyte extravasation, and degranulation.[34,36] PAF may also act on smooth muscle cells, and possibly pericytes, to induce vasoconstriction.[37]

V. RELATIVE SENSITIVITY OF ENDOTHELIAL CELLS TO PDT

Knowledge of relative sensitivities of endothelial and other cell types is of interest mechanistically, but is also of clinical relevance. As an example, PDT has been suggested as a method for treating large vessels stenosed by atherosclerotic plaques.[38] Current methods commonly have the drawback of restenosis through proliferation and migration of smooth muscle cells (SMC). For this to work, the PDT treatment should destroy already-migrated SMC and inhibit further migration from the medial layer without damage to the luminal endothelium.

A number of *in vitro* studies[39–43] have compared sensitivity of human and other normal and malignant cell lines (Table 12.2), measured by the survival of colony-forming cells. As a basis of comparison, we have abstracted from the published survival curves, an ED_{10} dose; that is, that light dose required to reduce the population to a surviving fraction of 10%. In all cases, endothelium (usually bovine aortic, but also including HUVEC) was the most sensitive cell type, whether compared to tumor cells or other normal tissues such as fibroblasts or SMC. The latter, of course, would be the preferred "target" in PDT for recanalization of (re)stenosing coronary arteries

Table 12.2 Relative Sensitivity to PDT of Endothelial and Other Normal and Malignant Cell Lines

Cell line	Sensitizer (left) or LDL receptor levels (right; mean log fluorescence) per cell			Sensitivity (ED10 dose)	Ref.
Human:	Photofrin II (pg/cell)			(J m^{-2})	39
Fibroblast	1.26			345	
Bovine:					
Aortic endothelium	1.08			255	
Smooth muscle	1.83			472	
Human:	ClAlPcS mM/10^7 cells			(kJ m^{-2})	
Fibroblast	nd			172	40
Hamster:					
Fibroblast	2			63	
Bovine:					
Endothelium	12			37	
Human:	HPD	LDL	AcLDL	(J cm^{-2})	
Colon ca	(ng/10^6 cells)				41, 42
	109 (log)	59	nd	0.193	
	98 (plat)	42	nd	3.20	
Bovine:					
Endothelium	287 (log)	68	148	0.717	
	204 (plat)	49	115	1.75	
		LDL		(J cm^{-2})	
Hamster:					
Fibroblast		nd		1.33	43
Human:					
Prostate ca		60		1.99	
Cervix ca		68		1.49	
Prostate ca		73		1.15	
Leukemia		nd		1.15	
Cervix ca		80		0.886	
HUVEC		nd		0.725	
Bovine:					
Aortic endothelium		101		0.604	

so that at first sight, the high sensitivity of endothelia would appear to contra-indicate this application. However, in a recent study, Dartsch et al.[38] find that within certain sensitizer dose limits, human endothelium has the same sensitivity as *normal* SMC but is significantly less sensitive than SMC isolated from atherosclerotic plaques, the population of therapeutic interest.

What is the basis for the differential sensitivity of endothelia compared to many other cells? The most obvious variable would be photosensitizer uptake into cells and in the series of Rosenthal and Ben-Hur[40] for ClAlPcS and West et al.[41,42] for HPD, the cellular sensitivities do correlate with the amount of drug per cell. It has been claimed that compounds such as HPD, a large proportion of which circulate around the body in association with lipoproteins and albumin, might be expected to be taken up to a greater degree by cells with high numbers of LDL (low-density lipoprotein) receptors on the cell surface (reviewed recently by Jori and Reddi[44]). Endothelia (and, particularly, proliferating endothelia, as may be the case in tumors) fulfill this criterion[42,43,45](Table 12.2). If this LDL pathway is proven to be of importance for sensitizer uptake, it could be accentuated even further by incorporating the photosensitizer into artificially modified LDL (e.g., acetylated), for which there is a highly expressed separate receptor confined to endothelia and macrophages (Table 12.2).[42,45] It should be acknowledged, however, that total photosensitizer concentration per cell is unlikely to be the sole parameter determining relative sensitivity (see the results of Gomer[39] in Table 12.2).

A second aspect of the expression of overall sensitivity, and a relatively unexplored area of PDT, is the role of posttreatment "repair" processes in different cell types. It might be that endothelia have a limited capacity for repair (note that here we are not necessarily equating "repair" with that of DNA, its usual sense). To test for repair is a more difficult exercise for PDT than for, say, ionizing radiation or hyperthermia, even *in vitro*. With the latter modalities, classical split dose experiments reveal recovery processes with time-scales typically of hours leading to, e.g., enhanced survival of clonogenic cells. In PDT, such an experiment is complicated by uncertainties as to how much photosensitizing drug is still present at the time of the second illumination; if this is less than before, then a given dose of light will be likely to have less effect, and such an outcome is not necessarily saying anything about repair. What other approaches are available? For other modalities, recovery processes are often quantitatively associated with the induction of specific genes and their protein products. Of known early-response proteins, transcriptionally induced up-regulation of the glucose-regulated proteins (GRP) did indeed confer resistance to subsequent PDT in fibrosarcoma cells.[46] However, using various cell lines and their thermotolerant variants (not endothelial), it was found that although the much-investigated 70-kDa family was expressed constitutively at much higher levels in the thermotolerant cells, the uptake of photosensitizer and the response to PDT were the same in the two groups.[47] The authors inferred from this that the classical "heat-shock" 70-kDa family may not be relevant to PDT response. The first results from our group, using BAE cells and two-dimensional electrophoresis, indicate that within 24 hr of sublethal HPD-mediated PDT, the synthesis of several proteins (41, 71, 73, 80, and 90 kDa) is greatly up-regulated or induced (Figure 12.2). At light doses of 0.5 J cm^{-2}, the overall degree of protein synthesis is reduced with the exception of these same up-regulated proteins and three additional proteins, at 62, 78, and 100 kDa. The 70- to 100-kDa proteins closely resemble the classical heat-shock proteins. However, the 41- and 62-kDa proteins more closely resemble those stress-proteins recently shown to be induced in BAE cultures by arsenite or hypoxia.[48,49] It is interesting to speculate whether, *in vivo*, the hypoxia known to be induced by vascular-mediated PDT treatments[50,51] might in turn induce stress-response molecules in "downstream" endothelial and other cell types. Ogawa et al. have recently examined the effect of hypoxia, heat shock, glucose deprivation, and other inducers of stress-proteins on BAE cultures. In common with the above studies, they found distinct variations in the pattern of stress proteins depending on the stimulus.[52] Such variations presumably reflect the involvement of several induction pathways, including heat-shock transcription factor,[53] protein kinase C,[54] and the recently reported stress-activated c-Jun kinases.[55] Ogawa et al. also describe an hypoxia-induced cell-surface factor X activator that would amplify the extrinsic coagulation pathway.[52] Implicit in this section has been the notion of functional

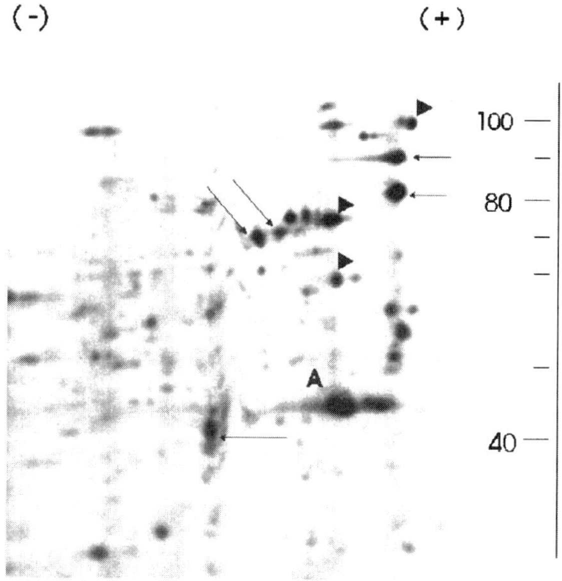

Figure 12.2. Two-dimensional protein profile of bovine aortic endothelial cells labeled with ^{35}S-methionine for 24 hr following PDT treatment by 10 μg ml^{-1} HPD and 0.5 J cm^{-2} of 630-nm laser light. Relative to controls (not shown), the vast majority of proteins are now absent. Of those remaining, five constitutive proteins (arrows) are up-regulated, and there are three induced proteins (triangles). Actin (A).

recovery of individual PDT-exposed cells. A further "recovery" pathway for the tissue may involve the elimination of damaged cells: local or systemic endothelial activators, such as bacterial lipopolysaccharide,[56] cytokines, and growth factors,[57] appear to interact with induced stress-proteins leading to cell death and, as such, may increase endothelial cell sensitivity to PDT.[58]

In vivo, determination of relative sensitivity is less straightforward. Undoubtedly, the endothelium of both normal and tumor tissues can be shown to be promptly damaged by a variety of photosensitizer-light combinations, and this promptness of effect is frequently assumed also to be an indicator of high sensitivity. It has been claimed for several sensitizers, on the basis of electron microscopy[59,60] or assays of clonogenicity,[61] that death of tumor cells is not due to direct cytotoxicity, but is a secondary consequence of the prompt vascular damage. Yoshida and colleagues have recently challenged this orthodoxy in the expression of Photofrin-mediated injury in the rodent brain, demonstrating by light-and electron-microscopy a more rapid response of astrocytes than endothelium.[62] One should recall, however, the dictum of radiobiology that radioresponsiveness and radiosensitivity are not necessarily synonymous. A final point to stress, is that "sensitivity" is strongly influenced by drug concentration at the time of illumination. Thus, Reed et al.[63] found pronounced damaging effects in bladder vasculature when DHE was administered 30 min before, when serum levels of DHE were high, and bladder tissue levels low, but very little vascular disturbance when the drug had been given 48 hr previously and when the bulk of the sensitizer had become localized in the bladder stroma. The consequences of these shifts in "sensitivity" are not trivial, and Reed et al.[63] suggest that they may have contributed to the severe bladder contraction in 4 of 5 patients treated with DHE 48 hr prior to therapy.

VI. SUMMARY AND CONCLUSIONS

Vascular endothelia, including human endothelia, are among the cell types most sensitive to the *in vitro* cytotoxic effects of photodynamic therapy. A widely accepted hypothesis is that this may be associated with high numbers of glycoprotein LDL receptors on the endothelial surface encouraging high uptake of LDL-incorporated photosensitizing drugs. Whether this association represents a correlation, in our view remains to be definitively established. Protein synthesis in endothelia treated by PDT is markedly reduced initially, with the exception of a number of constitutive and induced proteins, among which are recognized stress-response ("heat-shock") molecules. The precise physiological role of these proteins in PDT-mediated damage and recovery is an important area for investigation. *In vivo*, the combination of many (but not all?) photosensitizers and light leads to rapid physical destruction of the endothelium, often with exposure of the thrombogenic subendothelial surface, leading in turn to vessel occlusion and the secondary cell death of the dependent malignant cells of the tumor through sustained anoxia. Even relatively low doses of PDT may result in the release from endothelium of thrombogenic factors, most clearly von Willebrand's. Perhaps less clear, *in vivo* at least, is the origin of the predominant thrombogenic and vasoactive eicosanoids that are released in a variable cocktail from a number of PDT-treated cell types. Empirically, it can be shown that cyclooxygenase inhibitors, notably indomethacin, reduce the cytotoxic effects of PDT, including such critical parameters as tumor cure. The "target" cells in which this amelioration of cytotoxicity occurs may, or may not, be primarily endothelium. Given the demonstrated possibility of radically manipulating PDT outcome in tumors and normal tissues, a closer understanding of the precise biochemical events and their cell(s) of origin seems warranted.

REFERENCES

1. Meyer-Betz, F., Untersuchungen über die Biologische (photodynamische) Wirkung des hematoporphyrins und anderer Derivative des Blut-und-Gallenfarbstoffs, *Dtsch. Arch. Klin. Med.*, 112, 476, 1913.
2. Lim, H. W., Young, L., Hagan, M., and Gigli, I., Delayed phase of hematoporphyrin-induced phototoxicity: modulation by complement, leukocytes, and antihistamines, *J. Invest. Dermatol.*, 84, 114, 1985.
3. Moore, J. V., Keene, J. P., and Land, E. J., Dose-response relationships for photodynamic injury to murine skin, *Br. J. Radiol.*, 59, 257, 1986.
4. Fingar, V. H., Wieman, T. J., Wiehle, S. A., and Cerrito, P. B., The role of microvascular damage in photodynamic therapy: the effect of treatment on vessel constriction, permeability and leukocyte adhesion, *Cancer Res.*, 52, 4914, 1992.

5. Lim, H. W., Hagan, M., and Gigli, I., Phototoxicity induced by hematoporphyrin derivative in C5-deficient, mast-cell deficient and leukopenic mice, *Photochem. Photobiol.*, 44, 175, 1986.
6. Roberts, W. G. and Hasan, T., Role of neovasculature and vascular permeability on the tumor retention of photodynamic agents, *Cancer Res.*, 52, 924, 1992.
7. Roberts, W. G. and Hasan, T., Tumor-secreted vascular permeability factor/vascular endothelial growth factor influences photosensitiser uptake, *Cancer Res.*, 53, 153, 1993.
8. Sporn, L. A. and Foster, T. H., Photofrin and light induces microtubule depolymerization in cultured human endothelial cells, *Cancer Res.*, 52, 3443, 1992.
9. Foster, T. H., Primavera, M. C., Marder, V. J., Hilf, R., and Sporn, L. A., Photosensitised release of von Willebrand factor from cultured human endothelial cells, *Cancer Res.*, 51, 3261, 1991.
10. Reed, M. W. R., Wieman, T. J., Schuschke, D. A., Tseng, M. T., and Miller, F. N., A comparison of the effects of photodynamic therapy on normal and tumor blood vessels in the rat microcirculation, *Radiat. Res.*, 119, 542, 1989.
11. Miller, F. N., Sims, D. E., Schuschke, D. A., and Abney, D. L., Differentiation of light-dye effects in the microcirculation, *Microvasc. Res.*, 44, 166, 1992.
12. Gilissen, M. J., van de Merbel de Wit, E. A., Star, W. M., Koster, J. F., and Sluiter, W., Effect of photodynamic therapy on the endothelium-dependent relaxation of isolated rat aortas, *Cancer Res.*, 53, 2548, 1993.
13. Rosenblum, W. I., Nelson, G. H., and Povlishok, J. T., Laser-induced endothelial damage inhibits endothelium-dependent relaxation in the cerebral microcirculation of the mouse, *Circ. Res.*, 60, 169, 1987.
14. Chang, K. C., Chong, W. S., Park, B. W., Seung, B. W., Chun, G. W., Lee, I. J., and Park, P. S., NO- and NO_2-carrying molecules potentiate photorelaxation in rat trachea and aorta, *Biochem. Biophys. Res. Commun.,* 191, 509, 1993.
15. Deliconstantinos, G., Villiotou, V., and Fassitsas, C., Ultraviolet-irradiated human endothelial cells elaborate nitric oxide that may evoke vasodilatory response, *J. Cardiovasc. Pharmacol.*, 20(Suppl. 12), S63-65, 1992.
16. Warren, J. B., Loi, R. K., and Coughlan, M. L., Involvement of nitric oxide synthase in the delayed vasodilator response to ultraviolet light irradiation of rat skin *in vivo*, *Br. J. Pharmacol.*, 109, 802, 1993.
17. Nishimura, H., Rosenblum, W. I., Nelson, G. H., and Boynton, S., Agents that modify EDRF formation alter antiplatelet properties of brain arteriolar endothelium *in vivo*, *Am. J. Physiol.*, 261, H15, 1991.
18. Fingar, V. H., Wieman, T. J., and Doak, K. W., Changes in tumor interstitial pressure induced by photodynamic therapy, *Photochem. Photobiol.*, 53, 763, 1991.
19. Leunig, M., Goetz, A. E., Gamarra, F., Zetterer, G., Messmer, K., and Jain, R. K., Photodynamic therapy-induced alterations in interstitial fluid pressure, volume and water content of an amelanotic melanoma in the hamster, *Br. J. Cancer*, 69, 101, 1994.
20. Lim, H. W., Effects of porphyrins on skin, in *Photosensitising Compounds: Their Chemistry, Biology and Clinical Use*, Bock, G. and Harnett, S., Eds., John Wiley, Chichester, 1989, 148.
21. Henderson, B. W., Owczarczak, B., Sweeney, J., and Gessner, T., Effects of photodynamic treatment of platelets or endothelial cells *in vitro* on platelet aggregation, *Photochem. Photobiol.*, 56, 513, 1992.
22. Fingar, V. H., Wieman, T. J., and Doak, K. W., Mechanistic studies of PDT-induced vascular damage: evidence that eicosanoids mediate this process, *Int. J. Radiat. Biol.*, 60, 303, 1991.

23. Fingar, V. H., Wieman, T. J., Karavalos, P. S., Doak, K. W., Ouellet, R., and van Lier, J. E., The effects of photodynamic therapy using differently substituted zinc phthalocyanines on vessel constriction, vessel leakage and tumor response, *Photochem. Photobiol.*, 58, 251, 1993.
24. Herrmann, K. S., Platelet aggregation induced in the hamster cheek pouch by a photochemical process with excited fluorescein isothyocyanate-dextran, *Microvasc. Res.*, 26, 238, 1983.
25. Stern, S. J., Flock, S., Small, S., Thomsen, S., and Jacques, S., Chloroaluminum sulfonated phthalocyanine versus dihematoporphyrin ether: early vascular events in the rat chamber, *Laryngoscope*, 101, 1219, 1991.
26. Stern, S. J., Craig, J. R., Flock, S., and Small, S., Effect of aspirin on photodynamic therapy utilising chloroaluminum sulfonated phthalocyanine (CASP), *Lasers Surg. Med.*, 12, 494, 1992.
27. Rosenblum, W. I. and El-Sabban, F., Platelet aggregation in the cerebral microcirculation, *Circ. Res.*, 40, 320, 1977.
28. Reed, M. W. R., Miller, F. N., Wieman, T. J., Tseng, M. T., and Pietsch, C. G., The effect of photodynamic therapy on the microcirculation, *J. Surg. Res.*, 45, 452, 1988.
29. Levine, L., Radioimmunoassay of eicosanoids associated with tumor growth, *Biochemistry of Arachidonic Acid Metabolism*, Lands, W. E. M., Ed., Martinus Nijhoff, Boston, 405, 1985.
30. Ben-Hur, E., Heldman, E., Crane, S. W., and Rosenthal, I., Release of clotting factors from photosensitised endothelial cells: a possible trigger for blood vessel occlusion by photodynamic therapy, *FEBS Lett.*, 236, 105, 1988.
31. Sporn, L. A., Rubin, P., Marder, V. J., and Wagner, D. D., Irradiation induces release of von Willebrand protein from endothelial cells in culture, *Blood*, 64, 567, 1984.
32. McEver, R. P., Beckstead, J. H., Moore, K. L., Marshall-Carlson, L., and Bainton, D. F., GMP-140, a platelet alpha-granule membrane protein, is also synthesised by vascular endothelial cells and is localized in Wiebel-Palade bodies, *J. Clin. Invest.*, 84, 92, 1989.
33. Hattori, R., Hamilton, K. K., Fugate, R. D., McEver, R. P., and Sims, P. J., Stimulated secretion of endothelial von Willebrand factor is accompanied by rapid redistribution to the cell surface of the intracellular granule membrane protein GMP-140, *J. Biol. Chem.*, 264, 7768, 1989.
34. Koltai, K., Hosford, D., Guinot, P., Esanu, A., and Braquet, P., Platelet activating factor (PAF): A review of its effects, antagonists and possible future clinical implications, *Drugs*, 42, 9, 1991.
35. Patel, K. D., Zimmerman, G. A., Prescott, S. M., and McIntyre, T. M., Novel leukocyte agonists are released by endothelial cells exposed to peroxide, *J. Biol. Chem.*, 267, 15168, 1992.
36. Chao, W. and Olson, M. S., Platelet-activating factor: receptors and signal transduction, *Biochem J.*, 292, 617, 1993.
37. Camussi, G., Tetta, C., and Baglioni, C., The role of platelet-activating factor in inflammation, *Clin. Immunol. Immunopathol.*, 57, 331, 1990.
38. Dartsch, P. C., Coppenrath, E., Coppenrath, K., and Ischinger, T., Photodynamic therapy of vascular stenosis: results from cell culture studies on human endothelial cells, *Coron. Art. Dis.*, 4, 207, 1993.
39. Gomer, C. J., Rucker, N., and Murphree, A. L., Differential cell photosensitivity following porphyrin photodynamic therapy, *Cancer Res.*, 48, 4539, 1988.

40. Rosenthal, I. and Ben-Hur, E., Phthalocyanines in photobiology, in *Phthalocyanines: Properties and Applications*, Leznoff, C. C. and Lever, A. B. P., Eds., VCH Publishers, New York, 397, 1989.
41. West, C. M. L., West, D. C., Kumar, S., and Moore, J. V., A comparison of the sensitivity to photodynamic treatment of endothelial and tumor cells in different proliferative states, *Int. J. Radiat. Biol.*, 58, 145, 1990.
42. West, C. M. L., West, D. C., and Moore, J. V., Low-density lipoprotein uptake and the sensitivity to photodynamic treatment of cultured endothelial cells, *Int. J. Radiat. Biol.*, 60, 25, 1991.
43. West, C. M. L. and Moore, J. V., unpublished data, 1994.
44. Jori, G. and Reddi, E., The role of lipoproteins in the delivery of tumor-targeting photosensitisers, *Int. J. Biochem.*, 25, 1369, 1993.
45. Gaffney, J., West, D., Arnold, F., Sattar, A., and Kumar, S., Differences in the uptake of modified low density lipoproteins by tissue cultured endothelial cells, *J. Cell Sci.*, 79, 317, 1985.
46. Gomer, C. J., Ferrario, A., Rucker, N., Wong, S., and Lee, A.S., Glucose regulated protein induction and cellular resistance to oxidative stress mediated by porphyrin photosensitisation, *Cancer Res.*, 51, 6574, 1991.
47. Gomer, C. J., Rucker, N., and Wong, S., Porphyrin photosensitivity in cell lines expressing a heat-resistant phenotype, *Cancer Res.*, 50, 5365, 1990.
48. Zimmerman, L. H., Levine, R. A., and Farber, H. W., Hypoxia induces a specific set of stress proteins in cultured endothelial cells, *J. Clin. Invest.*, 87, 908, 1991.
49. Graven, K. K., Zimmerman, L. H., Dickson, E. W., Weinhouse, G. L., and Farber, H. W., Endothelial cell hypoxia associated proteins are cell and stress specific, *J. Cell. Physiol.*, 157, 544, 1993.
50. Reed, M. W. R., Mullins, A. P., Anderson, G. L., Miller, F. N., and Wieman, T. J., The effect of photodynamic therapy on tumor oxygenation, *Surgery*, 106, 94, 1989.
51. Tromberg, B. J., Orenstein, A., Kimel, S., Barker, S. J., Hyatt, J., Nelson, J. S., and Berns, M. W., In vivo tumor oxygen tension measurements for the evaluation of the efficiency of photodynamic therapy, *Photochem. Photobiol.*, 52, 375, 1990.
52. Ogawa, S., Clauss, M., Kuwabara, K., Shreeniwas, R., Butura, C., Koga, S., and Stern, D., Hypoxia induces endothelial cell synthesis of membrane-associated proteins, *Proc. Natl. Acad. Sci., U.S.A.*, 88, 9897, 1991.
53. Morimoto, R., Heat shock: the role of transient inducible responses in cell damage, transformation and differentiation, *Cancer Cells*, 3, 295, 1991.
54. Santell, L., Bartfeld, N. S., and Levin, E. G., Identification of a protein transiently phosphorylated by activators of endothelial cell function as the heat shock protein HSP27, *Biochem. J.*, 284, 705, 1992.
55. Kyriakis, J. M., Banerjee, P., Nikolakaki, E., Dai, T., Rubie, E. A., Ahmad, M. F., Avruch, J., and Woodgett, J. R., The stress-activated protein kinase sub-family of c-Jun kinases, *Nature*, 369, 156, 1994.
56. Buchman, T. G., Abello, P. A., Smith, E. H., and Bulkley, G. B., Induction of heat shock response leads to apoptosis in endothelial cells previously exposed to endotoxin, *Am. J. Physiol.*, 265, H165-70, 1993.
57. Saklatvala, J., Kaur, P., and Guesdon, F., Phosphorylation of the small heat-shock protein is regulated by interleukin 1, tumor necrosis factor, growth factors, bradykinin and ATP, *Biochem. J.*, 277, 635, 1991.
58. Breider, M. A., Lu, X., Panjehpour, M., and Frazier, D. L., Cytokine modulation of endothelial cell sensitivity to photodynamic therapy, *Lasers Surg. Med.*, 13, 305, 1993.

59. Berenbaum, M. C., Hall, G.W., and Hoyes, A. D., Cerebral photosensitisation by hematoporphyrin derivative. Evidence for an endothelial site of action, *Br. J. Cancer*, 53, 81, 1986.
60. Nelson, J. S., Liaw, L.-H., and Berns, M. W., Tumour destruction and kinetics of tumour cell death in two experimental mouse tumours following photodynamic therapy, *Photochem. Photobiol.*, 46, 829, 1987.
61. Henderson, B. W., Waldow, S. M., Mang, T. S., Potter, W. R., Malone, P. B., and Dougherty, T. J., *Cancer Res.*, 45, 572, 1985.
62. Yoshida, Y., Dereski, M. O., Garcia, J. H., Hetzel, F. W., Chopp, M., Photoactivated Photofrin II: astrocytic swelling precedes endothelial injury in rat brain, *J. Neuropathol. Exp. Neurol.*, 51, 91, 1992.
63. Reed, M. W. R., Schuschke, D. A., Ackermann, D. M., Harty, J. I., Wieman, T. J., and Miller, F. N., The response of the rat urinary bladder microcirculation to photodynamic therapy, J. *Urol.*, 142, 865, 1989.

Index

A

B

H

I

K

L

R

S

T

U

V

X

Z